Leonid Zhmud

The Origin of the History of Science in Classical Antiquity

Peripatoi

Philologisch-historische Studien zum Aristotelismus

Herausgegeben von

Wolfgang Kullmann,
Robert W. Sharples, Jürgen Wiesner

Band 19

2006
Walter de Gruyter · Berlin · New York

Leonid Zhmud

The Origin of the History of Science in Classical Antiquity

Translated from the Russian by

Alexander Chernoglazov

2006
Walter de Gruyter · Berlin · New York

♾ Printed on acid-free paper which falls within the guidelines
of the ANSI to ensure permanence and durability

ISSN: 1862-1465

ISBN-13: 978-3-11-017966-8
ISBN-10: 3-11-017966-0

Bibliographic Information published by Die Deutsche Bibliothek

Die Deutsche Bibliothek lists this publication in the Deutsche Nationalbibliografie;
detailed bibliographic data is available in the
Internet <http://dnb.dbb.de>.

Printed in Germany
Typesetting: Selignow Verlagsservice, Berlin
Printing and binding: Hubert & Co., Göttingen
Cover design: Christopher Schneider, Berlin

To the memory of my wife Irina
(1965–2002)

καινὰ δὲ ζητοῦντες ἐπιπόνως εὑρήσουσι

Isoc. *Antid.*, 83.

Preface

When writing a book on the Pythagorean school, I noticed that, although Aristotle's student Eudemus of Rhodes was regularly used as a source for Greek science, there was no scholarly treatment of him as the first historian of science. Indeed, as my further research showed, the whole area of ancient Greek historiography of science remained almost entirely unexplored. This prompted me to work first on Eudemus, and then on his predecessors, colleagues, and followers. The Center for Hellenic Studies, Washington (1995–1996); Maison des Sciences de l'Homme, Paris (1998); Institute for Advanced Study, Princeton (1998–1999); Alexander von Humboldt-Foundation, Bonn (2000, 2004); Wellcome Trust Centre for the History of Medicine, London (2000–2001); and Wissenschaftskolleg zu Berlin (2002–2003, 2005) have found my studies worthy of support. Without them this book would hardly have been written. The Alexander von Humboldt-Foundation and Wissenschaftskolleg zu Berlin generously sponsored the translation of this book into English. I express my deep gratitude to all these institutions.

Alexander Gavrilov, Elena Ermolaeva, Dmitri Panchenko, and Alexander Verlinsky have read the Russian version of the book, published in 2002 in St. Petersburg; their valuable suggestions have improved the text in many places. I am grateful to Gertrud Grünkorn for her offer to publish the English version of my book with the Walter de Gruyter Press. In preparing the new edition I have updated the bibliography and written an additional chapter, tracing the fate of the historiography of science after Eudemus.

István Bodnár, Carl Huffman, Charles Kahn, Paul Keyser, Colin G. King, and Henry Mendell have read and commented on separate chapters of the manuscript; their criticism has made it possible to eliminate many inaccuracies and to make essential improvements to the text. Lydia Goehr, John Hyman, Maria Michela Sassi, David Sider, and Heinrich von Staden have all been extremely helpful. I would particularly like to thank Geoffrey Lloyd, who has read a whole draft of the book and sent me his very helpful comments. The editors of the *Peripatoi* series, Wolfgang Kullmann, Robert Sharples, and Jürgen Wiesner, have also read the whole manuscript and provided their expert comments, which saved me from many mistakes. Those that still remain are my own responsibility.

I want also to express my special gratitude to Mitch Cohen at the Wissenschaftskolleg zu Berlin for his meticulous reading of my book in manuscript and for having greatly improved its English.

Several sections of this book (3.1–2, 4.2–3, 5.1-2, and 7.6) include revised versions of my earlier papers:

1) Plato as "architect of science", *Phronesis* 43 (1998) 211–244;

2) Eudemus' history of mathematics, *Eudemus of Rhodes*, ed. by I. Bodnár, W. W. Fortenbaugh, New Brunswick 2002, 263–306 (Rutgers University Studies in Classical Humanities, Vol. 11);

3) Historiographical project of the Lyceum: The peripatetic history of science, philosophy, and medicine, *Antike Naturwissenschaft und ihre Rezeption*, Vol. 13 (2003) 113–130;

4) "Saving the phenomena" between Eudoxus and Eudemus, *Homo Sapiens und Homo Faber. Festschrift für J. Mittelstraß*, ed. by G. Wolters, M. Carrier, Berlin 2005, 17-24.

I am grateful to the respective publishers and editors for their kind permission to use these papers.

St. Petersburg, January 2006 Leonid Zhmud

Contents

Works, quoted by the abbreviated title

AHM:	*Ancient histories of medicine. Essays in medical doxography and historiography in classical Antiquity*, ed. by Ph. J. van der Eijk, Leiden 1999.
Barker. *GMW* I:	Barker, A. *Greek musical writings*, Vol. 1: *The musician and his art*, Cambridge 1984.
Barker. *GMW* II:	Barker, A. *Greek musical writings*, Vol. 2: *Harmonic and acoustic theory*, Cambridge 1989.
Becker. *Denken*:	Becker, O. *Das mathematische Denken der Antike*, Göttingen 1957.
Burkert. *L & S*:	Burkert, W. *Lore and science in ancient Pythagoreanism*, Cambridge, Mass. 1972.
Dorandi. *Filodemo*:	Dorandi, T. *Platone e l'Academia (PHerc. 1021e 164)/ Filodemo*, Neapel 1991.
Dox.:	Diels, H. *Doxographi Graeci*, Berlin 1879.
Heath. *Elements*:	Heath, T. L. *The thirteen books of Euclid's Elements*, Vol. 1–3, Oxford 1927.
Knorr. *AT*:	Knorr, W. R. *The ancient tradition of geometric problems*, Boston 1986.
Knorr. *TS*:	Knorr, W. R. *Textual studies in ancient and medieval geometry*, Boston 1989.
Krafft. *Mechanik*:	Krafft, F. *Dynamische und statische Betrachtungsweise in der antiken Mechanik*, Wiesbaden 1970.
KRS:	Kirk, G. S., Raven, J. E., Schofield, M. *The Presocratic philosophers*, 2nd ed., Cambridge 1983.
Lasserre. *Léodamas*:	Lasserre, F. *De Léodamas de Thasos à Philippe d'Oponte*, Neapel 1987.
Neuenschwander. VB:	Neuenschwander, E. Die ersten vier Bücher der Elemente Euklids, *AHES* 9 (1973) 325–380.
Neugebauer. *ES*:	Neugebauer, O. *The exact sciences in Antiquity*, 2nd ed., New York 1962.
Neugebauer. *HAMA*:	Neugebauer, O. *A history of ancient mathematical astronomy*, Pt. 1–3, Berlin 1975.
van der Waerden. *EW*:	Waerden, B. L. van der. *Erwachende Wissenschaft*, Basel 1959.
Zhmud. Philolaus:	Zhmud, L. Some notes on Philolaus and the Pythagoreans, *Hyperboreus* 4 (1998) 243–270.

Introduction

Greek science and its historiography

1. The historiography of science in the 16th–18th centuries

Ancient Greek science has been studied for such a long time that the history of these studies themselves deserves an enquiry of its own. Like philology, which emerged in Europe as classical philology, the history of science was born as the history of ancient science. It is the theories and discoveries of Greek scientists that provided the material on which the methods of the history of science were worked out over the centuries. This process started much earlier and continued far longer than is commonly thought. Interest in the history of science appeared first in classical Antiquity and has experienced more than one rise and decline since then. The first rise came in the late fourth century BC, when the earliest works on the history of science were written. Then, after a long period of dwindling interest, medieval Arabic culture again focused attention on the history of Greek science, with later peaks of interest occurring during the Renaissance and the scientific revolution of the 17th century. The *modern* historiography of science, which takes contemporary science as its main reference point and has gradually mastered new methods of source criticism, emerged in the late 18th to early 19th centuries. This period coincided with a new infatuation with classical Antiquity, so that, ever since, the history of Greek science has remained a steadily growing field of study, combining classical philology with the history of science.

This is the history of studies in ancient science summarized in one paragraph. Those who seek a detailed history of the subject will be disappointed: none has ever been written. Generally speaking, historians of science, unlike classical philologists or historians of philosophy, have as yet shown no particular interest in the origin and the early stages of their discipline. In the few cases where these problems have appeared to draw attention, their examination proved superficially selective and seldom reached further back than the 18th century.[1] Apart from works on the ancient historiography of medicine[2] and

---

[1] Loria, G. *Guida allo studio della storia delle matematiche*, Milan 1946; Struik, D. J. Historiography of mathematics from Proklos to Cantor, *NTM Schriftenreihe für Geschichte der Naturwissenschaften, Technik und Medizin* 17 (1980) 1–22; Vogel, K. L'historiographie mathématique avant Montucla, *Kleine Schriften zur Geschichte der Mathematik*, Vol. 2, Stuttgart 1988, 556–562; Schneider, I. Hintergrund und Formen der Mathematikgeschichte des 18. Jahrhunderts, *AIHS* 42 (1992) 64–75; Laudan, R. Histories of the sciences and their uses: A review to 1913, *HS* 31 (1993) 1–33; Vitrac, B. Mythes (et realités) dans l'histoire des mathématiques grecques an-

Arabic historiography of science, the field remains almost untouched. The Renaissance historiography of science has only recently come to be studied.

As a matter of fact, there is nothing surprising about this. The object of the history of science is, in the first place, science itself. Historiography, whether rightfully or not, has always remained in the background. For a historian of science, the works of Euclid, Ptolemy, or Newton are of greater importance than the historico-scientific literature contemporary to them. To be sure, sometimes this literature may prove to be a valuable source, for example, when the original scientific writings have been lost. The first histories of science were written by the Peripatetic Eudemus of Rhodes even before Euclid's *Elements* summed up the first three centuries of Greek mathematics. Whereas from Euclid we learn *what* was discovered during this period, it is Eudemus who tells us *who* made these discoveries and *when*, also adding some material not included in the *Elements*. Similarly, the history of early Greek astronomy is known mainly from Eudemus and from the doxographical work of Theophrastus, his colleague in the Lyceum. This is what actually accounts for the pragmatic interest shown by historians of Greek science in the surviving fragments of Eudemus and other Peripatetics. Yet outside Antiquity and after the invention of printing in particular, the purely pragmatic approach to the historiography of science is hardly justified. Those who study the science of the 16th–18th centuries turn, as a rule, to primary sources, not to the historico-scientific literature of the epoch, which was mostly antiquarian in character and did not aim to cover the latest discoveries. As a result, the interest in this literature as a source is still smaller than that enjoyed by the historico-scientific tradition of Antiquity.

Our subject is the ancient historiography of science. 'Pre-modern' historiography of science interests us only insofar as it reveals a marked continuity with the ancient tradition, both on the formal and the thematic level. If the history of science revived in Europe as the history of Greek science, it was because the science of the 15th–17th centuries was itself oriented toward assimilating the classical heritage. In this period, the interests of scientists and historians of

---

ciennes, *L'Europe mathématique: histoires, mythes, identités*, ed. by C. Goldstein et al., Paris 1996, 31–51.

2 Smith, W. D. Notes on ancient medical historiography, *BHM* 63 (1989) 73–109; Staden, H. von. Galen as historian, *Galeno: Obra, pensamiento e influencia*, ed. by J. A. López Férez, Madrid 1991, 205–222; Pigeaud, J. La médicine et ses origins, *Canadian Bulletin of Medical History* 9 (1992) 219–240. The collection *Ancient histories of medicine. Essays in medical doxography and historiography in classical Antiquity*, ed. by Ph. J. van der Eijk, Leiden 1999, constitutes the first attempt at systematic approach to this subject. For the earlier literature, see Heischkel, E. Die Medizinhistoriographie im XVIII Jh., *Janus* 35 (1931) 67–105, 125–151; eadem. *Die Medizingeschichtsschreibung von ihren Anfängen bis zum Beginn des 16. Jh.s*, Berlin 1938; eadem. Die Geschichte der Medizingeschichtsschreibung, *Einführung in die Medizinhistorik*, ed. by W. Artelt, Stuttgart 1949, 202–237.

science converged to a much greater extent than they did, for example, in Antiquity, and this gave the historiography of science an important additional impetus. Such a convergence of interests is by no means common. As opposed to the history of philosophy – which is still an integral part of philosophy[3] – or the history of medicine – which remained an integral part of medicine up to the 19th century – the history of science usually focuses on tasks quite different from those of science itself. For most physicians of the late 18th century, Hippocrates and Galen remained topical,[4] just as the problems posed by Plato, Aristotle and Descartes remain topical for the greater part of modern Western philosophy. But the problems that occupied Archimedes, Ptolemy, or Copernicus are very far from those of modern science.[5] The history of science becomes really *necessary* for scientists only when, for whatever reason, the scientific or, in a more general sense, the cultural tradition, which normally ensures the transmission of knowledge from generation to generation, is disrupted. It is when foreign science is being assimilated that the main question of the history of science – 'who discovered what?' – arises in the process of scientific investigation itself. The absence of clear answers to this question can hinder research, for instance, by forcing scientists to spend time and energy proving what has already been proven or refuting what has already been refuted.

One such period was the 8th–10th centuries, when Greek science was appropriated by the Arabic-speaking world and became an integral part of Arabic science. Unlike medieval Europe and, in many ways, unlike Byzantium, Arabic culture borrowed, along with Greek science, both the ancient historico-scientific tradition[6] and its major methodological approaches to science.[7] It would

---

3 That is why its earlier stages are studied much more fully. See e.g. Braun, L. *Histoire de l'histoire de la philosophie*, Paris 1973; Del Torre, M. A. *Le origini moderne della storiografia filosofica*, Florence 1976; Piaia, G. "*Vestigia philosophorum*": *il Medioevo e la storiografia filosofica*, Rimini 1983; *Models of the history of philosophy*, ed. by G. Santinello, C.W. T. Blackwell, Vol. 1–3, Dordrecht 1993.

4 The classical history of medicine of the time, Sprengel, K. *Versuch einer pragmatischen Geschichte der Medizin*, T. 1–5, Halle 1792–1803, still regarded doctors' familiarity with ancient and Arabic medicine, which the author knew first-hand, as being of practical use.

5 On the ongoing 'dehistorisation' of mathematics since the 18th century, see Siegmund-Schultze, R. Über das Interesse der Mathematiker an der Geschichte ihrer Wissenschaft, *Amphora. Festschrift für H. Wussing*, ed. by S. Demidov et al., Basel 1992, 705–736.

6 On the Arabic historiography of science and medicine, see Meyerhof, M. Sultan Saladin's physician on the transmission of Greek medicine to the Arabs, *BHM* 18 (1945) 169–178; Rosenthal, F. Al-Asturlabi and as-Samaw'al on scientific progress, *Osiris* 9 (1945) 555–564; idem. Ishaq b. Hunayn Ta'rih al-attiba', *Oriens* 7 (1954) 55–80; idem. An ancient commentary on the Hippocratic Oath, *BHM* 30 (1956) 52–87; Plessner, M. M. Der Astronom und Historiker Ibn Sa'id al-Andalusi und seine Geschichte der Wissenschaften, *RSO* 31 (1956) 235–257; Hau, F. R. Die medizinische Geschichtsschreibung im islamischen Mittelalter, *Clio medica* 18 (1983);

hardly suffice to say that Muslim scientists had a lively interest in their Greek predecessors. They held them in the highest esteem, tried to find every last bit of information on them, made annotated catalogues of their works, translated the extant biographies of eminent scientists and physicians and compiled new ones.[8] Later, on the basis of all of this, a historiography of Arabic science and medicine arose, which in turn influenced both the Byzantine and the Western traditions.

In many ways, the situation in Europe in the 15th–17th centuries parallels the Arabic assimilation of Greek science. To return to ancient science after so many centuries; to edit and translate Euclid, Apollonius, Archimedes, Ptolemy, Diophantus, and Pappus; to understand 'who was who' in ancient science – all this urgently demanded at least a general historical picture of Greek mathematics and astronomy, which presented its achievements chronologically.[9] The absence of such a picture hampered, if it did not foreclose, progress to new discoveries. During the Renaissance, the historiography of science therefore remained as inseparable from the classical heritage as science itself;[10] the Middle Ages, apart from the Arabs, were usually ignored.

69–80; Brentjes, S. Historiographie der Mathematik im islamischen Mittelalter, *AIHS* 42 (1992) 27–63; Gutas, D. The 'Alexandria to Baghdad' complex of narratives, *Documenti e studi sulla tradizione filosofica medievale* 10 (1999) 155–193. See also below, 8.3.

7 On the methodology of science in the works of Arabic thinkers, see *Alfarabi. Über den Ursprung der Wissenschaften (De ortu scientiarum)*, ed. by C. Baeumker, Münster 1916; Wiedemann, E. Auszüge aus Ibn Sina's Teile der philosophischen Wissenschaften (mathematische Wissenschaften), *Aufsätze zur arabischen Wissenschaftsgeschichte*, Vol. 1, Hildesheim 1970, 146–154; idem. Definitionen verschiedener Wissenschaften und über diese verfaßte Werke, ibid., Vol. 2, 431–462; Maróth, M. Das System der Wissenschaften bei Ibn Sina, *Avicenna/Ibn Sina*, ed. by B. Brentjes, Vol. 2, Halle a. S. 1980, 27–34; Gutas, D. Paul the Persian on the classification of the parts of Aristotle's philosophy: A milestone between Alexandria and Baghdad, *Islam* 60 (1983) 231–267; Hein, C. *Definition und Einteilung der Philosophie. Von der spätantiken Einleitungsliteratur zur arabischen Enzyklopädie*, Frankfurt 1985; Daiber, H. Qosta ibn Luqa (9. Jh.) über die Einteilung der Wissenschaften, *ZGAIW* 6 (1990) 92–129.

8 Wiedemann, E. Einige Biographien von griechischen Gelehrten nach Qifti (1905), *Aufsätze zur arabischen Wissenschaftsgeschichte*, 86–96, cf. 62–77; *The Fihrist of al-Nadim*, transl. by B. Dodge, Vol. 2, New York 1970, 634ff., 673; Pinault, J. R. *Hippocratic lives and legends*, Leiden 1992.

9 Interestingly, until Commandino's edition (1572) Euclid was generally confused with Euclid of Megara, who lived hundred of years earlier.

10 Nutton, V. 'Prisci dissectionum professores': Greek texts and Renaissance anatomists, *The uses of Greek and Latin*, ed. by A. C. Dionisotti et al., London 1988, 111–126; idem. Greek science in the sixteenth-century Renaissance, *Renaissance and revolution*, ed. by J.V. Field, Cambridge 1993, 15–28 (with an extensive bibliography); Grafton, A. From apotheosis to analysis: Some late Renaissance histories of classical astronomy, *History and the disciplines*, ed. by D. R. Kelley, Rochester

Let us turn, for example, to one of the earliest works on the invention of sciences, arts, crafts, etc., the famous *De rerum inventoribus* by Polydore Vergil.[11] For Polydore, as a humanist, *scientiae et artes* were, in the first place, the classical arts and sciences. Of the more than hundred authors he cites, only a few belong to the medieval or to his own period, all the others being Greeks and Romans. The proportion of ancient to modern discoveries (particularly noted among the latter are printing and the invention of cannon) is roughly the same. In the arrangement of his material and his focus on the problem of 'who was the first to invent what?', Polydore follows heurematography, the ancient genre of writings on first discoverers known to him primarily through Pliny.[12] Polydore was one of the first to introduce to European historiography of science notions typical of the Jewish writers of Antiquity and the early Christian apologists who followed them: that the sciences were invented not by the Greeks, but by the Biblical Patriarchs, who lived at a much earlier date.

The famous encyclopaedist of the 16th century, Petrus Ramus, considered at length the history of mathematical sciences in the first book of his *Scholae mathematicae*, based exclusively on classical sources. Like Polydore, he followed Josephus Flavius on the origin of astronomy and arithmetic: both sciences were invented by the Chaldaeans; Abraham taught them to the Egyptians who, in their turn, handed them down to the Greeks.[13] Luckily, Ramus based his overview of Greek mathematics on Proclus' commentary to book I of the *Elements*, which in turn relied on Eudemus' *History of Geometry*.[14] Ramus' curious demand to free astronomy from hypotheses and return to the times when the Babylonians, the Egyptians and the Greeks before Eudoxus foretold celestial phenomena relying on observation and logic alone, goes back to Eudemus' *History of Astronomy*, where Eudoxus figures as the first Greek who ad-

---

1997, 261–276; Siraisi, N. Anatomizing the past: Physicians and history in Renaissance culture, *Renaissance Quarterly* 53 (2000) 1–30; Cifoletti, G. The creation of the history of algebra in the sixteenth century, *L'Europe mathématique*, 121–142.

11 Polydorus Vergilius. *De rerum inventoribus*, Venice 1499. In the course of three centuries, this book was translated into eight languages and ran to more than a hundred editions.

12 On Polydore's ancient sources and his predecessors, see Copenhaver, B. P. The historiography of discovery in the Renaissance: The sources and composition of Polydore Vergyl's *De inventoribus rerum*, Vol. 1–3, *J. of the Warburg and Courtauld Institutes* 41 (1971) 192–222; *Polydore Vergil. On discovery*, ed. and transl. by B. P. Copenhaver, Cambridge, Mass. 2002, vi–xxix.

13 Ramus, P. *Scholarum mathematicarum libri unus et triginta*, Basel 1569, 2. Ramus' four periods in the history of mathematics – Biblical (from Adam to Abraham), Egyptian, Greek, and Latin – soon become a common periodization. See below, 8.

14 Ramus' chronological table of eminent Greek mathematicians (ibid., 41) includes almost all the relevant figures from Thales to Theon of Alexandria. His main source for the pre-Euclidean period is Proclus (Eudemus), but he used also Diogenes Laertius, Iamblichus, and Eutocius (ibid., 6, 7, 9). Interestingly, Ramus refers to the problem of doubling the cube, initiated by Plato (ibid., 12). See below, 3.1.

vanced astronomical hypotheses to 'save the appearances'. Coming as it did in the midst of animated discussions on the status of astronomical hypotheses, this demand was echoed by the leading astronomers of the time, in particular Kepler, who sought confirmation of his views in the texts of the ancients while demonstrating his superiority to them.[15]

The first Renaissance history of medicine, *De medicina et medicis* by G. Tortelli, followed the way paved long before him by Celsus and Pliny.[16] Starting with *De antiquitate medicinae* by his contemporary Bartolotti, the historiography of medicine made increasing use of Galen's material as well as of his notions of medicine's past. Some biographies of scientists and physicians had already been known via Greek and Arabic sources and their Latin translations;[17] during the Renaissance this genre was revived. The first general history of mathematics, written by B. Baldi in 1580s, was a collection of 202 mathematicians' biographies, from Thales to Clavius, patterned after Diogenes Laertius and using a wealth of Greek, Latin and modern sources.[18] Antiquity occupies about two thirds of this voluminous work.

The works of the humanists did not so much investigate the origin and development of arts and sciences as illustrate them with biographies of scientists and doctors, supplementing the latter with bibliographical and doxographical evidence. Chronological outlines briefly describing the achievements of eminent scientists from Antiquity to the present were a fairly popular genre.[19] Among the important tasks of this antiquarian and genealogical history was the enhancement of the status of a given science by demonstrating its antiquity. Thus the majority of early histories of chemistry considered alchemy's claims

---

15 Jardine, N. *The birth of history and philosophy of science. Kepler's "A defence of Tycho against Ursus"*, Cambridge 1984; Jardine, N., Segonds, A. A challenge to the reader: Ramus on Astrologia without Hypotheses, *The influence of Petrus Ramus*, ed. by M. Feingold et al., Basel 2001, 248–266.

16 *Giovanni Tortelli on medicine and phycisians; Gian Giacomo Bartolotti on the antiquity of medicine: Two histories of medicine of the XVth century*, transl. by D. M. Schullian, L. Belloni, Milan 1954.

17 Musitelli, S. *Da Parmenide a Galeno. Tradizioni classiche e interpretazioni medievali nelle biografie dei grandi medici antichi*, Rome 1985 (*A. A. Lincei*, Vol. 28, fasc. 4); Pinault, *op. cit.*

18 Rose, P. L. *The Italian Renaissance of mathematics*, Geneva 1975, 243ff. A part of Baldi's learned work was published posthumously: Baldi, B. *Cronica de' matematici overo Epitome dell' istoria delle vite loro*, Urbino 1707. For biographies of the medieval and Renaissance mathematicians with a commentary and ample bibliography, see Baldi, B. *Le vite de' matematici*, ed. by E. Nenci, Milan 1998.

19 Champier, S. *De medicinae claris scriptoribus in quinque partibus tractatus*, Lyon 1506; Brunfels, O. *Catalogus illustrium medicorum sive de primis medicinae scriptoribus*, Strasbourg 1530; Gaurico, L. Oratio de inventoribus, utilitate et laudibus astronomiae, *C. Ptolemaei Centum sententiae*, ed. G. Trapezuntius, Rome 1540; Clavius, C. Inventores mathematicarum disciplinarum (1574), *Opera mathematica*, I, Mainz 1611.

to antiquity and refutations of them.[20] Hence, the ideas of the development of individual sciences from their legendary 'founding fathers' to the author's day and of the transmission of knowledge from one culture to another are characteristic, in various forms, of the historico-scientific literature of the Renaissance and distinguish it from the medieval Latin genealogies of sciences and arts.[21]

In the 17th century, the number and volume of works on the history of science increases, the range of problems widens, and the subject matter becomes more varied.[22] Voluminous works by such polymaths as Voss combine prodigious learning with uncritical retelling of old legends. It is only natural that, in the period when Hippocrates and Archimedes were topical as never before, a large number of historico-scientific works were directly related to Antiquity.[23] Even works of a more general character devoted the greater part of their attention to this period.[24] Daniel Le Clerc's fundamental *History of Medicine*, the first to go beyond the biographies of famous doctors and annotated lists of their works, finishes at Galen, so that it can rightfully be considered a history of ancient medicine.[25] In the historiography of medicine, this kind of proportion in the se-

---

20 Duval, R. *De veritate et antiquitate artis chemicae*, Paris 1561. On the continuation of this discussion in the 17th–18th centuries, see Weyer, J. *Chemiegeschichtsschreibung von Wiegleb (1790) bis Partington (1970)*, Hildesheim 1974, 17f.

21 Rose, *op. cit.*, 258.

22 See e.g. Moderus, S. J. *Disputatio de mathematicarum disciplinarum origine, Seu primis inventoribus etc.*, Helmstedt 1605; Biancani, G. *De natura mathematicarum scientiarum tractatio, atque clarorum mathematicorum chronologia*, Bologna 1615; Deusing, A. *De astronomiae origine, ejusdemque ad nostram usque aetatem progressu*, Hardwijk 1640; Voss, G. J. *De universae mathesios natura et constitutione liber, cui subjungitur chronologia mathematicorum*, Amsterdam 1650; Glanvill, J. *Plus ultra: or the progress of knowledge since the days of Aristotle*, London 1668; Borrichius, O. *De ortu et progressu chemiae dissertatio*, Copenhagen 1668; Dechales, C. F. M. *Cursus seu mundus mathematicus. Pars I. Tractatus prooemialis, de progressu matheseos et illustribus mathematicis*, T. 1, Leiden 1690, 1–108; Cassini, D. De l'origine et du progrès de l'astronomie (1693), *Mémoires de l'Académie Royale des Sciences* 8 (1730) 1–52.

23 Biancani, G. Aristotelis loca mathematica ... *atque Clarorum mathematicorum chronologia*, Bologna 1615; Molther, J. *Problema Deliacum, de cubi duplicatione*, Frankfurt 1619; Beverwyick, J. van. *Idea medicinae veterum*, Leiden 1637; Nottnagel, C. *De originibus astronomiae*, Wittenberg 1650; Schmidt, J. A. *Archytam Tarentinum dissertatione historica-mathematica*, Jena 1683; idem. *Archimedem mathematicorum principem dissertatione historico-mathematica*, Jena 1683; Valentini, M. B. *Medicina nov-antiqua, h.e. cursus artis medicae e fontibus Hippocratis*, Frankfurt 1698.

24 Riccioli, G. B. Chronicon duplex astronomorum, astrologorum, cosmographorum et polyhistorum, *Almagestum novum astronomiam veterem novamque complectens*, Bologna 1651; Boulliau, I. *Astronomia Philolaica ... Historia, ortus et progressus astronomiae in prolegomenis describitur*, Paris 1645.

25 Le Clerc, D. *Histoire de la médicine*, Geneva 1696. In the subsequent editions, Le

lection of material shows up quite frequently until the end of the 18th century. Thus, of the 33 chapters of Ackermann's history of medicine, 26 deal with Antiquity, 3 with the Arabs, 3 with the school of Salerno, and only one considers "the revival of Galen's and Hippocrates' medicine in Europe".[26]

The general history of a science, mathematics for example, was normally divided into the following periods: the mathematics of the Jews, starting with antediluvian times; the mathematics of the Egyptians and Babylonians, who received it from the Jews (an account of this was already based on Greek sources); the mathematics of the Greeks, who borrowed it from the Egyptians and Babylonians; the mathematics of the Arabs, who inherited it from the Greeks; etc.[27] As we have already noted, this perspective derives from early Christian writers and, in particular, from Clement of Alexandria and Eusebius, who tried, in the wake of such Jewish authors as Aristobulus, Philo, and especially Josephus Flavius, to combine the Bible with the doctrines of ancient philosophy.[28] After an account of fabulous discoveries made by Seth, Abraham, or Moses, the historians finally passed on to Thales and the Greek tradition, where they could rely on more dependable sources and demonstrate not only their learning, but their critical sense as well. With time, this perspective shifts progressively to the pagans, so that the biblical theme slowly but irrevocably disappears from works on the history of science.[29]

Yet within the limits of ancient Greek tradition, too, a clear boundary between mythologized and real history was lacking until the late 18th century. Le Clerc, following the authority of Celsus and Galen, started his history with Asclepius, not with Hippocrates. The solid *Historical Dictionary of Ancient and Modern Medicine* includes, along with the biographies of eminent doctors, articles on Asclepius and the centaur Chiron.[30] Even such an authority on the history of astronomy as Bailly still regarded Atlas, Zoroaster, and Uranus as the first astronomers.[31] To be sure, much depended on individual preferences. Thus

---

Clerc, influenced by the critics, added a brief survey of the history of medicine until the 16th century.

26 Ackermann, J. C. G. *Institutiones historiae medicinae*, Nuremberg 1792.

27 See e.g. Weidler, J. F. *Historia astronomiae, sive de ortu et progressu astronomiae liber singularis*, Wittenberg 1741. Cf. above, 5 n. 14.

28 Worstbrock, F. J. Translatio artium. Über Herkunft und Entwicklung einer kulturhistorischen Theorie, *ArKult* 47 (1965) 1–22. Cf. below, 8.3.

29 For growing criticism of the concept of *prisca sapientia* in the 18th-century historiography of philosophy, see Blackwell, C.W. T. Thales Philosophus: The beginning of philosophy as a discipline, *History and the disciplines*, 61–82.

30 Eloy, N. F. J. *Dictionnaire historique de la médicine ancienne et moderne*, T. 1–4, Liège 1755.

31 Bailly, J. S. *Histoire de l'astronomie ancienne depuis son origine jusqu'à l'établissement de l'École d'Alexandrie*, Paris 1775, 4. Bailly's curious idea of a source common to all the astronomies of Antiquity, which he identified with Atlantis (Pasini, M. *L'astronomie antédiluvienne*: Storia della scienza e origini della civiltà in J.-S. Bailly, *Studi settecenteschi* 11–12 [1988–89] 197–235), is similar to the thesis of a

Baldi (1589) opened his collection with Thales' biography, Biancani (1615) decided not to mention Atlas, Zoroaster, Orpheus, Linus, etc., because they were legendary figures impossible to date, while Montucla, even in the second edition of his famous *History of Mathematics* (1799), could not get rid of Thoth as the inventor of mathematics.

The 18th century, and its second half in particular, saw the rapid growth of literature on the history of science, which numbered hundreds of solid volumes.[32] As science itself developed and became more specialized, the chapters on ancient science in general treatises grew shorter, remaining, however, subject matter for most studies.[33] Moreover, the number of special works on ancient science grew at least as fast as that of writings based on the material of European science alone.[34] In the middle of the 18th century, a historian of mathematics could still allow himself to restrict his work to the biographies of ancient scientists.[35] Many writers continued to borrow from their Greek and Roman teachers not only evidence, but also the problems to be considered in the history of science. Of still greater importance than these particular borrowings was the perspective itself, in which ancient science continued to be an integral part of science as such, remaining, in this sense, *modern* until at least the end of the 18th century. Admittedly, the new type of historiography emerging at the threshold of the 19th century departs not from Greek science as such (the number of

---

well-known modern mathematician and historian of science who found the common ground of all the ancient mathematical traditions in the megalithic culture of the third to second millennium BC (Waerden, B. L. van der. *Geometry and algebra in ancient civilizations*, Berlin 1983).

32 The bibliography of works published from 1750 to 1800 on the history of mathematics alone includes about 200 titles (Cantor, M. *Vorlesungen über die Geschichte der Mathematik*, 2nd ed., Vol. 4, Leipzig 1901, 1–36).

33 See e.g. Heilbronner, J. C. *Historia matheseos universae a mundo condito ad seculum p. C. n. XVI*, Leipzig 1742; Weidler, *op. cit.*; idem. *Bibliographia astronomica*, Wittenberg 1755.

34 See e.g. Taelpo, S. *Scholium mathematicum de geometriae origine*, Aaboe 1700; Krebs, J. A. *Dissertatio de originibus et antiquitatibus mathematicis*, Jena 1727; Schulze, J. H. *Historia medicinae a rerum initio ad annum urbis Romae DXXXV deducta*, Leipzig 1728; Costard, G. *A letter concerning the rise and progress of astronomy amongst the antients*, London 1746; idem. *A further account of the rise and progress of astronomy amongst the antients*, Oxford 1748; Fabricius, J. A. Elenchus medicorum veterum, *Bibliotheca graeca*, Vol. 13, Hamburg 1746, 15–456; Neubronner, T. *Historiae zodiaci, sectio prima: de inventoribus zodiaci*, Göttingen 1754; Rogers, F. *Dissertation on the knowledge of the ancients in astronomy and optical instruments, etc.*, London 1755; Goguet, A. J. de. *De l'origine des loix, des arts et des sciences et de leurs progrès chez les anciens peuples*, Vol. 1–3, Paris 1758; Reimer, N. T. *Historia problematis de cubi duplicatione*, Göttingen 1798.

35 Frobesius, J. N. *Rudimenta biographiae mathematicae*, T. 1–3, Helmstedt 1751–1755. Though this series was interrupted by the author's death, an impartial opinion on another book of his (Cantor, *op. cit.*, Vol. 3, 499) shows that as a rule he hardly ever ventured beyond the Arabs.

works related to it continued to grow century after century),[36] but from ancient notions of its history that failed to stand up to critical examination. A case in point is the works of J. K. Schaubach and especially of C. L. Ideler,[37] which combine competent historico-philological criticism with professional expertise in astronomy. Due to such an approach, the history of ancient science finally managed to take a necessary distance from its subject.

Despite all the individual and typical features of the historico-scientific works of Antiquity, the Middle Ages, the Renaissance and the Enlightenment, they can be usefully regarded as in many ways a single tradition that paved the way for the modern historiography of science. The influence of classical patterns on the formation of Arabic and, later, European historiography of science is a subject of special studies. For our purposes, suffice it to state that the search for the beginnings of the historiography of science leads much further back than the Renaissance epoch. Even the texts of the compilers and commentators of late Antiquity that served for centuries as the main sources on the history of ancient science are but an intermediary instance. Five hundred years of studies in Greek science have not passed in vain. The historians of science have long been aware that the pioneer research in the history of knowledge was initiated by Aristotle and carried out by his pupils. It is from them, or, more precisely, from their sources that the study of the origin of the history of science should take its start.

2. The historiography of science in Antiquity

The history of science belongs to the series of historiographical genres that emerged at the Lyceum. Along with biography, whose first specimens were produced by Aristoxenus and Dicaearchus, still another genre popular in Antiquity was born here: the systematic account of doctrines on natural philosophy, known as doxography. Studies in doxography took their start from H. Usener's dissertation (1858) and the fundamental *Doxographi Graeci* (1879) by his pupil H. Diels. In recent decades they have been actively carried on by J. Mansfeld and D. Runia.[38] The history of science appeared to be eclipsed here as elsewhere by other historiographical genres; apart from occa-

---

36 For an overview of the achievements and tendencies in the historiography of the past two centuries, see Krafft, F. Der Wandel der Auffassung von der antiken Naturwissenschaft und ihres Bezuges zur modernen Naturforschung, *Les études classiques aux XIXe et XXe siècles: Leur place dans l'histoire des idées*, ed. by W. den Boer, Geneva 1980, 241–304 (*Entretiens Fondation Hardt*. T. 26).

37 Schaubach, J. K. *Geschichte der griechischen Astronomie bis auf Eratosthenes*, Göttingen 1802; Ideler, C. L. *Historische Untersuchungen über die astronomischen Beobachtungen der Alten*, Berlin 1806.

38 See e.g. Mansfeld, J. Aristotle, Plato, and the Preplatonic doxography and chronography, *Studies in historiography of Greek philosophy*, Assen 1990, 22–83; Mansfeld, J., Runia, D. *Aëtiana: The method and intellectual context of a doxographer*, Vol. 1, Dordrecht 1997.

sional notes scattered through the works on Greek astronomy and mathematics, not a single serious study has so far been written on it.

The reasons for this have already been suggested above. In Antiquity, the history of philosophy and the history of medicine were parts of philosophy and medicine respectively. The problems posed by Plato and Hippocrates continued to preoccupy philosophers and physicians until Greek philosophy and medicine ceased to exist – hence the large number of writings on these subjects, some of which are still extant. The works directly related to the genre of the history of science were obviously much fewer in number, and very few of them survive in fragments. Apart from them, at our disposal is the vast historico-scientific *material* found in texts of different genres. Sundry as our sources are, they are certainly not scarce and the evidence they bring shows many features in common. Though the historiography of science has not taken shape as a viable genre, the existence of a historico-scientific *tradition* lasting from the classical period until the last centuries of Antiquity is beyond doubt.

The Peripatetic works related to the history of science have been studied since the mid-19th century, with a focus on important testimonies they contain. Indeed, those who accept what Eudemus reports on Thales' geometry, or Theophrastus on Anaximander's astronomy, or Aristoxenus on Pythagoras' arithmetic have quite a different view of the early Greek science from those who reject this evidence. But the problem lies not so much in the assessment of separate fragments or individual authors as in the general approach to the Peripatetic historiography and its separate branches – doxography, the history of science, biography, etc. When reconstructing early Greek science, we are compelled to rely not on original sources, but on preserved historico-scientific evidence. As a result, our knowledge of it remains largely dependent on what was regarded as science by the Peripatetics themselves, what, where, and in what way they actually recorded, and what they neglected. The main conceptual approaches to science, which predetermined for many centuries to come the comprehension of this phenomenon in Antiquity and in the modern period, were established in the fourth century BC. The comparison of Plato's and Aristotle's views on science with modern conceptions of it has repeatedly proved to be fruitful: the differences between them allow us to grasp the specificity of the approach to science at different times, while the common features demonstrate the invariable nature of the phenomenon itself. It is important, however, to consider Plato's and Aristotle's positions in the context of the discordant opinions that existed in Antiquity, particularly the opinions of those who created the science of the time first-hand.

A terminological remark is needed here. 'Greek science' in this book is mostly confined to the exact sciences – geometry, arithmetic, astronomy, and harmonics, though in some contexts 'science' inevitably takes on a broader meaning. It is in the realm of the exact sciences that we find the closest possible match between ancient and modern *concepts* of what science is as well as between ancient and modern *practice* of scientific research. From the late fifth century BC on, the Greeks called the exact sciences by a special term μαθήματα and clearly distinguished them from the other intellectual pursuits, e.g. from physics, which they considered a part of philosophy. Unlike many other disciplines practiced by the Greeks, the exact sciences, joined in the fourth century BC by optics and mechanics, reached a truly scientific level in Antiquity. Their special status is confirmed by the fact that Greek historiography of science deals only with *mathēmata*; no other scientific discipline became a subject of a historical work, though histories of medicine were written. Thus, Greek historiography of science gives further justification for our rather restrictive treatment of Greek science, which proceeds from the modern concepts but tries to pay due attention to the ancient ones.

The idea that the history of science allows us to trace the development of the human mind in a more reliable and spectacular way than any other kind of history was repeatedly expressed in the age of Enlightenment.[39] The 20th century provided a corrective to this idea, giving it more precision: the progress of knowledge is best studied by tracing the growth of scientific knowledge.[40] If science constitutes the best embodiment of the progress of knowledge, its historiography can be usefully considered an example of changing notions of knowledge, science, and progress, an integral part of intellectual and cultural history.

Thus, our research aims not only at collecting the most important evidence related to the origins of the historiography and methodology of science in Antiquity, but also at answering the following questions. What was the socio-cultural context in which the history of science emerged? What do the main approaches to science that found expression in the Peripatetic historiography stem from? Did classical Antiquity comprehend science as a special form of cognitive activity, and did this comprehension find its expression in the historiography of science? To what extent did the Greek historiography of science constitute a historical analysis of the development of knowledge? Did it proceed from philosophical premises, or remain purely descriptive? What was the fate of the historiography of science in the Hellenistic period? Why did it fail where doxography succeeded in creating a stable and popular genre?

Partly anticipating the analysis of the aforementioned problems, let us give a general overview of the tradition under study. Among its first landmarks was the trend in Greek thought that sought an answer to the popular question of 'who discovered what'. By this trend, I mean the early heurematography of the sixth and fifth centuries BC, which treated most different elements of culture as discoveries (εὑρήματα) and showed interest in their first discoverers (πρῶτοι εὑρεταί). At the beginning of the fourth century, it gave birth to a special genre, a sort of 'catalogues of discoveries', which survived until the very end of Antiquity and later provided a model for Arabic and European writers.[41]

---

39 See e.g. Montucla, J.-E. *Histoire des mathématiques*, Vol. 1, Paris 1758, viii.

40 Popper, K. R. *The logic of scientific discovery*, London 1959, 15.

41 Kleingünther, A. ΠΡΩΤΟΣ ΕΥΡΕΤΗΣ, Leipzig 1933; Wiedemann, E. Über Er-

Originally the quest for *prōtoi heuretai* as a sort of intermediary link between the past and the present had little to do with history. It can rather be termed a rationalization of the mythical past, the more so because the *heuretai* themselves were often legendary and mythical figures. Still, the tradition, though connected with myth, was nourished by genuine interest in the real authors of cultural innovations – poets, musicians, inventors, sages. It is due to these innovations, whose fame started to spread throughout the Greek world from the seventh century on, that every element of culture came with time to be regarded as someone's discovery. Since the mid-sixth century, the various discoveries mentioned in heurematography include scientific ones as well. The attention the Greeks paid to questions of priority, which had so large a part in the formation of Greek science in general, helped to save the memory of such important discoveries as, for example, Thales' prediction of the solar eclipse.[42] Admittedly, the Peripatetic history of science is linked with heurematography more intimately than by just employing it as a source of information on scientific discoveries. In Eudemus' history of the exact sciences, the traditional question of 'who discovered what' remains among the most important. As the earliest forerunner of the Peripatetic history of science, heurematography undoubtedly deserves to be considered in detail.

One of the characteristic features of the search for *prōtoi heuretai* consisted in ruling out the possibility that the same discovery had been made twice.[43] Although heurematography did often mention several authors for one and the same discovery, it implied that only one of these versions was true. Astronomy was discovered either by the Egyptians, or by the Babylonians, or by the Phoenicians, but it could not emerge independently in several cultures at once. The emergence and spread of cultural phenomena was conceived of within the narrow framework of the 'learning (imitation) – discovery' formula: the new could either be learned from another, or found independently. Any thing that showed a superficial similarity with another, earlier one, could be declared a borrowing. This 'naive diffusionism' resulted in a bias toward according priority in the invention of sciences to the Orient, especially since the Greeks were

---

finder nach arabischen Angaben, *Gesammelte Schriften zur arabisch-islamischen Wissenschaftsgeschichte*, Vol. 2, Frankfurt 1984, 848–850. – The history of technology continued to exist in the form of the catalogues of discoveries until the late 18th century: Beckmann, J. *Beyträge zur Geschichte der Erfindungen*, Vol. 1–5, Leipzig 1783–1805; Busch, G. C. *Versuch eines Handbuchs der Erfindungen*, Vol. 1–8, Eisenach 1790–1798.

42 The first to mention Thales' discoveries was Xenophanes (21 B 19 = Eud. fr. 144). Part of the evidence on Thales as *prōtos heuretēs* (D. L. I, 23–27) goes back to the oral tradition of the sixth century.

43 Πρῶτος εὑρετής was always μόνος, the possibility of the existence of δεύτερος seems never to have been seriously considered (Kleingünther, *op. cit.*, 57 f.). Emerging later, however, was the motif of bringing the first discoveries to perfection (see below, 2.5).

well aware of the youth of their culture in comparison with the Egyptian or Babylonian. Strengthened by Jewish and early Christian authors, who derived Greek philosophy and science from the Pentateuch, this tendency not only prevailed in early modern historiography, but repeatedly came into the foreground even in the 19th–20th centuries.[44] Taking this tendency into account will make the analysis of the ancient evidence on the Oriental origins of sciences the more instructive.

In the second half of the fifth century, interest shifts gradually from individual discoveries to the emergence of whole branches of knowledge and skills (τέχναι) and, later, to the origin and development of culture as a whole. The history of individual τέχναι (for example, music and poetry) and philosophical doctrines on the origin of culture as the sum total of τέχναι were directly influenced by heurematography. Still more important was the Sophistic theory of τέχνη: undertaken within its framework were the first attempts at analyzing scientific knowledge, such as the Hippocratic treatise *On Ancient Medicine* and Archytas' work *On Mathematical Sciences*, which considered scientific knowledge from both methodological and historical points of view.

In this period, the exact sciences (geometry, arithmetic, astronomy, and harmonics), though comprising a separate group among other τέχναι, had not yet become a model of science conceived of as ἐπιστήμη. The transition from science-τέχνη to science-ἐπιστήμη is largely associated with Plato, who created a theory of knowledge modeled on mathematics. According to Plato, the chief aim of ἐπιστήμη consists, not in serving society's practical needs, but rather in knowledge as such, which is the worthiest occupation of a free man. The paradigmatic character of mathematics in Plato's teaching left little place for interest in its history, and most of Plato's mathematical passages important for the history of science do not yield to simple interpretation. The numerous works on exact sciences written by the Academics Speusippus, Xenocrates, Philip of Opus, etc., were also oriented toward the systematic account of scientific knowledge, rather than its history. At the same time, the Platonists showed an interest in tracing the effect of their teacher on the development of science: an Academic legend assigned to Plato the role of an 'architect of science' who posed the main problems for mathematicians and defined the methods they should use.

The central part of this book is concerned with the first generation of Aristotle's pupils – Eudemus, Theophrastus, Aristoxenus, and Meno. Particularly interesting is the Peripatetic historiographical project, which aimed at the collection, systematization, and preliminary analysis of material related to the

---

44 Zhmud, L. *Wissenschaft, Philosophie und Religion im frühen Pythagoreismus*, Berlin 1997, 141ff., 202ff. This tendency is also visible in the recent discussions of the origins of Greek culture: Bernal, M. *Black Athena. The Afroasiatic roots of classical civilization*, Vol. 1–2, New Brunswick 1987–1991. Cf. Palter, R. Black Athena, Afro-Centrism, and the history of science, *HS* 31 (1993) 227–267.

knowledge accumulated by the Greeks. The exploration of different kinds of knowledge was distributed among the Peripatetics in accordance with Aristotle's division of theoretical sciences into mathematics, physics, and theology. The methods of organizing, describing, and analyzing the material used within the framework of the project were different for each particular science. Particularly important for us are geometry, arithmetic, and astronomy, whose histories Eudemus considered in three special treatises. He placed the discoveries of Greek mathematicians in chronological sequence, starting with Thales and ending with Eudoxus' pupils, who were his own contemporaries. In many of its aspects, the plan of Eudemus' histories closely followed Aristotle's favorite idea of all arts and sciences as gradually approximating to perfection. His description of scientific discoveries and methods was, however, based on the intrinsic criteria of exact sciences rather than on philosophical premises.

Whereas Eudemus wrote of mathematicians and their discoveries, the doctrines of physicists were treated in Theophrastus' fundamental work *Opinions of the Natural Philosophers* (Φυσικῶν δόξαι). Along with purely philosophical problems, this treatise included mainly those we associate with natural sciences (cosmology, physics, meteorology, physiology, etc.). Meno's *Medical Collection* ('Ιατρικὴ συναγωγή), dealing with medical theories of the fifth and fourth centuries, is linked to physical doxography and followed its methods of organizing material. This work was concerned, not with discoveries in the field of medicine, but with the theories of doctors and certain physicists on the causes of diseases. Extant from Aristoxenus' work *On Arithmetic* is a single fragment, which holds some interest for the history of science.

The historiography of science flourished only for a short period. With the decline of the Lyceum in the third century BC, the development of the genre seems to have come to a standstill. Let us give a brief outline of some other genres. While the biographies of philosophers who also pursued science do occasionally include some evidence of their discoveries, biographies of 'pure' scientists are practically unknown to us. Eratosthenes' introduction to his *Geography* includes a short historical overview of this science, and his *Platonicus* is a literary version of the history of solving the problem of doubling the cube. In his *Theory of Mathematical Sciences*, Geminus (first century BC), who is traditionally considered an intermediary between Eudemus and late Antiquity, paid principal attention to the methodology and philosophy of mathematics; his evidence on individual mathematicians lacks historical context. Pappus limited his voluminous *Collectio* (ca. 320 AD) to purely professional tasks, but dealt with mathematical problems of the past as an anthologist, not as a historian. A commentary on Euclid's book I by Proclus (fifth century AD), concerned as it was with mathematics as such and, to a smaller extent, its history, paid particular attention to the philosophy and the theology of mathematics. Commentaries on Archimedes' works by Eutocius (sixth century AD) include selected solutions to the famous geometrical problems of Antiquity.

That the history of Hellenistic astronomy or mathematics, which could have been of inestimable help to us, has never been written is, naturally, disappointing. Our problems, however, did not concern the ancients. They were writing for themselves, their disciples and contemporaries and could hardly imagine that, of the total scientific literature, only one-tenth at best – if not one-fiftieth – would ultimately survive. With time, the growing awareness of science's gradual decay spurred them to find and collect as much evidence of ancient science as possible. Since the first century AD, we find in many authors (Dercyllides, Theon of Smyrna, Anatolius, Porphyry, Proclus, etc.) lists of mathematicians and astronomers with their discoveries, which were borrowed from Eudemus. The most extensive excerpts from Eudemus' writings are found in Simplicius, the Neoplatonic commentator of the sixth century AD. This antiquarian trend, though helpful in salvaging what would otherwise have been irretrievably lost, did not bring about the revival of the history of science. Eudemus had to wait for his followers for many hundreds of years.

3. Greek notions of science and progress

Eudemus' *History of Geometry* and *History of Astronomy* show approaches to science and to the selection of material rather close to serious studies of modern times.[45] We do not find in them either legends and anecdotes, or a particular interest in the philosophy and theology of mathematics, or any inclination to the number mysticism characteristic of the Platonists, for example. They are concerned exclusively with scientific discoveries, with the development of new theories and methods carried out within the framework of the professional community – *mathemata mathematicis scribuntur*. This trait of the Peripatetic historiography of science is determined, ultimately, by the fact that the conception of exact sciences formed in the Lyceum appeared to be very close to the views of the mathematicians themselves. Since the dependence of historiography on the general notions of science is quite obvious, our book also considers such problems as the comprehension of science by scientists themselves, notions of boundaries between the exact sciences and natural philosophy, the philosophy and methodology of science in the Academy and the Lyceum, the classification of sciences, etc.

Outlining briefly this range of problems, let us touch upon the quite animated discussion several decades ago of whether there was in classical Greece a notion of progress, and specifically of scientific progress, i.e., the idea of a

45 Eudemus understood the history of mathematics as a chain of discoveries that links scientists to each other; cf. "Geschichte einer Wissenschaft ist meines Erachtens: wie ihre Lehren sind entdeckt, bekannt gemacht, bestimmt, berichtiget, dargetan, erläutert, angewandt worden." (Kästner, A. G. *Geschichte der Mathematik seit der Wiederherstellung der Wissenschaften bis an das Ende des 18.Jh.s*, Vol. 1, Göttingen 1796, 13).

steady growth of knowledge. L. Edelstein demonstrated quite convincingly that classical Antiquity was not unaware of the idea of progress.[46] One should beware, however, of taking an idea for an ideology. The popular 19th-century conviction that in the *future* we will experience *constant improvement in all spheres of human life* is not to be found in Antiquity. The Greek notion of progress was based preeminently, though not exclusively, on achievements in human knowledge and technology and hence proved much more limited than the 19th-century one.[47] Even granting that some notions of progress current at the time did include the idea of steady social and moral improvement, it was the real achievements of the *past* and the *present*, not imaginary future prospects, that the Greeks were concerned with.[48] Such a view, free of the 'totality' of the 19th-century progressivist ideology and its eager anticipation of the future,[49] could comfortably coexist with a cyclic conception of history as, for example, in Aristotle or, later, in Jean Bodin.[50]

The limited or, rather, realistic character of the classical idea of progress is due, first of all, to the difference in scale between the actual changes that took place in ancient Greece and in Western Europe respectively in the eighth–fourth centuries BC and in the 15th–19th centuries AD.[51] We should keep in mind that the idea of progress made its first appearance only three hundred years after the emergence of writing in Greece and less than one hundred years after the origin of philosophy and science. In Europe, which had infinitely more opportunities to ascertain the steady character of progress, this idea took root only after the French Revolution and the beginning of the Industrial Revolution. The limited character of ancient notions of progress underscores their scientific and, in a larger sense, cognitive component, which was not questioned even by those who, on the whole, denied the existence of such notions in classical Greece.[52] Without the idea of the progressive growth of knowledge, the history of science would have hardly come about, and we have abundant evidence that the science of the past and the present was indeed described by the Greeks in the terms of 'progress'. In the fifth–fourth centuries BC, the idea

---

46 Edelstein, L. *The idea of progress in classical Antiquity*, Baltimore 1967.

47 The notions of progress that reappeared in the 16th–17th centuries were chiefly based on the same two components (Edelstein, *op. cit.*, XIX n. 24; Koselleck, R. Fortschritt, *Geschichtliche Grundbegriffe*, ed. by O. Brunner et al., Vol. 2, Stuttgart 1975, 392).

48 Thraede, K. Fortschritt, *RLAC* 7 (1965) 162; Meier, C. 'Fortschritt' in der Antike, *Geschichtliche Grundbegriffe*, 354.

49 The 20th century is characterized by a notable decline of the progressivist ideology: Nisbet, R. *History of the idea of progress*, New York 1980, 317ff.

50 Bodin, J. *Method for the easy comprehension of history*, New York 1945, 296ff.

51 Meier, C. Ein antikes Äquivalent des Fortschrittsgedankens: das "Könnens-Bewusstsein" des 5. Jh.s v. Chr., *HZ* 226 (1978) 265–316.

52 See Edelstein, *op. cit.*, XX n. 27; Boer, W. den. Progress in the Greece of Thucydides, *MKNAdW* 40.2 (1977).

of progress was most often denoted by the word ἐπίδοσις; emerging later was the notion προκοπή, whose Latin analogue, *progressus*, has entered all modern European languages.[53]

The idea of the progressive growth of knowledge (as well as many others) can well be expressed without being labeled with a specific term.[54] The lack in Greek of a special term for science as a whole, as distinguished from its individual branches, is hardly crucial, either. Considering this fact, some scholars still argue that, in Antiquity, science in the modern sense of the word did not exist; others, that, in the early period at least, it was not distinguished from philosophy, both having been denoted by the same term, ἐπιστήμη. Even after their separation, which is believed to have taken place at the end of the classical period or even later, philosophy continued to exert on science, including mathematics, a much greater influence than it has in modern times, and (according to this view) the differences between them went unnoticed by the Greeks.

An ancient language's possession of a term denoting a field of creative activity as precisely as a modern term is hardly indispensable for the flourishing of this field. The Greeks did not have such terms for, say, art and literature. The absence of minimally elaborated terminology could, indeed, constitute a serious obstacle for the *analysis* of science, for its methodology and historiography. Yet the corresponding Greek terms for the well-ordered areas of knowledge appeared by the early fourth century at the latest;[55] some of them are of much earlier date. In the second half of the fifth century, the educational curriculum in mathematics starts to include arithmetic, geometry, astronomy, and harmonics, which were termed μαθήματα, branches of learning or areas of knowledge. In the late fifth to early fourth centuries this term came to mean mathematics as such. That ἐπιστήμη could denote both *mathēmata* and philosophy did not in the least identify the latter with mathematics.

The idea of the original syncretism of philosophy and science stems partly from terminological confusion: physics, pursued, according to Aristotle, by the Presocratics, is indiscriminately termed both *natural philosophy* and *natural science(s)*. But this confusion apart, such syncretism seems to me hardly plausible because of the fundamental epistemological heterogeneity between philosophy and science, which in the final analysis can be reduced to the following.[56]

---

53 Edelstein, *op. cit.*, 146; Thraede. Fortschritt, 141ff.; Meier. 'Fortschritt', 353.

54 See e.g. Xenophanes (21 B 18) and below, 1.3.

55 γεωμετρία denoted geometry, ἀστρολογία and ἀστρονομία astronomy, λογισμός and λογιστική arithmetic, ἁρμονική harmonics, μηχανική mechanics, ὀπτική optics, ἰατρική medicine, περὶ φύσεως ἱστορία and φυσιολογία natural philosophy or natural science.

56 See Fritz, K. von. *Grundprobleme der Geschichte der antiken Wissenschaft*, Berlin 1971, 3ff.; Zaicev, A. The interrelationship of science and philosophy in Antiquity, *Selected papers*, ed. by N. Almazova, L. Zhmud, St. Petersburg 2002, 403f. (*in Russian*). See also Zhmud, L. Die Beziehungen zwischen Philosophie und Wissenschaft in der Antike, *Sudhoffs Archiv* 78 (1994) 1–13.

Scientific problems are sooner or later solved, if they are correctly posed, or withdrawn, if they are incorrectly posed, while genuinely philosophical problems (not technical ones, like those of logic) have never received generally recognized and irrefutable solutions. Since there are no *a priori* reasons to believe that the differences between philosophy and science change with time, we may expect them to have manifested themselves already at the earliest period of their synchronic development in ancient Greece. The development of mathematics and astronomy would, indeed, have been impossible had each of them not singled out a special class of problems solvable by specific methods, i.e., the axiomatico-deductive and the hypothetico-deductive method respectively. The achievements made by these sciences by the end of the fifth century show that Greek scientists succeeded very early in isolating solvable problems and developing adequate methods of dealing with them. Owing to this, the exact sciences, unlike the natural sciences, appeared to be independent of contemporary philosophy and were not perceived as a part of it.[57] Astronomy, which originally included a natural-philosophic component, was divided by the end of the fifth century *de facto* into cosmology, pursued by philosophers, and the mathematical theory of the motions of heavenly bodies, which was the domain of trained specialists, μαθηματικοί. By the mid-fifth century, professionalisation becomes quite pronounced: the mathematicians Hippocrates of Chios, Theodorus, Theaetetus, and the astronomers Oenopides, Meton, and Euctemon have left practically no traces of philosophical preoccupations. In the fourth century, the same can be said of Eudoxus' numerous pupils; Eudoxus himself participated in some Academic philosophical discussions, but left no works on these subjects. It is from this situation that Aristotle and his students proceeded, clearly formulating the difference between *mathēmata* and physics and consistently following the distinction in their works. Thus, even the first two centuries of the development of philosophy and the exact sciences do not confirm the idea of their original syncretism. Neither is this idea proven by cases in which science and philosophy come to be joined in one person (Thales, Pythagoras, Archytas); modern history provides still more examples of such 'personal union' (Pascal, Descartes, Leibnitz, Russell, etc.).

Presocratic natural philosophy did, in fact, study problems that we regard as related to physics, meteorology, or biology, while the medicine, botany, and zoology of the classical period stayed, in their turn, under the strong (though not equally intense) influence of philosophical doctrines. Ancient physics remained part of philosophy to the very end – but that is precisely why it never

---

[57] "I am convinced that the mathematical studies were autonomous, almost completely so, while the philosophical debate, developing within its own tradition, frequently drew support and clarification from mathematical work ... My view conforms to what one may observe as the usual relation between mathematics and philosophy throughout history and especially recently." (Knorr, W. R. Infinity and continuity: The interaction of mathematics and philosophy in Antiquity, *Infinity and continuity in ancient and medieval thought*, ed. by N. Kretzmann, Ithaca 1982, 112).

became a science. Nevertheless, in some fields of physics the Greeks succeeded in isolating the particular problems that they were equipped to solve and raised their research to the scientific level. As a rule, these were fields in which experimentation proved comparatively simple and its results could be expressed in mathematical form: acoustics, optics, mechanics, statics, and hydrostatics.[58] Interestingly, the Greeks themselves related these fields not to physics (≈ philosophy), but to *mathēmata* (≈ science). In spite of the great number of discoveries and the wealth of accumulated material, other branches of natural science were not able to cross the boundary between pre-science and science until the modern epoch.

These remarks are not meant to deny the obvious fact that the ancient division of the cognitive and – in a larger sense – the cultural space is often remarkably different from that accepted at present. In the classical period that particularly concerns us now, culture was usually understood as the sum total of πᾶσαι τέχναι, while the word τέχνη itself could equally refer to mathematics and poetry, medicine and pottery. The term ἐπιστήμη meant 'firm knowledge' and was, hence, the closest to the notion of science in the modern sense. It was far, however, from embracing all kinds of scientific knowledge: according to Plato, it did not include Presocratic φυσιολογία and μετεωρολογία. The term ἐπιστήμη, on the other hand, could denote not only astronomy, but also rhetoric and even ironwork. According to Aristotle, theoretical sciences included theology (first philosophy), physics, and mathematics, and each of them could be indiscriminately referred to as ἐπιστήμη or φιλοσοφία. *Mathēmata*, which numbered originally among τέχναι, in the fourth century came to include mechanics and optics (which passed in the modern period into the domain of physics) and were normally referred to as ἐπιστῆμαι. At the same time, the four *mathēmata* entered the educational canon (ἐγκύκλιος παιδεία, *artes liberales*), formed by the time of Hellenism, the other three parts of which – rhetoric, grammar, and dialectic – were usually related to as τέχναι.[59]

There is no need to multiply these examples. It is obvious enough that the problem cannot be reduced to a trivial terminological discrepancy, for instance, that in the early period astronomy bore the name of ἀστρολογία, while the astrologers of late Antiquity were called μαθηματικοί. What we face here is a different configuration of forms and results of creative and, in particular, cognitive activity deeply rooted in linguistic, cultural, and philosophical tradition. Having assimilated and modified this tradition, the Academics and later the Peripatetics failed to eliminate most of the contradictions inherent in it. As a result, they often indiscriminately applied the same notion to different fields and denoted the same field by different notions, and the fields themselves tended to

58 Lloyd, G. E. R. *Early Greek science: Thales to Aristotle*, London 1970, 30f., 139f.

59 Fuchs, H. Enkyklios paideia, *RLAC* 5 (1962) 365–398; Hadot, I. *Arts libéraux et philosophie dans la pensée antique*, Paris 1984. Hadot's dating of this canon in the Imperial age seems too late to me.

overlap. But here, as in the case of the syncretism of philosophy and science, we should not exaggerate the importance of differences between ancient and modern terminology and the classification of sciences, reducing the history of ideas to a superficially understood history of terms. The fact that in the epoch of Plato and Aristotle and much later, geometry and ironwork were both denoted by the word τέχνη, does not at all mean that the Greeks had difficulties distinguishing between them. When necessary, language always finds means to distinguish between things called by the same words: thus, ironwork was related to βάναυσοι τέχναι and geometry to λογικαὶ τέχναι.

Comparing the ancient classification of arts and sciences (ἐπιστήμη–τέχνη, *scientia–ars*) with the modern one, we should bear in mind that the latter took its final shape only during the 19th century, after more than three centuries of rapid scientific progress. Earlier it was the ancient, basically Aristotelian, canon that everywhere remained in use. It is to this canon that we owe much of the confusion, both in ancient and modern languages, about what belongs to the 'sciences' and what to the 'arts'. Reflections on the general category under which sciences ought to be considered, as well as on distinctions between sciences, arts, and philosophy, fell considerably behind the progress of science itself and even tended to slow it down.

Zabarella, a Paduan philosopher of the 16th century, like most of his contemporaries, based his classification on Aristotle: heading the list of sciences are metaphysics, natural philosophy, and mathematics. Looking closer, however, at what at that time was regarded as related to *artes* and what to *scientiae*, we find, rather than a clearly defined hierarchy, a field of overlapping meanings.[60] *Scientia*, in the largest sense of the word, comprised every kind of knowledge, including all practical fields, for example, medicine, which more properly should be considered an *ars* (though many physicians objected to this). *Ars*, on the other hand, could denote both the trades and theoretical philosophy. The analysis of more than a hundred university textbooks shows that this situation lasted throughout the 16th–17th centuries.[61] As a rule, theoretical philosophy was subdivided into metaphysics, physics, and mathematics, so that the sciences of the quadrivium, regarded as *scientiae*, were part of philosophy and *artes liberales* at the same time.

At the end of the 17th century, Newton revealed the fundamental laws of the new physics in his *Philosophiae naturalis principia mathematica*. It took one and a half centuries for the *philosophia naturalis* to transform itself into the *science* of the 19th century. Throughout the whole of the 18th century, no one in England or in Europe managed to understand clearly to which of the two

---

60 Mikkeli, H. The foundation of an autonomous natural philosophy: Zabarella on the classification of arts and sciences, *Method and order in Renaissance philosophy of nature*, ed. by D. A. Di Lischia, Aldershot 1997, 211–228.

61 Freedman, J. S. Classifications of philosophy, the sciences and the arts in sixteenth- and seventeenth-century Europe, *The Modern Schoolman* 72 (1994) 37–65.

groups – *arts* or *sciences* – each particular discipline belonged.[62] In his article 'Art' for the *Encyclopaedia*, Diderot still followed the Aristotelian division between 'active' art and 'contemplative' science, thus leaving out of consideration the growing number of applied sciences that did not fit into either category. Some contemporary dictionaries noted that the notions of 'art' and 'science' were often used indiscriminately, the same discipline figuring among *liberal arts* as well as among *liberal sciences*; others identified science with "any art or kind of knowledge".

The German word *Wissenschaft* also acquired its modern meaning on the threshold of the 19th century, while some of its earlier meanings remained very close to τέχνη.[63] Thus, *Kunst* and *Wissenschaft* were, as a rule, used as synonyms; to specify the particular field in question, one needed to have recourse to adjectives: *schöne, nützliche, ernste Wissenschaften. Schöne Wissenschaften* denoted letters, while *schöne Künste* meant fine arts. It is only by the end of the 18th century that *Wissenschaft* (in the singular) began to embrace the sum total of sciences; at the same time it separated from philosophy, with which it was initially identified.

It is obvious that discrepancies and contradictions between the actual configuration of sciences at a given epoch and the way it is comprehended by contemporaries is no less characteristic of the modern period than it was of classical Antiquity.[64] In the course of our study, we will try to record these contradictions and trace the fate of certain ancient classifications of sciences. Let us note finally that interest in the classification of various fields of knowledge was growing at the close of the Hellenistic period, when the development of Greek science slowed down and eventually came to a standstill. Commentators eagerly studied classifications of sciences and their philosophical foundations in late Antiquity. Inherited by medieval encyclopaedias,[65] these classifications remained among the few vestiges of ancient science, which had by that time long ceased to exist.

---

62 Spadafora, D. *The idea of progress in England*, New Haven 1990, 29ff.

63 Thus, *Wissenschaft* in the subjective sense used to mean "persönliche Fähigkeit, Fertigkeit, Geschichtlichkeit", in the objective sense "jeder Wissenszweig samt der praktisch-nützlichen Anwendung" (Bumann, W. Der Begriff der Wissenschaft im deutschen Sprach- und Denkraum, *Der Wissenschaftsbegriff. Historische und systematische Untersuchungen*, ed. by A. Diemer, Meisenheim am Glan 1970, 64–75).

64 The future researcher may well find similar contradictions in what seems logically obvious to us.

65 *A source book in Medieval science*, ed. by E. Grant, Cambridge, Mass. 1974, 3f., 53ff.

Chapter 1

In search of the first discoverers: Greek heurematography and the origin of the history of science

1. Πρῶτοι εὑρεταί: gods, heroes, men

In theory, a study of the origins of the history of science in Antiquity should start from the point where history and science first intersect, i.e., from a historical overview of the scientific discoveries of the past. The problem, however, is that such overviews are unknown before the second half of the fourth century BC, whereas the sporadic mentions that historians, for example Herodotus, make of scientific discoveries belong not so much to history as to heurematography. Yet this is not the only reason to regard heurematography, an utterly unscientific genre with apparently little to offer history, as one of the forerunners of the history of science. Heurematography raised the question of how knowledge and skill are originated and transmitted long before the history of science appeared, and various answers to this question are part of the latter's prehistory. Which is why the common origins of the interest in *prōtoi heuretai* shared by both genres can best be traced in this 'prehistoric' material.

* * *

Interest in the past is inherent, to different extents, in all societies, including preliterate societies. The forms of its manifestation in ancient time are quite various, but generally they fit into the long worked out typology of folklore and early literary genres. Among the folklore genres, cosmogonic and etiological myths are to be mentioned first, then the heroic epic, which in many though not all cultures becomes the earliest literary genre. Another early literary genre worth noting is the historical chronicle, characteristic of the Chinese and, to a lesser extent, the Jewish tradition. This list does not, of course, exhaust the variety of questions the ancients asked about their past. It simply reduces our analysis to a number of definite themes that aroused constant interest and led to the formation of stable genres. Thus, a cosmogonic myth answered the question of the origin of the universe, an etiological myth explained the origin of particular elements of the civilization, say, a craft or a product important in a given culture, such as beer in Sumer or wine in Greece. A heroic epic and, later, a chronicle, told of things of still greater interest: ancestors' glorious feats.

Ancient Greece, whose literary and cultural history begins with Homeric and Hesiodean epics, manifests the same tendencies. The *Iliad* tells of the heroic deeds of the Achaeans and the Trojans, the *Theogony*, with its peculiar 'genealogical' attitude, depicts the origin of the world inhabited by gods and men. But the interest in the past characteristic of the epic is not identical to historical interest as such. The first is satisfied with legends about gods and ancient heroes; the second, oriented primarily toward men and their accomplishments, seeks to explain the present by linking it with the past. For all the uniqueness of the Homeric and Hesiodean epics, they have very little about them to suggest that their authors had a properly historical interest. It is only natural, therefore, that we do not find in either Homer or Hesiod any traces of a tradition on the *prōtoi heuretai* and their inventions.[1]

The first surviving evidence on the *prōtoi heuretai* is found in a fragment of *Phoronis*, an epic poem of the first third of the sixth century.[2] It mentions the Idaean Dactyls, mythical creatures named after the mountain range Ida in Troas.[3] Originally, the Dactyls were represented as dwarfish smiths, yet in *Phoronis* they take quite a different shape. The author calls them Phrygian sorcerers (γόητες Ἰδαῖοι Φρύγες ἄνδρες), the first to have invented blacksmith's work (οἳ πρῶτοι τέχνην πολυμήτιος Ἡφαίστοιο εὗρον). Though the question of *prōtos heuretēs* is in itself new,[4] it is applied to the traditional, albeit somewhat transformed material. From the traditional dwarfish blacksmiths, the Idaean Dactyls turn here into Phrygian sorcerers who discovered the art reputed to be under the patronage of Hephaestus. Later Hephaestus himself will turn from patron of the blacksmith's work into its first discoverer, in accordance with the pattern applied to most of the gods. But the author of *Phoronis*, though well aware that ironwork constitutes "the art of the wise Hephaestus", assigns its discovery not to the divine patron, but (using the modern idiom) to foreign specialists endowed with supernatural qualities. The discovery is thereby transferred from the divine sphere into the human one, unusual as these people appear to be,[5] and attributed to the neighbors of the Greeks.

---

1 On Homer's and Hesiod's treatment of different τέχναι and their role in human life, see Erren, M. Die Geschichte der Technik bei Hesiod, *Gnomosyne*, ed. by G. Kurtz, Munich 1981, 155–166; Schneider, H. *Das griechische Technikverständnis*, Darmstadt 1989, 11ff., 31ff.

2 *Schol. Apoll. Rhod.* I, 1129f. See Kleingünther, *op. cit.*, 26ff.

3 For material on the Dactyls, see Hemberg, B. Die Idaiischen Daktylen, *Eranos* 50 (1952) 41–59.

4 Referring to the fragment of Pseudo-Hesiodean *On the Idaean Dactyls* (fr. 282 Merkelbach–West), Schneider, *op. cit.*, 46, attributes the tradition of the invention of iron by the Dactyls to Hesiod. Meanwhile, fr. 282 merely repeats what is said in *Phoronis*, and the work *On the Idaean Dactyls* is a result of ancient philologists' combinations: Rzach, A. Hesiod, *RE* 8 (1912) 1223; Schwartz, J. *Pseudo-Hesiodea*, Leiden 1960, 246f.

5 In Greek mythology, the Dactyls figure along with other fabulous dwarfs, the Cabiri and Telchines, who are also credited with the invention of metalwork (Hemberg, B. *Die Kabiren*, Uppsala 1950; Dasen, V. *Dwarfs in Ancient Egypt and Greece*, Oxford 1993). Though the tradition of gnome-blacksmiths connected one way or another with Hephaestus is very old, the author of *Phoronis* has in mind people rather than

Interpreting this evidence in the manner of Euhemeristic rationalization of myth, one could find in it a reminiscence of the real history, namely, how iron smelting, discovered by the Hittites, spread from Asia Minor to Greece. Yet it would be unfounded to suppose that the author of *Phoronis*, active in Argos in the early sixth century, had heard anything of this history or taken interest in it. The early searches for the *prōtoi heuretai* focused, characteristically, not so much on their identities and historical background as on the technical and cultural achievements as such.[6] "X discovered y" is a classic formula of heurematography, featuring simply the name of the author and, in a very few cases, his origin. The time of the discovery is hardly ever recorded, let alone the circumstances, and the discoverer himself was far from being a real figure. In the heurematographic tradition that comes to us through the catalogues of discoveries of the Imperial age,[7] god-inventors (Athena, Demeter, Apollo) and cultural heroes (Triptolemus, Palamedes, Daedalus) are virtually matched in number by the other two major groups: historical personalities (Pheidon, Stesichorus, Thales) and Greek or 'barbarian' cities and nations.[8]

That heurematography's shift from mythography to real events was gradual and remained unfinished is not surprising. Greek historiography, as represented by Hecataeus, Herodotus, and Hellanicus of Lesbos followed the same path. In the absence of written evidence and adequate methods for the analysis of sources, heurematography (as well as history) could become *historical* only by turning to recent or contemporary developments. When trying to 'reconstruct' the distant past as recorded, if at all, in oral tradition, it resorted to the most fan-

---

gods, for the gods could hardly be called γόητες Ἰδαῖοι Φρύγες ἄνδρες. For more detail, see Zhmud, L. ΠΡΩΤΟΙ ΕΥΡΕΤΑΙ – Götter oder Menschen?, *Antike Naturwissenschaft und ihre Rezeption*, Vol. 11 (2001) 9–21.

6 Thraede, K. Erfinder, *RLAC* 5 (1962) 1192.

7 For material on ancient heurematography, see Brusskern, J. C. *De rerum inventarum scriptoribus Graecis*, Bonn 1864; Eichholtz, P. *De scriptoribus* Περὶ εὑρημάτων (Diss.), Halle 1867; Kremmer, M. *De catalogis heurematum* (Diss.), Leipzig 1890; Wendling, E. *De Peplo Aristotelico* (Diss.), Strasbourg 1891; Kleingünther, *op. cit.*, passim; Kienzle, E. *Der Lobpreis von Städten und Ländern in der älteren griechischen Dichtung* (Diss.), Kallmünz 1936; Thraede. Erfinder, 1191 ff.; idem. Das Lob des Erfinders. Bemerkungen zur Analyse der Heuremata-Kataloge, *RhM* 105 (1962) 158–186.

8 My calculations, based on the alphabetical index of inventors in Kremmer (*op. cit.*, 113 f.), give the following numbers: men – 56; cities and peoples – 43; gods – 33; heroes – 56. This data is certainly very approximate, because: 1) Kremmer's catalogue is selective and based mainly on late sources, where many historical figures are lacking; 2) I omit almost all names that cannot be related to any group; 3) the gods include Dactyls, Kouretes, Centaurs, Moirae, Cyclops, etc.; 4) the heroes include not only Roman kings (Numa Pompilius, etc.), but also a great number of etymological fictions, such as Iambe, the inventor of iambus; 5) on the other hand, numbered among the men are such doubtfully historical personalities as Anacharsis and King Midas, who could not, anyway, count as heroes.

tastic combinations. "The more arbitrary the first suggestion was, the better chances it had to be taken up."[9] Hence, the value of the evidence on the Idaean Dactyls is not that it could (or was meant to) point out the real inventors of the blacksmith's work. Apart from marking the lower limit of the period when interest in *prōtoi heuretai* arose, it contains the germs of two important tendencies that were to be developed later. I mean, first, the gradual and incomplete replacement of gods by semi-divine/heroic figures and next by people, and second, the Greeks' proclivity to assign inventions, including their own, to Oriental neighbors.

Let me stress again that this was not a linear process; sometimes the changes were of an alternative character. Depending on the public mood, the peculiarities of each particular work, the goals and attitudes of its author, and, last but not least, the character of the invention itself, different figures came to occupy the foreground.[10] A character from an earlier tradition who had receded into the background could reappear side by side with 'new' inventors. If heurematography records, on the whole, hardly more 'human' discoveries than those assigned to gods and heroes, this is rooted in the natural inclination to associate the beginnings of civilization with divine assistance and in the obscurity and anonymity of the real inventors of old. A tendency, peculiar to the epideictic literature, to honor divine inventors by crediting them with as many discoveries as possible also has to be taken into account. In the late catalogues of discoveries, it resulted in ascribing the same invention to several gods and heroes, usually without any attempt to reconcile the mutually exclusive versions.[11]

From the late fifth century on, professional literature dealing first with the history of poetry and music and then with that of philosophy, science, and medicine gradually reduces to a minimum the divine and heroic share in discoveries. While the history of music, in particular that of its earlier stages, still features such names as Orpheus, Musaeus, or Marsyas, the histories of philosophy, astronomy, and geometry include only real historical characters. In this respect, Peripatetic historiography is more critical than many historical works of the 17th and even 18th centuries, whose accounts of Greek astronomy start with Atlas, Uranus, and other mythological figures. Admittedly, in Antiquity the historicity in the treatment of material depended not so much on when a given work was written as on its genre. The author of an encomium, a hymn, a tragedy or a work *On Discoveries* would hardly be seriously concerned with the

---

9 Thraede. Erfinder, 1207.

10 In the course of the sixth–fourth centuries BC, the invention of writing was successively attributed to Cadmus, Danaus, Palamedes, Prometheus, Actaeon, and the Egyptian god Thoth (see e.g. *FGrHist* 1 F 20, 10 F 9, 476 F 3). On the 'secondary sacralization' of the *prōtoi heuretai*, see below, 37.

11 The bulk of the catalogues of discoveries actually derives from the epideictic literature: Thraede. Lob des Erfinders; Cole, T. *Democritus and the sources of Greek anthropology*, Ann Arbour 1967, 6f.

reliability of the reported information.[12] In such genres as doxography or history of science, the writers usually avoided making up obvious inventions of histories, even though they repeated some inventions made by the others.

There is one more reason why the succession 'gods – heroes – men' was not strictly linear. In Homer and Hesiod and, naturally, before them, the Greek gods were represented not as the first discoverers but as the 'donors of goods' and as the patrons of crafts that they had *taught* to men.[13] They turn into *prōtoi heuretai* only after the fame of the *human* inventors had spread throughout the Greek world. Interest in first discoverers in the absolute sense, i.e., in those who invented metallurgy, agriculture, writing, or music, awakens gradually, stimulated by growing attention to innovations as such and to the question of priority in their creation. Though the rapid social and cultural development of Greece about 800–600 BC led to a lot of discoveries in all spheres of life, a certain space of time was needed for specific interest in them to arise and take root. To judge by the available evidence, the real creators of technical and cultural innovations – inventors, poets, musicians, painters, sculptors – commanded public attention in the early seventh century.

Revealing in this respect is a fragment of one of the early lyric poets, Alcman, in which he professes his admiration for his predecessors, who "taught people wonderful, soft and new sounds".[14] The vocabulary of this fragment, and the expression ἀνθρώποις … ἔδειξαν in particular, is very close to that used in the tradition on *prōtoi heuretai*,[15] even though the motif of a first discoverer is only implicit here. Although poets did teach people new sounds, it is their relative rather than absolute novelty that Alcman must have had in mind: the key notions πρῶτοι and εὗρον are still lacking here. Yet in another fragment, we find ϝέπη τάδε καὶ μέλος Ἀλκμὰν εὗρε (fr. 39 Page), whereby the poet appears to claim the status of first discoverer for himself. A Homeric hymn to Hermes ascribes to him the invention of the seven-string lyre (IV, 24–61). Meanwhile, by that time there undoubtedly existed a tradition crediting Terpander with this discovery,[16] Hermes himself hardly ever having been associated with music before.[17] The gradual character of the transformation of gods

[12] This is true of Peripatetic heurematography as well; see below, 43.

[13] See e.g. *Od.* VI, 232f. on "a cunning workman whom Hephaestus and Pallas Athena have taught all manner of craft" (ὃν Ἥφαιστος δέδαεν καὶ Παλλὰς Ἀθήνη τέχνην παντοίην). Cf. *Od.* XX, 72.

[14] θαυμαστὰ δ' ἀνθρώποις … γαρύματα μαλσακά … νεόχμ' ἔδειξαν … (fr. 4.1 Page).

[15] Davies, M. The motif of the πρῶτος εὑρετής in Alcman, *ZPE* 65 (1986) 25–27.

[16] This hymn is usually dated in the sixth century (Schmid, W., Stählin, O. *Geschichte der griechischen Literatur*, Vol. 1, Munich 1974, 236f.; Janko, R. *Homer, Hesiod and the Hymns*, Cambridge 1982, 140f.).

[17] Kleingünther, *op. cit.*, 22, 29; Terpander as *prōtos heuretēs* was first mentioned by Pindar (fr. 125 Snell), but this tradition certainly goes back to the seventh century.

into first discoverers is confirmed by the Homeric hymn to Aphrodite.[18] Athena is called here the first to have taught (πρώτη ἐδίδαξε) craftsmen the art of making chariots and carriages and maids that of handiwork (weaving, probably). Though Athena is described as πρώτη, her merit here, as in Homer, is not the invention of handicrafts, but their instruction.[19] If later the Greek cities renowned for their crafts were credited with the invention of things formerly considered to be under the patronage of gods,[20] this does not mean at all that initially the *prōtos heuretēs* model was created on the mythological material and applied to gods alone.[21]

Since Greek literature before the sixth century happens to be represented only by poetic genres, what we know best are the innovators in music and poetry. When Glaucus of Rhegium (late fifth century) undertook in his *On the Ancient Poets and Musicians* one of the first attempts to systematize the early history of Greek poetry and music, he wrote mainly of who invented what, who borrowed what from whom, etc., relying in the first place on references made by poets themselves.[22] Still, the oral and epigraphic tradition that survived until later times shows that inventions in other spheres were being recorded as well.

The fame of the Argivan king Pheidon (first half of the seventh century), who was regarded as the inventor of an improved system of measures, the so-called μέτρα Φειδώνια,[23] obviously preceded the renown of Palamedes as the

---

18 V, 12–15. The hymn is dated within the period of the eighth through fifth centuries BC, most often the seventh (Janko. *Homer*, 180).

19 For the notion of gods who instructed people in handicrafts, see also *Hymn. Hom.* XX, 2f. (Hermes), Solon. fr. 13, 49 (Athena and Hermes), Pind. *Ol.* VIII, 50f. (Athena). In the Orphic theogony, Athena and Hermes turn unexpectedly from teachers into pupils: πρῶτοι τεκτονόχειρες, οἳ Ἥφαιστον καὶ Ἀθήνην δαίδαλα πάντ' ἐδίδαξαν (fr. 178–179 Kern).

20 Thebes becomes the inventor of the chariot, Athens of ceramics (*DK* 88 B 1.10, 12), Corinth of horse gear and the dithyramb (Pind. *Ol.* XIII, 18; cf. Hdt. I, 23). See Kienzle, *op. cit.*, 72ff.

21 So Schneider, *op. cit.*, 103. Interestingly, the Muses, while remaining patrons of τέχναι, never turned into their inventors. Kremmer, *op. cit.*, 111, adduces a list of the Muses along with the 'historical' inventors of arts (*Schol. in Oppian. halieutica* I, 78): Clio – history (Herodotus), Thalia – comedy (Menander), Melpomene – tragedy (Euripides), Euterpe – auletics (Stesichorus), Terpsichore – lyre (Pindar), Erato – cymbals (Hermes!), Calliope – poetry (Homer), Urania – astronomy (Aratus), Polyhymnia – geometry (Euclid).

22 See below, 2.1. On the references of the early Greek lyric poets to their predecessors, see Janko, R. Schield of Heracles, *CQ* 36 (1986) 41 n. 18. On the polemics among poets, see Zaicev, A. *Das Griechische Wunder. Die Entstehung der griechischen Zivilisation*, Konstanz 1993, 146f.

23 Hdt. VI, 127; Her. Pont. fr. 152; Arist. *Pol.* 1310b 19f.; Ephor. *FGrHist* 70 F 115, 176; Schwabacher, W. Pheidonischer Münzfuß, *RE* 19 (1938) 1946ff.; Andrewes, A. The Corinthian Actaeon and Pheidon of Argos, *CQ* 43 (1949) 74ff.

inventor of weights and measures.[24] The shipbuilder Ameinocles of Corinth, who was invited to build ships on Samos, worked in the mid-seventh century or even earlier.[25] From the early seventh century on, vase painters, potters, and, later, sculptors considered it natural to sign their works,[26] so that the names of the early *prōtoi heuretai* in this field go back to signatures left by artists themselves.[27] These include, for example, Butades of Sicyon (seventh century), the legendary inventor of ceroplastics (the art of modeling in wax), whose works, signed and dedicated to the temple, were preserved in Corinth until the Hellenistic epoch.[28] Glaucus of Chios (early sixth century), a renowned master whom Herodotus calls the inventor of iron soldering (σιδήρου κόλλησις), produced and signed a silver crater on an iron stand that the Lydian king Alyattes later dedicated to the temple in Delphi (Hdt. I, 25; cf. Paus. X, 2–3). The architect Mandrocles of Samos, who built the bridge across the Bosporus for Darius' expedition against the Scythians (513 BC), spent part of his generous reward to commission a picture of the bridge. He dedicated it to the temple of Hera, supplying it with an epigram that mentioned his name (Hdt. IV, 87–89).

Even this fragmentary evidence of the archaic epoch testifies that the search for *prōtoi heuretai* reflected contention for priority, typical of Greek culture on the whole. Thus, the tradition on first discoverers leads us to the problem of priority for all sorts of social and cultural innovations, a problem much broader than both heurematography and the history of science.

2. Heurematography and the 'Greek miracle'

The investigation of the 'Greek miracle', the unique complex of qualities that distinguishes Greek culture from everything that preceded it, brings the problem of the authorship of cultural achievements into particularly sharp focus.

---

24 Kleingünther, *op. cit.*, 82. The tradition on the Lydian invention of the golden coins (Xenoph. 21 B 4; Hdt. I, 94) also goes back to the seventh century.

25 Thuc. I,13.3. According to Thucydides, whose information derives from written sources of the fifth century, Ameinocles was invited to Samos "300 years before the beginning of the Peloponnesian war". See Hornblower, S. *A commentary on Thucydides*, Vol. 1, Oxford 1991, 42f.

26 Jeffery, L. H. *The local scripts of archaic Greece*, 2nd ed., Oxford 1990, 62, 83, 230f.; Philipp, H. *Tektonon Daidala*, Berlin 1968, 77f.; Walter-Karydi, E. Die Entstehung des beschrifteten Bildwerks, *Gymnasium* 106 (1999) 289–317.

27 Thraede. Erfinder, 1181.

28 Robert, C. Butades, *RE* 3 (1897) 1079; Fuchs, W., Floren, J. *Die griechische Plastik*, Vol. 1, Munich 1987, 197 (cited here are the names of other early masters from Corinth). – Renowned in the mid-sixth century was the family of Chian sculptors and architects, Archermus and his sons Bupalos and Athenis; their signed works survived until the time of Augustus (Plin. *HN* 36, 5; Paus. IV,30.5; Svenson-Evers, H. *Die griechischen Architekten archaischer und klassischer Zeit*, Frankfurt 1996, 108f.).

Characteristic of this complex is, first and foremost, the emergence of literature that is no longer anonymous. This implies that poets asserted the importance of their work and hoped for its public acknowledgement, for which they sought to link their names inextricably with their creations. Hand in hand with this go criticism and praise of contemporaries and forerunners and a striving after thematic and formal innovations.[29] These features, as typical of contemporary literature as they are alien to the anonymous and traditional writings of the Near East,[30] appeared in Greek poetry in the course of the several generations that followed Homer and Hesiod. Still more important, Greek society of the archaic epoch acknowledged these notions of the literary process as a norm.

It would be far-fetched to say that it was poets and musicians who taught the Greeks to appreciate authorship and individual efforts as such. Parallel processes were at work in the visual arts. From the seventh century on, artists' claims to authorship are reflected in the signatures left on ceramics and sculptures.[31] Names of the prominent architects are attested since the first third of the sixth century, not least due to their own efforts. Chersiphron and Metagenes, the builders of the famous temple of Artemis at Ephesus, invented a new method of transporting stone columns with the help of special wooden rollers. Not satisfied to be known as the authors of a famous edifice, they wrote a technical treatise that announced this and probably many other inventions as well.[32] The architect and sculptor Theodorus, who built the temple of Hera on Samos, also wrote a book about his work (Vitr. VII, praef. 12). According to Pliny (*HN* 7, 198), Theodorus' inventions include the setsquare (*norma*), the water gauge (*libella*), and even the key (*clavis*). This information is likely, in part at least, to go back to Theodorus' book.[33] If the evidence that Theodorus executed a sculptural 'self-portrait' (*HN* 34, 83) is true, we are justified in comparing this remarkable artist of the archaic epoch with the masters of the Renaissance, who were well aware of the value of their artistic genius. Thus, we see that by personifying the anonymous discoverers of the past, heurematography was reproducing an attitude already predominant in the contemporary society.[34]

---

29 *Zaicev. Griechisches Wunder*, ch. 4.

30 "Literary works from the ancient Middle East are generally completely anonymous, but sometimes the attempt is made to attribute them to some authoritative thinker or other. " (ibid., 128 n. 96). A few names of the Babylonian authors adduced by M. L. West (*The east face of Helicon. West Asiatic elements in Greek poetry and myth*, Oxford 1997, 63, 65, 68, 81) confirm the general rule. The only exception seems to be the Jewish prophetic literature.

31 Philipp, *op. cit.*, 77, stresses the kinship of motives that were at work in poetry and in the visual arts.

32 Vitr. VII, praef. 12; X,2.11–12; Fabricius, K. Chersiphron, *RE* 3 (1899) 2241–2242.

33 Svenson-Evers, *op. cit.*, 40f.

34 "The already established practice of claiming the authorship and, accordingly, acknowledging it with regard to a large variety of cultural products was projected into

The same claims to authorship and the fame that goes with it, the same contention for priority, mark the beginnings of Greek science and philosophy. It is no wonder that Thales, their common founder, was credited with a dictum that the best reward for his mathematical discovery would be to link it permanently with his name (11 A 19). Apocryphal as this saying is, there is no doubt that the problem of authorship was a matter of serious concern for both Thales and his contemporaries. Otherwise, his theorems would not have reached us under his name, for Thales himself did not write anything. Criticism of predecessors for their poor understanding of the subject is a constant motif in most branches of learning from history and geography to medicine and philosophy, a motif designed to set off the remarkable novelty of one's own theories and achievements.[35] Accusations of plagiarism, aimed at undermining others' claims to priority, are also a very early phenomenon.[36] Hence, it seems fair to say that the sharpened interest in priority and consequently in the authorship of *any* achievements in *every* kind of creative activity, being the motive behind the search for first discoverers, was in turn itself the product of forces that created Greek literature, art, philosophy, and science.

In his pithy article on the *prōtoi heuretai*, Thraede names some of the "sociological conditions" under which this tradition emerged.[37] Yet the majority of the factors he adduces – colonization, "genetic world outlook", the proliferation of real discoveries and the growing importance of personality in culture – belong rather to the historico-cultural prerequisites than to sociological conditions. When considering the history of discoveries from a contemporary rather than an ancient point of view, the theory of the Greek 'cultural upheaval' developed by A. Zaicev seems to be the most fruitful attempt to answer the question of the sociological components of the 'Greek miracle'.

It has been repeatedly noted that the behavior of a Greek of that epoch was regulated to a large extent by the appraisal of his or her social group.[38] The orientation toward the approval of others, toward public acknowledgement of one's merits, and the aspiration to fame and honors were among the most important motives for individual behavior. This attitude, by no means unique in a socio-psychological typology of societies, was strengthened by an additional tendency. The early Greek polis was a highly competitive society. The orientation toward success, toward surpassing others in the achievement of one's life goals, played a tremendous role. Especially important is that the competitive spirit animated not only such spheres of conflicting practical interests as econ-

---

the past: one tried to find an inventor, often a mythical one, for nearly every achievement of human civilization." (*Zaicev. Griechisches Wunder*, 129).

35 Ibid., 122ff.

36 Stemplinger, E. *Das Plagiat in der griechischen Literatur*, Berlin 1919.

37 Thraede. Erfinder, 1192.

38 Dodds, E. *The Greeks and the irrational*, Berkeley 1951, 18f. For important reservations, see Cairns, D. L. *Aidōs. The psychology and ethics of honour and shame in ancient Greek literature*, Oxford 1993, 27ff., 43f.

omy and politics, which is quite often the case, but also those where a victory brought no, or hardly any, practical benefits at all, for example in athletic games. The agonistic spirit took root in the Greek society already in the preliterate epoch; no wonder the 'institution' of the Olympic games in 776 is the first dated event in Greek history and the lists of Olympic victors remain among its earliest documents.[39] Priority in sport, particularly a victory at the Olympics,[40] which required great energy, wealth, and leisure, brought athletes the kind of glory earlier bestowed only on kings or warlords. Quite often it was followed by heroic honors and a cult. In the eighth to sixth centuries, the epoch of the disintegration of traditional norms and values, the growth of private initiatives, and economic and territorial expansion, the Greek agonistic spirit contributed to establishing a new value orientation toward *priority as such*, independent of whether the victor himself or his polis benefited from it materially. This emerging anti-utilitarian socio-psychological orientation contributed in turn to the creation of a social climate in which any person of remarkable attainments in a cultural sphere could enjoy wide public acknowledgement. Creative achievements of all sorts were stimulated, irrespective of their practical utility; the pressure of tradition was substantially decreased.[41]

Already in the early seventh century, fame could be achieved by accomplishments whose practical significance for society was far from evident: victory in a stadium race, poetic endowment, or invention of a musical instrument. Greek athletics, poetry, and music manifested themselves earlier than other activities simply because they were rooted in the traditions of the preliterate epoch. In the first part of the sixth century, to which our earliest mention of *prōtos heuretēs* belongs, fame could be claimed by those who proved a geometrical theorem, drew a geographical map, or imparted a new philosophical theory to his fellow citizens.

This anti-utilitarian orientation should not, however, be overestimated. The social weight of practical inventions did not become any smaller, otherwise they would not have been assigned to gods, heroes, and, later, famous philosophers.[42] In the mid-fifth century, the renowned architect Hippodamus of Miletus proposed a law calling for honors to be bestowed on inventors of things useful to the state.[43] In the classical epoch, not only τέχναι, but also philosophy and theoretical sciences made claims to practical importance and social utility

39 See Moretti, L. *I vincitori negli antichi agoni olimpici*, Rome 1957.

40 Let us note that Greek athletics knew nothing of second and third places; only victory counted.

41 Zaicev. *Griechisches Wunder*, esp. ch. 2–3.

42 See below, 35 n. 60. On the positive attitudes of the Greeks to technology, see Schneider, *op. cit.*, 52ff.; Schürmann, A. *Griechische Mechanik und antike Gesellschaft*, Stuttgart 1991.

43 Arist. *Pol.* 1268a 6f., b 23f. = *DK* 39 A 1. Aristotle approved of Hippodamus' idea, particularly with regard to arts and sciences, objecting only to too-frequent changes of laws.

(2.1). In the third century, Eratosthenes was so proud of having invented a new device for drawing curves that he dedicated to King Ptolemy III a bronze model of it, adding a fine epigram that emphasized the practical utility of the device.[44] Archimedes' biographer Heraclides pointed out that his book *On Measuring the Circle* is useful for the necessities of life.[45]

In the sixth and the fifth centuries, almost all of civilization was regarded as a sum total of various τέχναι,[46] among which μουσική, ποιητική, ἰατρική, and λογιστική (τέχνη) figured along with other crafts and arts. Until the first part of the fourth century, the sciences as we understand them, i.e., *mathēmata*, still were related to τέχναι, only gradually being set apart as a separate group. Aeschylean Prometheus, speaking of his services to civilization, calls himself the inventor of πᾶσαι τέχναι (506), numbering among them building, astronomy, arithmetic, writing, shipbuilding, medicine, divination, and metallurgy (450 ff.) – in a word, every sphere through which social life is civilized. Greek thought of that time does not seem to make any fundamental distinction between practical and unpractical discoveries. Xenophanes mentions the Lydians' invention of the coin (21 B 4) and Thales' prediction of a solar eclipse (21 B 19), and Pindar refers to Corinth's invention of the dithyramb and horse gear (*Ol.* XIII, 18). In the archaic epoch, Palamedes was credited with the invention of measures, weights, and the alphabet,[47] Hermes with the invention of the lyre and the art of making fire (*Hymn. Hom.* IV, 24 f., 108 f.). Athena was associated with the appearance of the chariot, flute playing, the cultivation of olive trees, etc.[48]

The ancient Oriental tradition has also brought us the names of many gods and cultural heroes associated with the beginnings of human civilization. Among their gifts to mankind, alongside purely practical things like agriculture, the plow, or beer, there are such socially important inventions as writing and music.[49] Yet we do not find here the names of those who invented new genres in poetry, new styles in architecture, new trends in music, or new methods in mathematics and astronomy, though such people undoubtedly existed.[50] The lack of interest in human *prōtoi heuretai* correlates with the fact

---

44 Eutoc. *In Archim. De sphaer.*, 88.3–96.9; Knorr. *TS*, 131 ff.

45 Eutoc. *In Apollon. con.*, 168.5 f. = *FGrHist* 1108 F 1–2: πρὸς τὰς τοῦ βίου χρείας ἀναγκαῖον.

46 See Joos, P. ΤΥΧΗ, ΦΥΣΙΣ, ΤΕΧΝΗ: *Studien zur Thematik frühgriechischer Lebensbetrachtung* (Diss.), Winterthur 1955, 31 f.; Thraede. Fortschritt, 145, 152. These, however, did not include laws and regulations, νόμοι.

47 The alphabet is first mentioned in Stesichorus (fr. 213 Page).

48 Kleingünther, *op. cit.*, 28 f.

49 See e.g. the Sumerian myth of the origin of agriculture and cattle breeding (Kramer, S. N. *Sumerian mythology*, Philadelphia 1944, 53 f.).

50 The sole exception seems to be the Egyptian tradition on Imhotep (later ranked among gods) as the inventor of the pyramid: Wildung, D. Imhotep, *Lexikon der Ägyptologie* 3 (1980) 145 f.

that the gods of Sumer, Egypt, and Babylon, just like the gods in Homer, were thought of as the donors or teachers of crafts, not as their inventors. In the myth of the goddess Inanna, for instance, she wheedles from Enki, the god of wisdom, more than a hundred divine institutions, including various crafts, language, writing, and music, and imparts them to people.[51] It does not follow from the myth that Enki invented all these things. The Egyptian gods responsible for various crafts (*Berufsgötter*) also were not regarded as their inventors, with the possible exception of Thoth.[52]

3. Inventors and imitators. Greece and the Orient

Both tendencies indicated above – the willingness of the Greeks to attribute their own inventions to their Oriental neighbors and the secularization of the notion of *prōtoi heuretai* – were developed by Hecataeus and, later, by Herodotus, whose influence on later literature proved decisive.[53] Hecataeus revealingly corrects the generally accepted Greek tradition of assigning the invention of wine to Dionysus and attributes it to the Aetolian king Oresteus, the son of Deucalion (*FGrHist* 1 F 15). He names Danaus as the inventor of the Greek alphabet (F 20), thus pointing to the Egyptian origin of this discovery.[54] In Herodotus, Greek gods and heroes never figure as first discoverers at all, discoveries usually being attributed by him to 'barbarian' nations, first of all to Egyptians. We shall return later to this *egyptophilia*, which in certain authors will grow into a genuine *egyptomania*.[55] Meanwhile, I would like to point out still another important source of the rapidly changing concept of *prōtoi heuretai*.

In the late sixth to early fifth centuries, we encounter in Greek philosophy the first theoretical reflections on the origin and development of culture. The notions ζήτησις and εὕρεσις play the key role in them. Depending on context, they can be understood to mean 'search – find' or 'research – discovery', the latter more often in philosophical and scientific writings. A well-known fragment of Xenophanes (21 B 18), where this pair of notions first occurs, marks an important stage in secularizing the search for first discoverers:

> The gods did not reveal to men all things in the beginning,
> but in the course of time, by searching, they find out better.[56]

---

51 Kramer, *op. cit.*, 61 ff.

52 Hieck, W. Berufsgötter, *Lexikon der Ägyptologie* 2 (1974) 641 f. On Thoth as the inventor of writing, see below, 6.3.

53 Vogt, J. Herodot in Ägypten, *Genethliakon W. Schmid*, Stuttgart 1929, 97–137.

54 Hellanicus of Lesbos (*FGrHist* 4 F 175) also pointed out that the vine was first discovered in Egypt; he ascribes the invention of iron weapons to the Scythians (F 189).

55 For an introduction to this topic, see Froidefond, C. *Le Mirage égyptienne dans la littérature grecque d'Homère à Aristote*, Paris 1971.

56 οὔτοι ἀπ' ἀρχῆς πάντα θεοὶ θνητοῖσ' ὑπέδειξαν, / ἀλλὰ χρόνωι ζητοῦντες ἐφευρίσκουσιν ἄμεινον (transl. by W. Guthrie).

Without totally denying the part of gods in creating civilization, Xenophanes strongly emphasizes the independent efforts of people leading in time to new discoveries and inventions.[57] Viewed from such a perspective, man is no longer the beneficiary of divine care, but rather the subject of a civilization whose progress is due, first and foremost, to his own efforts. These two verses are often justly regarded as the first clear formulation of the notion of progress in human society.[58] To judge from the importance Xenophanes accorded to σοφίη (21 B 2), the agent of progress was not the ordinary man but the sage.

At the end of the fifth century, Archytas, echoing Xenophanes, begins his account of the discovery of the art of calculation with a discourse based on the same companion notions 'research – discovery' (47 B 3):

> To know what was heretofore unknown, one has either to learn it from another, or to discover himself... Discovering without research is difficult and (happens) seldom, by research it is easy and practicable, but without knowing (how) to research it is impossible to research.

Only the knowledgeable man (ἐπιστάμενος), the specialist in his field, or, still better, a sage can find something new. It is in the person of the sage (ἀνὴρ σοφός) that philosophy (and later the Sophistic) finds a new cultural hero.[59] No wonder the biographies of the Seven Sages and the first philosophers contain such a wealth of references to their εὑρήματα, related to the sphere of culture in the largest sense: from discoveries in astronomy and geography to the introduction of weights and measures and the invention of the anchor.[60] Revealingly,

---

57 For the motif of the gradual increase of discoveries in classical literature, see Aesch. *Prom.* 447f.; Eur. *Suppl.* 201f., fr. 60, 236, 542, 771, 813, 931 Nauck; *VM* 3; Isoc. *Paneg.* 32; Pl. *Leg.* 678b 9–10; Chairem. *TrGF* 71 F 21; Arist. *SE* 183b 20f.; *Met.* 982b 13–15 (later evidence: Thraede. Fortschritt, 148). In the fourth century BC it will be contrasted with a different motif – the speed at which the inventions follow each other in recent times (Arist. fr. 52–53 Rose; *De an.* 417b; *EN* 1098a 22f.); see also a passage from an early Academic work (below, 87f.).

58 Edelstein, *op. cit.*, 3f.; Thraede. Fortschritt, 142; Babut, F. L'idée de progrès et la relativité du savoir humain selon Xenophane (Fr. 18 et 38 D–K), *RPhil.* 51 (1977) 217–228; Schneider, *op. cit.*, 60f.

59 In Isocrates (*Paneg.* 32; *Panath.* 48; *Nic.* 8–9) and Aristotle (*Protr.* fr. 8 Ross), φιλοσοφία figures as a (co)inventor of all the τέχναι. See also Posidonius (fr. 284 E.-K.) and objections by Seneca (*Ep.* 90).

60 Thales was the first to study astronomy, to discuss physical problems, to maintain the immortality of the soul, to inscribe a triangle in a circle, to divide the year into 365 days, and to estimate the sizes of the sun and the moon (D. L. I, 23–27); Solon was the first to institute the nine archons and to call the thirtieth day of the month the Old-and-New day (I, 58); Chilon was the first to propose the office of ephores (I, 68); Periander was the first who had a bodyguard and who established a tyranny (I, 98); Anacharsis invented the anchor and the potter's wheel (I, 105); Pherecydes was the first to write of nature and (the origin of) gods (I, 116); Anaximander was the first to invent the gnomon, the geographical map, and the celestial globe (II, 1–2); Pythagoras coined the word 'philosophy', discovered the monochord, was the first to intro-

the names of famous philosophers and scientists later continue to be associated with inventions that stand very far from the properly mental sphere. Democritus invented the arch (Sen. *Ep*. 90, 32), Protagoras the shoulder pad for carrying burdens (D. L. IX, 53, cf. IV, 2), Archytas the children's rattle and the mechanical dove (47 A 10), and Plato the water alarm clock (Athen. IV, 75). Eudoxus was the first to arrange the couches at a banquet in a semicircle (D. L. VIII, 88). Some of these testimonies come from people who are hardly to be suspected of idle talk, for example, Aristotle (*Pol*. 1340b 25f.). But what is important for us now is not so much their historical reliability as the tendency, carried to extremes, to regard any, even the most insignificant element of civilization as the result of somebody's research and discovery. Because of the growing number of new *prōtoi heuretai*, gods and cultural heroes are pushed gradually into the background, particularly with regard to discoveries made within the horizon of history. Finally, religion itself is declared to be man's own creation.[61]

Prometheus, who figures in Aeschylus as the inventor of writing, medicine, astronomy, and arithmetic (457f., 478f.), was later superseded by human discoverers, a separate one for each of these τέχναι.[62] To be sure, Prometheus himself is far from being a traditional divine *prōtos heuretēs*. Let us have a closer look at the Titan's own description of his gifts to mankind, particularly because this seems to be the first mention of mathematics and astronomy as discoveries:

> ... I taught them to discern the rising of the stars
> and their settings, ere this ill distinguishable.
> Aye, and numbers, too, chiefest of sciences,
> I invented for them, and the combining of letters ...[63]

Though the names of sciences themselves do not occur in Aeschylus, his words make perfectly clear what branches of learning he means: the knowledge of risings and settings refers to astronomy, ἀριθμός (called ἔξοχον σοφισμάτων) to arithmetic, and γραμμάτων συνθέσεις to writing (grammar). The order in which Aeschylus lists different τέχναι and σοφίσματα is arbitrary enough,[64]

---

duce the meat diet for athletes and measures and weights into Greece, to identify the Morning and the Evening Star with Venus, to call the heaven the cosmos and the earth spherical, etc. (VIII, 12, 14, 48).

61 Prodicus (84 B 5), Critias (88 B 25); Democritus (68 A 77–79, B 166, 297); Thraede. Erfinder, 1218f.

62 Theophrastus, discussing the problem of the emergence of arts and sciences in the context of the argument on the age of humankind, claims that people who discovered these things lived only a thousand years before (fr. 184.125f. FHSG). The participation of gods is not even mentioned here.

63 ... ἔστε δή σφιν ἀντολὰς ἐγὼ / ἄστρων ἔδειξα τάς τε δυσκρίτους δύσεις. / καὶ μὴν ἀριθμόν, ἔξοχον σοφισμάτων / ἐξηῦρον αὐτοῖς, γραμμάτων τε συνθέσεις (457–460, transl. by H. Smyth).

64 Hardly convincing is an attempt to find an 'evolutionary' sequence in the list of crafts in *Prometheus* and thus to postulate a philosophical or Sophistic source that he

but the fact that arithmetic immediately follows astronomy may show his awareness of an intrinsic kinship between these sciences.[65] He was unlikely to single them out as abstract sciences, contrasting them with such practical arts as agriculture, cattle breeding, and ship-building: astronomy remained for Aeschylus a practical discipline indispensable to farmers and navigators.[66] We should recall, however, that research in astronomy and mathematics was joined from the time of Thales. In Anaximander's or, to a still greater extent, Pythagorean teaching, numbers and heavenly bodies hold a special place. Pythagorean arithmetic was even parodied in Epicharmus' comedy (23 B 2). It is also revealing that Prometheus speaks not of separate discoveries, but of already mature τέχναι, which occupy a deserved place among crafts traditional from Homeric times, like medicine and shipbuilding.

To what extent do Prometheus' words reflect the attitude of Aeschylus himself? Did he really associate the emergence of all the arts and sciences, as well as of civilized life as a whole, with the activities of a philanthropic Titan, thus debarring humanity from any participation in their discovery?[67] Prometheus as *prōtos heuretēs* is undoubtedly Aeschylus' own creation, for the earlier tradition does not know of such a figure. In the tragedy featuring the inventor Palamedes, who was renowned already in the archaic epoch, Aeschylus credits him with the discovery of writing, astronomy, and mathematics (fr. 303a Mette), in accord with most of the fifth-century writers.[68] The version of the divine origin of τέχναι seems then to be connected with artistic ends Aeschylus pursued when writing the *Prometheus*, rather than with his views on the origin of culture. As for Palamedes, he was, unlike Prometheus, a hero, i.e., a mortal, not a god; besides, the inventions originally ascribed to him (weights and measures) reflected the glory conferred on human discoverers.[69]

It is hard to say whether the transference of Palamedes' discoveries to Prometheus was part of the growing tendency toward the 'secondary sacralization' of inventors. Anyway, by the end of the fifth century, the tendency had

---

might have used: Conacher, D. Prometheus as founder of the arts, *GRBS* 18 (1977) 189–206.

65 Joos, *op. cit.*, 34. Further in the same play, medicine adjoins mantic, in keeping with current opinion on the close kinship between these τέχναι.

66 For his attitude to the value of knowledge, see: ὁ χρήσιμ' εἰδώς, οὐχ ὁ πόλλ' εἰδὼς σοφός (fr. 390 Nauck).

67 Cf. Guthrie, W. K. C. *In the beginning*, London 1957, 83f.; Boer, W. den. Prometheus and progress, *Miscellanea tragica in honorem J. C. Kamerbeek*, ed. by J. M. Bremmer et al., Amsterdam 1976, 17–27. By contrast, Joos, *op. cit.*, 35, regards Aeschylus' Prometheus as embodying the idea that culture can exist against the will of gods, which foreshadows the Sophistic notion of culture without gods.

68 Kleingünther, *op. cit.*, 78f. See Sophocles (fr. 399 Nauck), Euripides (fr. 578 Nauck), Gorgias (76 B 11a, c. 30), Alcidamas (*Od.* 22); cf. Pl. *Res.* 522d 1f. The invention of writing by Palamedes is first mentioned by Stesichorus (fr. 213 Page), so that this version is obviously older than Aeschylus.

69 See above, 28 n. 23.

become quite pronounced. Besides a protest against attributing to people the discovery of all the important τέχναι, it reflected the attempts to justify the activity of gods from the standpoint of new notions of culture, namely, by showing them as inventors.[70] In his well-known list of human accomplishments, Sophocles does not make direct reference to the participation of the gods in the progress of civilization (*Ant.* 332–375). A similar list in Euripides' *Suppliant Women* takes the form of Theseus' praise of the divinity (195–213) and concludes by reproaching people who strive to surpass the gods intellectually (216–217). A reaction to the onset of rationalism and agnosticism is particularly manifest in the polemical verses from the pseudo-Epicharmean comedy *Politeia* (late fifth century):

> The λόγος steers mankind aright and ever preserves them.
> Man has calculations (λογισμός), but there is also the divine Logos.
> But the human Logos is sprung from the divine Logos,
> And it brings to each man his means of life, and his maintenance.
> The divine Logos accomplishes all the τέχναι,
> Itself teaching men what they must do for their advantage;
> For no man has discovered any τέχνη, but it is always God.[71]

One of those who may have caused such a reaction was Herodotus, notable for his rationalistic views on the origin of culture. Gods and heroes do not figure in him as *prōtoi heuretai* at all. He either names the author of an invention or assigns it to a particular nation.[72] Most of the inventions mentioned by Herodotus occur in book II and are associated with the Egyptians. Thus, at the very beginning of the book (II, 4) he asserts, referring to local priests, that the Egyptians were the first to establish the (exact) length of the year by dividing it into 12 months with 30 days plus 5 additional days. It follows, accordingly, that the basis for astronomy, or at least calendar astronomy, was laid in Egypt. Elsewhere, speaking of the Egyptian origin of geometry (i.e., the art of land surveying), Herodotus makes a reservation: two important astronomical instruments, namely the gnomon and polos, as well as the division of the day into 12 hours, come from Babylon, not Egypt (II, 109).

In the first passage, Herodotus explicitly refers to Egyptians priests as his immediate source. In the second, he seems to voice his own opinion on the origin of geometry (δοκέει δέ μοι). The source of information on Babylonian as-

---

70 This was pointed out by Prodicus (84 B 5); Thraede. Erfinder, 1219f.

71 23 B 57, transl. by K. Freeman. According to Aristoxenus (fr. 45), *Politeia* was written by a certain Chrysogonus. The author of the Hippocratic treatise *On Diet* (11) also maintained that the divine mind taught men the crafts, which are, according to him, the imitation of the divine nature. On the 'secondary sacralization' of inventors in Plato, see below, 6.3. The Epicureans, by contrast, and Diogenes of Oenoanda in particular, contended that τέχναι do not owe their origin to gods, but arise because of circumstances and needs (fr. 12 II, 4–11 Smith).

72 For material, see Kleingünther, *op. cit.*, 47ff.

tronomy is not indicated, but judging by the historian's polemical tone, he was not ready to accept that all astronomical knowledge derives from Egypt. (This may have been an idea of his precursor Hecataeus.) Considering other assertions by priests, namely, that the Egyptians were one of the most ancient peoples on earth (II, 2), the first to erect altars, statues, and temples to gods, and further, that the names of the twelve principal gods of the Greeks come from Egypt (II, 4, 43, 50), we ought to admit that we are dealing not so much with Herodotus' own guesses as with the purposeful propaganda of Egyptian priests suggesting to Greek travelers the idea of the superiority of local culture and religion in particular.[73] Most of the 'discoveries' mentioned in book II belong to religion, which obviously concerned priests first of all. In late Egypt, calendar astronomy was in the hands of priests, so their claim appears to be quite logical. Land surveying, the domain of trained specialists, the so-called *harpedonaptai*,[74] apparently did not evoke such claims. The Egyptian origin of geometry seems then to be Herodotus' own conclusion, a quite natural one, taking into account how old Egyptian civilization was.[75] If geometry appeared first in Egypt and then in Greece, any other conclusion was ruled out. The possibility of two independent discoveries was not even considered.[76] Herodotus' inference must have seemed all the more conclusive to him, since he says that the Egyptians avoid adopting not only Hellenic customs but anything foreign at all (II, 91).

In view of Herodotus' general tendency, it appears strange that while repeatedly praising Egyptian medicine he does not say that it was borrowed by the Greeks. The Egyptians are the healthiest people in the world, with the exception of the Libyans. They live healthily (II, 77) and their medicine is at such a high level that the whole country is full of doctors, each specializing in a particular kind of ailment, for example, ocular, dental, internal, etc. (II, 84). Nevertheless, neither in book II nor in the passage about Greek physicians (III, 125, 129–137) is there any hint that medicine originated in Egypt. Furthermore, the story of the Crotonian physician Democedes, who cured the Persian king Darius after the best Egyptian doctors failed in the case (III, 129), seems to demonstrate the superiority of Greek over Egyptian medicine. The historian's reserve in this matter may well be one of the reasons why the idea of the Oriental origin of medicine never enjoyed particular popularity in later literature.[77]

---

73 Vogt, *op. cit.*

74 See Democr. 68 B 299; Gands, S. Die Harpedonapten oder Seilspanner und Seilknüpfer, *Q & St* 1 (1930) 255–277.

75 Lloyd, A. B. *Herodotus Book II. Commentary 1–98*, Leiden 1976, 34.

76 Edelstein's objections (*op. cit.*, 88) are based on a misinterpretation of an Aristotelian passage (*Pol.* 1329b 25f.) that says that the same things are invented in different *successive* civilizations and then get lost because of catastrophes. See *Aristoteles. Politik. Buch II*, transl. by E. Schütrumpf, Berlin 1993, 205f.; cf. *Cael.* 270b 19f., *Met.* 1074b 10f.

77 Cf., however, Isoc. *Bus.* 22 and below, 8.4.

At one point Herodotus corrects the notion of Egyptian superiority in astronomy, ascribing the invention of the gnomon and polos and the division of the day into 12 hours to the Babylonians.[78] He may not have seen a gnomon in Egypt and may have noticed it only in Babylon, though this instrument was known, in fact, in both cultures. As for the polos, Herodotus could not have seen it outside Greece, for its hemispherical form implies a notion of the heavenly sphere that was foreign to both the Babylonians and the Egyptians. The division of the day into 12 parts (by analogy, probably, with the division of the year into 12 months) was known in Egypt already in the second millennium, and it is from Egypt, not Babylon, that the Greeks must have borrowed it. Thus, neither confidence in priests nor independent reasoning could protect Herodotus from mistakes, despite all his aspiration to truth.

While the Egyptian and Phoenician discoveries were mentioned in the earlier Greek literature, Herodotus was the first to refer to borrowings from Babylon. Though he does not call Babylonians the *prōtoi heuretai*, this is clearly suggested by the context of all his accounts of Greek borrowings from them. Another way Herodotus differs from the preceding heurematography is that he writes of scientific discoveries, or at least of the things that came to be understood as such later. After Herodotus, the idea that geometry originated in Egypt and astronomy in Babylon (or in Egypt) became commonplace and survived until the end of Antiquity; later it passed into medieval and early modern historiography. Since the fourth century, Egypt and Babylon are joined by Phoenicia as the motherland of arithmetic (Eud. fr. 133; cf. Pl. *Leg.* 747 a–c). Herodotus does not assert this directly, but in his account of the origin of the Greek alphabet in Phoenicia he notes: “These Phoenicians who came with Cadmus… among many other kinds of learning (ἄλλα τε πολλὰ διδασκάλια) brought into Hellas the alphabet” (V, 58). It is hard to say definitely whether these “kinds of learning” included the art of calculation, but this conjecture seems to me quite plausible.

What can explain Herodotus’ persistent efforts to emphasize the non-Greek origin of many discoveries and, further, to interpret typically Greek customs as borrowings? The reason is hardly his individuality as a historian, nor the *barbarophilia* Plutarch imputed to him. Isocrates, Plato, and Aristotle, who were not noted for sympathy with the ‘barbarians’, admitted as well that they were the teachers of many discoveries. To answer this question, we have to understand a quite peculiar way the Oriental borrowings were reflected in Greek tradition. Archytas’ fragment on the two ways of acquiring new knowledge (47 B 3), mentioned above, reveals an important and widely employed pattern of Greek thought. One can either learn something new from another and with another’s assistance or discover it himself and by his own means. As a matter of fact, the companion notions of ‘learning/imitation – discovery’ (μάθησις/μίμησις – εὕρεσις) were one of the few available instruments for explaining

[78] On this subject, see Zhmud. *Wissenschaft*, 206 f.

the origin of cultural phenomena.[79] All 'discoveries' (i.e., any element of civilization) not attributed to gods and in want of authors may be said to have fallen into two groups: indigenous and borrowed. The foreign inventors, apart from such 'personalities' as Busiris, Cadmus, or Anacharsis, remained as a rule anonymous, their names being of little interest,[80] while the Greek *prōtoi heuretai* tended to be personified. This scheme may account for the importance of travel to the Orient as one of the major means of learning and transmitting knowledge. Such a voyage was traditionally assigned to nearly every one of the famous thinkers from Thales and Pythagoras to Democritus and Plato. Before setting to the task of invention himself, a sage had to study with his teacher and then make a journey to Egypt, Babylon, or, at least, to the Persian Magi; this was considered an indispensable part of his education.

This prevailing scheme implies that the Greeks were ready to admit their substantial debt to their Oriental neighbors. The problem is that they did it in a very inadequate way. A contemporary list of the safely attested Oriental borrowings made ca. 850–500 BC is quite impressive; it includes dozens of things in the most diverse areas of technology and culture;[81] a lot more things are still disputed. But this list owes very little to direct references in the Greek sources. Furthermore, it is for the most part incompatible with the respective Greek lists of the Oriental 'discoveries'. With a few notable exceptions, the Greek tradition either passes over in silence things that were really taken over or attributes them to its own cultural heroes.[82] On the other hand, as a kind of compensation for this, it persistently ascribes to the most of its neighbors a lot of things they did not invent or did not have at all. The tradition on the borrowings in philosophy is totally fictitious. In the case of *mathēmata* it is distorted, widely exaggerated, and manifestly imprecise. Among the hundreds of references to Egyptian and Chaldaean mathematics and astronomy that fill the Greek literature, at most a few can be accepted as historically correct. The real picture of the Babylonian influence on Greek astronomy would later be reconstructed on the basis of the cuneiform studies and is still a matter of continual debate.[83]

---

79 For more details, see below, 2.3.

80 Kremmer, *op. cit.*, 113f., lists a prodigious number of nation-inventors: Africans, Arabs, Assyrians, Babylonians, Cappadocians, Carians, Chaldaeans, Egyptians, Etruscans, Gauls, Illyrians, Isaurians, Jews, Libyans, Lydians, Memphites, Mysians, Pelasgians, Persians, Samnites, Sicilians, Syrians, Telchites, Troglodytes, Phrygians, Phoenicians (and, separately, Carthaginians, Sidonians, and Tyrians), Thracians. Individuals among the foreign inventors are represented only sporadically.

81 Burkert, W. *The orientalizing revolution*, Cambridge, Mass. 1992; West, *op. cit.*, ch. 1.

82 E.g. weights, measures, the chariot, the olive tree, musical instruments, etc. Writing was widely but by no means universally acknowledged to have been taken from Phoenicians. See above, 26 n. 10.

83 Neugebauer. *HAMA* II, 589ff.

Thus, the model 'invention – imitation' was a rational, though unsuccessful attempt to account for such wide range of phenomena as the unprecedented social and cultural creativity of the archaic and classical epochs, the character and paths of influences from the Orient, the correlation of indigenous and borrowed in Greek culture. If even modern scholarship still fails to explain some of these things adequately, the ancient Greek scheme obviously could offer, at best, only a very approximate answer to the question of 'who discovered what'.

On the whole, by the turn of the fifth century BC, when heurematography created its own particular genre, a kind of catalogue of achievements under the standard title Περὶ εὑρημάτων,[84] Greek thought had already acquired a persistent tendency to associate a considerable share of its own civilization with the influence of neighbors, especially Oriental neighbors.[85] It is not always possible to figure out in each particular case why Greek authors gave preference to foreign *prōtoi heuretai*. While the version of the invention of the alphabet by Cadmus the Phoenician leaned, in the final analysis, on historical tradition,[86] Ephorus' idea of the invention of the anchor and potter's wheel by Anacharsis seems to be utterly absurd.[87] Admittedly, Ephorus, known for his extreme idealization of the Scythians, may have chosen Anacharsis because the legendary Scythian was an itinerant sage, i.e., a person who combined 'imitation' and 'invention' and thus exemplified the ideal of a *Kulturträger*. Oriental sages did not visit Greece often, so more often than not the role of a *Kulturträger* was performed by a Greek sage: after traveling to the Orient, he brought some discoveries home, adding to them some of his own. Eudemus' *History of Geometry*, for instance, attributes its discovery to the Egyptians, in full agreement with Herodotus and Aristotle (*Met.* 921b 23). It is said, further, that Thales, having traveled to Egypt, was the first to bring this science to Greece and to discover in it some important things (fr. 133–135). Repeatedly reiterated in the later tradition, such constructions reveal their genetic kinship with heurematography; at the same time they demonstrate how narrow was the range of means available to Greek historiography of culture.

The fact that the history of science was practiced in the Lyceum along with heurematography testifies that, by the fourth century, interest in *prōtoi heuretai* had grown more differentiated and the paths of the two genres had parted.[88]

---

84 Simonides of Keos the younger (second half of the fifth century) is mentioned as the author of Εὑρήματα (*FGrHist* 8 T 1). Scamon of Mytilene (the son of Hellanicus of Lesbos) is considered one of the earliest authors of Περὶ εὑρημάτων (Athen. XIV, 637b; *FGrHist* 476 F 4). See Jacoby, F. Skamon von Mytilene, *RE* 3 AI (1927) 437.

85 Kleingünther, *op. cit.*, 151.

86 Edwards, R. B. *Kadmos the Phoenician*, Amsterdam 1979, 174f.

87 *FGrHist* 70 F 42. No wonder this idea was soon contested by Theophrastus (fr. 734 FHSG).

88 Heraclides Ponticus (fr. 152), Theophrastus (fr. 728–734 FHSG), and Strato (fr. 144–147) were the authors of works Περὶ εὑρημάτων. Aristotle also wrote on

Theophrastus' and Strato's heurematography is little different from the earlier specimens of the genre; its preoccupation with the prehistoric past precluded in itself a critical approach to facts, so that the names of first discoverers and thought patterns remain the same. In Theophrastus' *On Discoveries*, Prometheus figured as the founder of philosophy (fr. 729 FHSG), whereas in his doxography this name was, of course, lacking. Properly speaking, the history of science and heurematography coincide thematically only in regard to the initial, 'prehistoric' period to which the emergence of science was related and on which little definite was to be said. Eudemus, agreeing with Herodotus on the Egyptian origin of geometry, immediately leaves this topic and proceeds to describe particular discoveries made by Greek mathematicians. Philip of Opus, having admitted the Oriental origin of astronomy (*Epin.* 986e – 987 A), concludes with the famous remark: the Greeks bring to perfection what they borrow from the 'barbarians' (987d–e). It is the course of these documentarily attested improvements that the history of science actually dealt with, while heurematography remained, as a rule, on the level of 'initial', often fictitious inventions and borrowings.

The obvious continuity between heurematography and a number of trends in Peripatetic historiography should not be regarded as one stream flowing smoothly into another. Among the important intermediate links between them are the theories of the origin of culture that emerged in the second half of the fifth century. They connected notions of ζήτησις and εὕρεσις with the new concepts of τέχνη put forward by the Sophists and gave powerful stimulus to the study of culture in all its aspects. Most of these theories are known to us in fragments and paraphrases; the only one that has survived in full is found in the Hippocratic treatise *On Ancient Medicine* (*VM*). Though its author is only a generation younger than Herodotus, his views on the development of his τέχνη, medicine, seem much more mature than the naïve genealogical constructions of the historian. Going far beyond the cursory mentions of 'discoveries' that are unrelated to each other, his original and integral conception considers the invention of medicine against a background of the progress made by human civilization as a whole. In the late fifth century, the search for *prōtoi heuretai* obviously acquires a new dimension, which is reflected in systematic attempts to create both a general theory of the origin of culture and the history of individual τέχναι. These trends, independent of their kinship with heurematography, deserve special consideration – both in themselves and as forerunners of the historiography of science. It is, accordingly, time to draw a preliminary conclusion from our survey of the early heurematographic tradition.

* * *

this subject (Plin. *HN* VII, 194–209 = fr. 924 Gigon); Eichholtz, *op. cit.*, 24f.; Wendling, *op. cit.*, passim.

Behind the variety of answers given by writers of the sixth and fifth centuries to the question of 'who discovered what?' one can see the obvious tendency toward the growing complexity of these answers. The question itself often appears to be merely a pretext to expound one's views on the particular phenomenon or to explain the dynamics of evolving civilization in general. This tendency derives, clearly, not so much from heurematography as such – after the early fourth century it had hardly undergone serious change – but rather from the general process of rationalization, the rapid development of philosophy and science, and the growing self-awareness of their proponents. By the late fifth century, Greek culture had actually acquired a history of its own. It is from this time that we have the first reconstructions of mankind's distant past and the first attempts to devise a thorough and systematic account of the history of music and poetry and to explore the origins of medical, philosophical and, probably, scientific thought.

Little more than a hundred and fifty years passed between the earliest surviving mention of *prōtoi heuretai* and the emergence of philosophical theories on the origin of culture. Less than a century separates Xenophanes, who first openly questioned the divine origin of discoveries, from the *VM* that presents original and profound insights into the history and methodology of science. The changes in the search for first discoverers are as obvious as they are drastic. But are we justified to consider all of them changes in *one and the same* phenomenon, i.e., in heurematography? Such an approach seems to me legitimate only to the degree that the question of 'who discovered what?', formulated in early heurematography, remains valid both for the author of *VM* and for those who created the history of science in the fourth century. Still, we should keep in mind that the evidence of *Phoronis* on the Idaean Dactyls lies on the same plane as the material of both the earliest and the latest Greek catalogues of discoveries.[89] The character of the questions raised and the answers offered within heurematography practically did not change through the centuries of its existence. Meanwhile, in the history and theory of culture and, later, in the Peripatetic history of science, each of the initial question's elements – the 'who', the 'what' and the 'discovered' – becomes part of new conceptual fields, serves to express new ideas, and, accordingly, takes on different meanings. Each of these elements could, depending on the context, become prominent or sink into the background. The personality of *prōtos heuretēs* was of far greater importance for the history of poetry than for theory on the origin of culture, preoccupied as the latter was with the driving forces of civilization. The history of science discarded some of the variants of the answer to 'who?' and added a new question – 'how?' – to the traditional 'what?' Thus, in a wider perspective, the history of science proves to be not the direct descendant of the tradition on first discoverers, but rather the offspring of several trends of Greek thought working together.

---

89 On Roman and Christian catalogues of discoveries, see Thraede. Erfinder, 1232f., 1247f.

Chapter 2

Science as τέχνη: theory and history

1. The invention of τέχνη

In the second half of the fifth century, most activities involving skills based on knowledge and experience were subsumed under the notion of τέχνη. Initially a term used in handicraft, this notion was thoroughly practical. The purpose of τέχνη was to help people, either by improving their life (agriculture, medicine, house building) or by embellishing it (music, poetry).[1] With the Sophists, too, who appeared on the intellectual scene at this period, knowledge was, as a rule, looked upon as purely utilitarian. Apart from rare exceptions – for example, Hippias of Elis, who taught mathematical sciences – most Sophists taught things presumed to be helpful in making a career. For σοφιστικὴ τέχνη that aspired to make people wise and happy, such disciplines as geometry and astronomy were of no use.[2] The same common-sense attitude toward theoretical mathematics was characteristic both of Socrates, who did not much differ in this respect from the majority of the Sophists,[3] and of Isocrates, their rightful successor.[4] Similar, though more differentiated, was the Sophists' attitude to natural philosophy. For those who conceived of their occupation as τέχνη, the μετεωρολογία of the Presocratics was synonymous with fruitless discussions on idle subjects (ἀδολεσχία) that do not result in any kind of firm knowledge.[5]

---

1 The division of arts into 'useful' and 'pleasurable' is first found in Democritus (68 B 144).

2 Protagoras criticized the proposition that a tangent touches a circle at one point (80 B 7). Antiphon and Bryson, using sophistic rather than geometrical methods, attempted unsuccessfully to solve the problem of squaring the circle (Arist. *Cat.* 7b 27f., *APo* 75b 37f., *SE* 171b 12f., *Phys.* 185a 14f.; Eud. fr. 139–140, cf. 59 A 38 on Anaxagoras). That is about all we know of Sophists' preoccupation with the exact sciences.

3 On the utilitarian attitude of Socrates toward the exact sciences, see Xen. *Mem.* IV,7.1–8 (in this case Xenophon is more reliable than Plato). The Socratics Antisthenes and Aristippus also took a negative view of science. See also Olson, R. Science, scientism and anti-science in Hellenic Athens: A new Whig interpretation, *HS* 16 (1976) 179–199.

4 *Antid.* 261–266, *Panath.* 26–29. See his programmatic statement: "It is much better to have an approximate idea of useful things than the exact knowledge of useless ones." Isocrates does not deny, however, the pedagogical importance of mathematics (see below, 74).

5 Eur. fr. 913 Nauck; Ar. *Nub.* 1480f.; Gorg. *Hel.* 13 = 82 B 11; *VM* 1; *De aere* 2; Isoc. *Antid.* 268; cf. also μετεωρολογία = ἀδολεσχία in Plato (*Crat.* 404b 7, *Res.* 488e

With time, this initially practical conception of τέχνη, opposed both to natural philosophy and to *mathēmata*, took a more and more intellectual turn, until it finally served as an interpretative model of science itself. To a considerable degree, this change can be accounted for by the fact that the circle of disciplines taught by the Sophists included subjects related to intellectual activities that, though practically oriented, had little to do with traditional handicrafts. The very novelty of their pedagogical practice made it necessary for the Sophists to explain and justify it by arguing that the subjects they taught qualified as τέχνη, since they involved both skill and knowledge. As a result, the Sophists considerably enriched and developed the notion of τέχνη. Their methodological research on the problem of what τέχνη is and under what conditions it actually emerges laid the basis for the history of culture, understood as the sum total of different τέχναι.[6]

Compared to modern views on handicraft, art, and even technology – the notions we now use to render τέχνη – the intellectualism of the Sophistic and, on the whole, of the classical understanding of τέχνη seems highly unusual. It can, at best, be compared to the intellectualism, no less remote from us, of Greek ethics, which made it possible for Socrates to use the notion of τέχνη in creating the new science of philosophical ethics. The Socrates of Plato's dialogues quite naturally uses the word, which originally referred to the art of a cook or a stonemason, in discussing intellectual and moral problems. This seems to indicate that he relied not only on the common use of the word, but also on the *theory* of τέχνη that had already been developed by the Sophists. F. Heinimann, who studied this theory, gives these as the common characteristics of a τέχνη: 1) τέχνη is meant to be useful; 2) each τέχνη serves a definite purpose: medicine keeps one healthy, agriculture provides one with food, etc.; 3) τέχνη is based on the knowledge of specialists who are in command of all means necessary to their ends; and 4) each τέχνη can be transferred by teaching; only that which can be transferred by teaching is entitled to be called a τέχνη.[7] It is obvious that these characteristics are applicable to more than art or handicraft.

8, *Phdr.* 270a 4, *Polit.* 299b 7). See Capelle, W. ΜΕΤΕΩΡΟΛΟΓΙΑ, *Philologus* 71 (1912) 414–448. Admittedly, some of the Sophists show a certain interest in natural philosophy.

6 The problem of the origin of culture interested Archelaus as well (60 A 4), but his ideas do not appear original, compared with those of his Sophist contemporaries. We know little of the views of his teacher Anaxagoras on this subject (59 B 4, 21); their reconstruction (Uxkull-Gyllenband, W. *Griechische Kultur-Entstehungslehren*, Berlin 1924, 6ff.) is highly hypothetical. One of the earliest theories of the origin of culture belongs to Democritus (for its reconstruction based on later texts, see Cole, *op. cit.*), but even that is believed to have been influenced by his older contemporary Protagoras (Uxkull-Gyllenband, *op. cit.*, 32; Emsbach, M. *Sophistik als Aufklärung: Untersuchungen zu Wissenschaftsbegriff und Geschichtsauffassung bei Protagoras*, Würzburg 1980, 202ff.).

7 Heinimann, F. Eine vorplatonische Theorie der τέχνη, *Mus. Helv.* 18 (1961) 105f.

Taken together, "they form a genuine theory of science (*Wissenschaftslehre*) – science understood as τέχνη, whose final aim is practical use rather than theoretical knowledge."[8]

In the traditional pairing of skill and knowledge, it is knowledge that gradually comes into the foreground. Particular attention is paid to its origin, acquisition, and application. In the course of the fifth and the greater part of the fourth centuries, the notion of ἐπιστήμη – which originally meant 'knowledge' and later came to mean 'science' as well – is used as a synonym for τέχνη.[9] The newborn scientific disciplines, such as mathematics, are also treated within the framework of the same model. It is revealing that unlike such old Ionic terms as ἀστρονομία/ἀστρολογία and γεωμετρία, the names of the scientific disciplines born in the fifth century – ἀριθμητική, λογιστική, and ἁρμονική[10] – are all focused on the notion of τέχνη; the fourth century added such terms as μηχανική, ὀπτική, and some others.[11]

The treatment of scientific disciplines on the model of τέχνη was strongly marked by a certain dynamic quality peculiar to this notion. This quality finds its best expression in the concepts of ζήτησις and εὕρεσις.[12] Τέχναι are understood as systematic 'research', 'discovery', or 'invention' of new things, new knowledge or new skills. Everything that is known and accessible today results either from the 'discoveries' made in the course of constant 'research' by our predecessors, or from μίμησις, imitation.[13] That τέχνη could be transferred by

---

8 Ibid., 106.

9 Snell, B. *Die Ausdrücke für den Begriff des Wissens in der vorplatonischen Philosophie*, Berlin 1924, 86f.; Schaerer, R. ΕΠΙΣΤΗΜΗ et ΤΕΧΝΗ. *Etude sur les notions de connaissance et d'art d'Homère à Platon*, Mâcon 1930; Isnardi Parente, M. *Techne. Momenti del pensiero greco da Platone ad Epicuro*, Florence 1966. Used in this meaning, ἐπιστήμη generally designated the part of τέχνη related to knowledge and cognition, rather than to practical skills. At the same time, ἐπιστήμη could refer to merely practical abilities, as well (Xen. *Oecon.* I, 1; VI, 8; Isoc. *Antid.* 213, 252).

10 Μουσική (Pind. *Ol.* I, 15) and ἰατρική (Hdt. II, 84; III, 129) are soon followed by ἀριθμητική and ἁρμονική (Archytas, B 1–3), λογιστική (Archytas, B 3; Xen. *Mem.* I,1.7). The widespread use of words with the suffix -ικος is often associated with the Sophists, particularly with their attempts at classifying the new τέχναι (Ammann, A. N. -ικος *bei Platon*, Freiburg 1953, 267f.). Cf. γεωδαισία (Arist. *Met.* 987b 26) and στερεωμετρία ([Pl.] *Epin.* 990d 8; Arist. *APo* 78b 38), formed after the model of γεωμετρία.

11 Μηχανική and ὀπτική are first mentioned in Aristotle (*APo* 75b 16, 76a 24, 77b 2, 78b 37; *Met.* 997b 20, 1078a 14–16) and in a quotation from an Academic treatise (Dorandi. *Filodemo*, 127.5; see below, 87), figuring in the *Second Analytics* as accomplished scientific disciplines. Aristotle sometimes referred mechanics to the τέχναι (*Mech.* 847a 18f.) and sometimes to the ἐπιστῆμαι (*APo* 78b 37); in Archytas, μαθήματα are still a part of τέχναι (see below, 61f.).

12 See above, 34f.

13 According to Democritus, people learned weaving from spiders, house building from swallows, and singing from songbirds (68 B 154). Aristotle wrote that the best

teaching ensured the handing on of knowledge (μάθησις) from teachers to their disciples. This model allowed a move from sundry mentions of the *prōtoi heuretai* to the systematic analysis of the origin and development of arts and sciences, both being conceived as a history of discoveries. Later, when τέχνη and ἐπιστήμη gradually separate and ἐπιστήμη turns from 'knowledge' functioning as the cognitive aspect of τέχνη into independent theoretical 'science',[14] some of the characteristics peculiar to τέχνη are transferred to ἐπιστήμη. Aristotle, for example, considered it characteristic of ἐπιστήμη that it can be transferred by teaching (*EN* 1139b 25, cf. *Met.* 981b 7–10). Another important feature of ἐπιστήμη, its usefulness (χρήσιμον, ὠφέλιμον), figured as a standard rubric not only in manuals of rhetoric, medicine, and tactics,[15] but also in the 'introductions' to exact sciences, in mathematical and astronomical treatises, in commentaries on them, etc.[16]

The development of historical views on τέχνη follows two main courses. The interest in how the τέχναι, emerging in succession have fashioned modern civilization gives birth to a teaching on the origin of culture (*Kulturentstehungslehre*). This teaching, dealing as a rule with the remote past, i.e., with the preliterate and hence prehistoric period, is not to be identified with the history of culture (*Kulturgeschichte*), the latter starting with the first dated events and (quasi-)historic personages. To be sure, the chronology of the 'heroic' epoch was quite artificial and its heroes belonged to the realm of legend, but since the Greeks themselves had always considered the events of the Trojan War to be their early history, this period can be regarded as a conventional chronological boundary between *Kulturentstehungslehre* and *Kulturgeschichte*. Democritus' theory of the origins of culture, for example, ended with the epoch preceding the Trojan War. The invention of writing marks the boundary between history and prehistory, so that the further development of music and other τέχναι be-

---

tools, such as the compass and the ruler, were discovered by observing nature and imitating it (*Protr.* fr. 47–48 Düring). In the treatise *On Diet* (11–24), this theory is brought to the point of absurdity: each τέχνη appeared in imitation of human nature (Joly, R. *Recherches sur le traité pseudo-hippocratique Du régime*, Paris 1960, 52f.).

14 In Plato the two notions are still, as a rule, synonymous (see below, 125).

15 See Heinimann, *op. cit.*, 117 n. 58.

16 Archimedes' book *On Measuring the Circle*, as his biographer Heraclides claims, is "indispensable for the necessities of life" (*FGrHist* 1108 F 1); see below, 294. For a vivid defense of the utility of *mathēmata*, see e.g. Iambl. *De comm. math. sc.*, 79.1 ff. Utility was often understood not in a practical sense, as e.g. the usefulness of mechanics (Papp. *Coll.* VIII, 1022.1 ff.), but in a merely formal one: such a text is useful for understanding the theory of conic sections. Ptolemy considered mathematics to be useful for studying theology and physics (*Alm.*, 7.4f.); this opinion was shared by Proclus (*In Eucl.*, 21.25f.). For further material, see Mansfeld, J. *Prolegomena mathematica: From Apollonius of Perga to the late Neoplatonists*, Leiden 1998, 4, 20f., 66 n. 229, etc. and the *utility* rubric in the index, 173.

longs to the subject of the history of culture.[17] This boundary is to some extent conventional, for *Kulturentstehungslehre* did not always end where *Kulturgeschichte* had already stepped in. Thucydides in the *Archaeology* and Dicaearchus in the *Life of Hellas* pass from the prehistoric state to historical events. Nevertheless, this criterion allows us to classify the theories of Archelaus, Protagoras, Democritus, and the author of *VM* as *Kulturentstehungslehren*, since their heroes were anonymous first discoverers of prehistoric times, and not historical or quasi-historical figures like Linus or Cadmus.

As for the history of culture itself, one of its early forms was the history of separate τέχναι. Poetry, for example, was considered in Glaucus of Rhegium's work Περὶ τῶν ἀρχαίων ποιητῶν καὶ μουσικῶν.[18] Starting with the legendary inventors of music, Orpheus and Musaeus – with whom Democritus (68 B 16) apparently ended – Glaucus passed on to Homer and, after him, to the poets of the archaic period (Terpander, Stesichorus, and Archilochus) to end, probably, with his own time (fr. 1–6 Lanata).[19] Glaucus paid particular attention to two interrelated problems: priority in musical discoveries and the relative chronology of the musicians, which made it possible to establish who influenced whom.[20] Glaucus was, to all appearances, the first who endeavored to order the historical material according to the principle of *prōtos heuretēs*.[21] The same problems are even more closely connected in Hellanicus of Lesbos' Καρνεονίκαι, a work devoted to the winners of the musical agones at the Carnea festivals in Sparta. The fragments of this chronicle of musical events say that Terpander was older than Anacreon, being "the first of all" to win at the Carnea (*FGrHist* 4 F 85 a) and that Lasus of Hermione was "the first to introduce κύκλιοι χοροί"

---

17 Cole, *op. cit.*, 41 f., 57. Democritus dated his Μικρὸς διάκοσμος to 730 after the fall of Troy (68 B 5) and, consequently, might have taken interest in the chronology of that epoch. According to Theophrastus, the discoverers of arts and sciences had lived about a thousand years earlier (fr. 184.125 f. FHSG). Proclus, referring to the history of discoveries (ἱστορία περὶ εὑρημάτων), claimed that γράμματα καὶ τέχναι had been discovered not a very long time before (*In Tim.*, 125.11 f.). See already Pl. *Leg.* 677 c–d.

18 Glaucus' fragments preserved, for the most part, in Ps.-Plutarch's *De musica* (Barker. *GMW* I, 205 ff.) are collected in: Lanata, G. *Poetica pre-platonica*, Florence 1963, 270–281; see also Huxley, G. Glaukos of Rhegion, *GRBS* 9 (1968) 47–54; Fornaro, S. Glaukos von Rhegion, *DNP* 4 (1998) 1093 f.

19 Fr. 6 mentions Empedocles, fr. 5 Glaucus' contemporary Democritus. The latter, though not known as a poet, wrote on music and poetry a great deal (68 A 33, X–XI).

20 Jacoby, F. Glaukos von Rhegion, *RE* 7 (1910) 1417–1420. We know practically nothing of the book Περὶ ποιητῶν καὶ σοφιστῶν by his contemporary Damastes of Sigeum (*FGrHist* 5 T 1), though it follows from some evidence that εὑρήματα (F 6) and the problems of chronology (F 11) also preoccupied him. Writings dealing with individual poets, e.g. *On Homer* by Stesimbrotus of Thasos (*FGrHist* 107) or *On Theognis* by Antisthenes (VA 41 Giannantoni), fall outside our subject.

21 Blum, R. *Kallimachos. The Alexandrian library and the origin of bibliography*, Madison 1991, 19 f.

(F 86). Another chronological writing of Hellanicus is the list of the priestesses of the sanctuary of Hera in Argos (*FGrHist* 4 F 74–84). Still more important for the later chronology is the list of Olympic victors compiled by Hippias of Elis (*FGrHist* 6 F 2). Another work by Hippias, Συναγωγή, can be regarded as the first treatise on the history of ideas, the predecessor of the Peripatetic doxography.[22] In this book Hippias tried to find similarities between the ideas of the ancient poets, on the one hand, and the prose writers, mainly philosophers, on the other.

These early researches, continued by Aristotle,[23] served as a model for the Peripatetic histories of music, including such works as Συναγωγὴ τῶν ἐν μουσικῇ <διαλαμψάντων> by Heraclides of Pontus,[24] Περὶ μουσικῆς by Aristoxenus (fr. 71–89), whose first book deals with the history of musical 'inventions',[25] and Dicaearchus' treatise on musical agones (fr. 75–76, 85). Hippias played an important role in Peripatetic doxography as the author who conveyed information about Thales and the other early figures;[26] his book was one of the sources for Eudemus' *History of Theology* and *History of Geometry*. The problems raised by Glaucus and the chronological principle of organizing cultural discoveries introduced by him also exerted strong influence on Aristotle's and Eudemus' works.

More often than not, the genealogy of different τέχναι was considered not in special writings, but in prefaces or introductions to the systematic treatises.[27]

---

22 Patzer, A. *Der Sophist Hippias als Philosophiehistoriker*, Freiburg 1986.

23 See e.g. the lists he compiled of the winners at the Dionysiac agones (D. L. V, 26 No. 135) and the Olympian and Pythian games (D. L. V, 26, No. 130–131, fr. 615–617 Rose = fr. 408–414 Gigon; Blum, *op. cit.*, 20ff.), the dialogue *On Poets* (fr. 70–77 Rose = fr. 14–22 Gigon), where much attention is paid to the founders of various genres, and, in particular, the history of rhetoric in Τεχνῶν συναγωγή (see below, 4.3).

24 Fr. 157–163. Lasserre completed the title of this work as Συναγωγὴ τῶν <εὑρημάτων> ἐν μουσικῇ. Wehrli, however, considered it to be identical to Heraclides' Περὶ μουσικῆς (cf. D. L. V, 87). In the latter case, its historical part might have been limited to the first two books, and the title Συναγωγή may have appeared at a later date (Wehrli. *Herakleides*, 112); cf., however, Hippias' Συναγωγή and Aristotle's Τεχνῶν συναγωγή (see below, 51 n. 31).

25 Wehrli. *Aristoxenos*, 69f. The historical evidence in Ps.-Plutarch's *De musica* goes back mainly to Glaucus, Heraclides, and Aristoxenus.

26 On the influence of Hippias on Aristotle's doxography, see Mansfeld. Aristotle, 28ff.; Patzer, *op. cit.*

27 This genre of 'manual' goes back to the epoch of the Sophists (Fuhrmann, M. *Das systematische Lehrbuch*, Göttingen 1960, 122ff.). Among such manuals are works on medicine ([Hipp.] *De arte*), horse breeding (see Xen. *Eq.* I, 1 on his precursor Simon), gymnastics (Iccus of Tarentum, *DK* 25), architecture (Hippodamus, *DK* 39), scenography (Agatharchides, Vitr. VIII, praef. 11), sculpture (Polyclitus *DK* 40), music (Damon, *DK* 37), rhetoric (Tisias, Protagoras, Gorgias, and Critias), mathematics (*Elements* by Hippocrates of Chios, 42 A 1), and harmonics (Archytas, 47 B 1).

From the late fifth century on, the topic *origo artis* becomes a regular part of the 'introductions' to various τέχναι and ἐπιστῆμαι[28] and continues to be reproduced in various forms until the end of Antiquity (8.4). Depending on the availability of the respective material and the interests of the author, this topic could be limited to a eulogy of a legendary or real founder of a particular branch of knowledge[29] or to a brief survey of precursors' achievements;[30] it could take the form of a *Kulturentstehungslehre*, as in the case of *VM*, or that of a short history of the τέχνη in question, as in Aristotle's Τεχνῶν συναγωγή,[31] Aristoxenus' *On music*, or Celsus' *De medicina*. The rich legendary and historical tradition concerning the inventors of music and poetry gave to these τέχναι an advantage over such disciplines as, for example, medicine. Even if he had attempted to, the author of *VM* would hardly be able to write the history of his τέχνη starting, say, with the time of Homer. No wonder his 'history' took the shape of the *Kulturentstehungslehre* and did not mention a single name. By the time of Celsus (first century AD), Greek medicine had a long and glorious history marked by a line of famous physicians starting with Asclepius and his sons Podalirius and Machaon.

The synchronization, already traditional by the fifth century, of the origin of music and poetry with the several generations immediately preceding and following Homer may have helped to form an opinion of the order in which various τέχναι emerged. According to Democritus, it is the necessary τέχναι that, under the pressure of need, emerge first. Later, when a surplus of wealth has been gained, these are followed by τέχναι that serve pleasure – namely, music (68 B 144). Since Democritus considers Musaeus to be the inventor of the most ancient hexameter poetry (68 B 16),[32] he must have dated the invention of music and poetry to the time of Homer's predecessors, i.e., at the very end of the 'prehistoric' epoch. According to Cole's reconstruction of Democritus' theory, the emergence of astronomy should be dated to the same period.[33] Thus, surplus of wealth and the leisure that results from it were put at the service not only of pleasure, but also of knowledge. The simultaneous emergence of arts

---

28 Heinimann, *op. cit.*, 117 n. 55.

29 Radermacher, L. *Artium scriptores* (*Reste der voraristotelischen Rhetorik*), Vienna 1951, 1ff.

30 E.g. Isoc. *Antid.* 180–181; Arist. *SE* 183b 29f. See Mansfeld. *Prolegomena*, the rubric *historical note/overview* in the index, 173.

31 Fr. 136–141 Rose = fr. 123–134 Gigon. Much historical evidence on the first steps of rhetoric contained in the late introductions into this discipline might be traced to this treatise (Rabe, H., ed. *Prolegomenon Sylloge*, Leipzig 1931, VIIIf.; Radermacher, *op. cit.*, 11ff.).

32 Kleingünther, *op. cit.*, 107f.; Luria, S. *Democritea*, Leningrad 1970, 568f.; Cole, *op. cit.*, 42f., 57.

33 Cole, *op. cit.*, 42f. This idea, though not occurring as such in Democritus' fragments, is found in the theories of Diodorus (I,16.1) and Lucretius (V, 1437f.), which probably go back to Democritus.

and sciences distinguishes Democritus' theory from that of Aristotle, which included three stages: 1) necessary τέχναι; 2) arts (music, in particular); 3) sciences and philosophy aimed at pure knowledge.[34] We do not know whether the notion of σχολή, so important for Aristotle, figured in Democritus, but it is this notion that, from the early fourth century, marked the transition from the invention of necessary crafts to arts and sciences.

In *Busiris,* for example, one of Isocrates' earlier speeches (ca. 390), this legendary legislator is said, first, to have provided the Egyptians a sufficiency (ἀναγκαῖα) and even abundance of goods (εὐπορία, περιουσία, 12–15). Afterward, he divided the people into three classes: priests, warriors, and craftsmen and peasants (15). The priests, spared the necessity of toil and war and enjoying affluence and leisure (εὐπορία, σχολή, 21), invented (ἐξεῦρον) medicine and philosophy (22) and also went in for astronomy, geometry, and arithmetic (23). This theory, like that of Democritus, implies two stages in the development of τέχναι, with the difference that Isocrates makes no mention of music and other arts that serve pleasure and seemed irrelevant in the context of Egyptian history, but passes immediately on to medicine, sciences, and philosophy.[35]

The frequency with which the subject of inventors crops up in Isocrates' speeches, including the political ones, shows that interest in the development of culture and in the past of humankind was not the prerogative of a narrow circle of intellectuals. Isocrates' importance for us lies not only in his ability "to formulate best what most of his educated contemporaries felt and wanted to say".[36] The main cluster of his ideas and interests was formed on the threshold of the fourth century and underwent no further essential changes, although there are variations depending on the subject and audience. As a result, his later works, along with his earliest ones, can provide material for the analysis of conceptions current at the turn of the century. In *Panegyricus* (ca. 380), which extols the role of Athens in the development of Greek culture, Isocrates combines

---

34 Met. 981b 13–22, 982b 22f.; fr. 53 Rose = *Protr.* fr. 8 Ross = fr. 74.1 Gigon (p. 314b 12f.). Some assert that this fragment comes from *Protrepticus* (Flashar, H. Platon und Aristoteles im *Protrepticos* des Jamblichos, *AGPh* 47 [1965] 66ff.), others that it is from *On Philosophy* (Düring, I. *Aristotle's Protrepticus. An attempt at reconstruction*, Göteborg 1961, 227f.; Effe, *op. cit.*, 68ff.). Spoerri, W. Kulturgeschichtliches im Alpha der aristotelischen "Metaphysik", *Catalepton. Festschrift B. Wyss*, Basel 1985, 45–68, believed that the theory on the origin of culture put forward in *Met.* 981b 13–22, 982b 22f. goes back to both these works. The similar theory, found in *De philos.* fr. 8 B Ross, belongs not to Aristotle, but to Aristocles of Messina, a Peripatetic of the first century AD. See Haase, W. Ein vermeintliches Aristoteles-Fragment bei Johannes Philoponos, *Synusia. Festgabe für W. Schadewaldt*, ed. by H. Flashar, Pfullingen 1965, 323–354; Tarán, L. Rec., *AJP* 87 (1966) 467–468; Moraux, P. *Der Aristotelismus bei den Griechen*, Vol. 2, Berlin 1984, 83ff., 92ff.

35 On the sources of *Busiris*, see Froidefond, *op. cit.*, 246f. (cf. below, 226). Leisure is also mentioned by Plato (*Crit.* 110a).

36 Meyer, E. *Geschichte des Altertums*, 4th ed., Vol. 5, Stuttgart 1958, 329.

a theory of the emergence of culture with a socio-political history. First of all, he writes, Athens took care to provide its citizens with food and the necessities of life (28). Citing as a proof the myth of Demeter and Persephone, the rhetorician turns to more serious arguments:

> But apart from these considerations, if we waive all this and carry our inquiry back to the beginning, we shall find that those who first appeared upon the earth did not at the outset find the kind of life which we enjoy today, but that they procured it little by little through their own joint efforts (κατὰ μικρὸν αὐτοὶ συνεπορίσαντο). Who, then, must we think the most likely either to have received this better life as a gift from the gods or to have hit upon it through their own search (ζητοῦντας αὐτοὺς ἐντυχεῖν)? Would it not be those who are admitted by all men to have been first to exist, to be endowed with the greatest capacity for τέχναι, and to be the most devoted to the worship of gods? (32–33).[37]

Immediately after his praise for the first discoverers in the prehistoric epoch, Isocrates turns to the role of Athens in the colonization of Ionia and the struggle with the 'barbarians' (34–37). Having succeeded and provided for the necessities of life, Athens did not stop there. This city was the first to lay down laws, invent various necessary as well as pleasurable τέχναι, and teach them to others (38–40).[38] The 'co-inventor and co-organizer' of all these wonderful things was rhetoric (47–48), i.e., the kind of φιλοσοφία practiced by Isocrates himself.[39]

In Isocrates' later works, the theory of the origin of culture is even more closely interwoven with the history of rhetoric. Man, he taught, is naturally inferior to many animals in strength, agility, and other qualities. He makes up for it, however, by his innate art of convincing others by means of speech. Owing to this art, people not only abandoned the beast-like way of life, but also, coming together, founded cities, laid down laws, and invented τέχναι.[40] At first, while starting to socialize, all people sought more or less the same things. Since then, however, we have made such progress (ἐπειδὴ δ' ἐνταῦθα προεληλύθαμεν) that both laws and discourses have become innumerable (*Antid.* 81–82). But while laws are respected when old, discourses are held in esteem when new, and those who seek for what is new, will have great difficulty in finding it (καινὰ δὲ ζητοῦντες ἐπιπόνως εὑρήσουσι, 83). Returning to the history of eloquence, or philosophy, Isocrates remarks: some of our ancestors, seeing that many τέχναι had been devised for other things, while none had been prescribed for the body and the soul, invented physical training and philosophy for them (181).

---

37 Transl. by G. Norlin.

38 On the invention of laws, τέχναι, and philosophy in Athens, see also *Hel.* 67; *Panath.* 119, 148.

39 Cf. *Antid.* 181. According to Thraede (Fortschritt, 145), for Isocrates "language and eloquence are the source, the culmination and the guarantee of progress".

40 *Nic.* 5–7, *Antid.* 254–255. See θηριωδῶς already in *Bus.* 25.

Some of the elements of the *Kulturentstehungslehre* presented in Isocrates may well go back to Democritus or one of the Sophists contemporary with him.[41] We need not, however, regard Democritus as *the* creator of study on the origin of culture. He was, rather, the author of one of a number of such theories, whose close affinity to each other need not be explained by direct influence. The most interesting of them from the historico-scientific perspective was a theory of the origin of medicine found in the *VM*. It is worth detailed consideration.

2. The theory of the origin of medicine

Many Hippocratic treatises, which were addressed to a larger public as well as to specialists, were written, in particular, to defend medicine from criticism that its methods were arbitrary and ineffectual.[42] When attempting to explain the nature of the medical art, a Hippocratic physician had the opportunity of stating his views not only of medicine and its method, but also of the differences and affinities between medicine, other τέχναι, and philosophy. Sometimes he spoke of the origin of medicine as well. The *VM,* dated to the last quarter of the fifth century,[43] is the only work of the Hippocratic corpus in which the problem of the origin of medicine is discussed in detail. Other Hippocratics touched upon the topic but slightly. It has been noted above that the author of *VM* did not aim to write a history of medicine as a history of individuals; it is primarily his own understanding of the medical art that he wanted to expound. On the other hand, his own theory of the origin and the development of medicine was not a formal tribute to the subject of *origo artis*, but an integral part of his conception of medicine as a profession. It is the history of medicine that proves the medical art to possess every quality that makes it a proper τέχνη. Since he is familiar with the principal intellectual trends of his time, the author of *VM* manages to show us the attitude of his contemporaries toward cognitive activity and scientific progress, the way they accounted for the origin of τέχναι and the growth of knowledge, and the significance they attached to the scientific method.

He begins his work with criticism of the natural philosophers who regard health as dependent on the excess of a certain quality (cold, hot, etc.). Such the-

---

41 The significant role of lawgiving in Isocrates reminds us of Protagoras' πολιτικὴ τέχνη (Pl. *Prot.* 322f.). Cf. also Critias (88 B 25.5f.).

42 See e.g. *De arte* 1, 4. Also important were the polemics among the physicians themselves: Ducatillon, J. *Polémiques dans la Collection hippocratique*, Lille 1977, 96f.

43 On the whole, I follow the text and the interpretations of Jouanna, J. *Hippocrate. L'ancienne médicine*, Paris 1990. See also Wanner, H. *Studien zu περὶ ἀρχαίης ἰατρικῆς* (Diss.), Zurich 1939; *Hippocrate. L'ancienne médicine*, ed. by A.-J. Festugière, Paris 1948.

ories, he believes, introduce into medicine unverifiable hypotheses of their own, ignoring the results medicine has already achieved independently.

> But medicine has long had all its means to hand, and has discovered both a principle and a method, through which the discoveries made during a long period are many and excellent, while full discovery will be made, if the discoverer be competent, conduct his researches with knowledge of the discoveries already made, and make them his starting-point. But anyone who, casting aside and rejecting all these means, attempts to conduct research in any other way or after another fashion, and asserts that he has found out anything, is and has been the victim of deception. His assertion is impossible (2).[44]

It is significant that the verb εὑρίσκειν and its derivatives occur five times in this passage alone. In the whole of the treatise, comprising about twenty pages, the word εὑρίσκειν is used twenty-three times, the word ἐξευρίσκειν five times, and the noun εὕρημα three times. For the Hippocratic corpus, such frequent use is unique.[45] It is revealing also that the verb ζητεῖν, which forms part of the well-known pair of notions ζήτησις–εὕρεσις, occurs in this work seven times, along with the noun ζήτημα, which is not found elsewhere in the Hippocratic corpus.

The author of *VM* is not only enthusiastic about the progress in investigations and discoveries that are enriching medicine with new knowledge,[46] but also believes medicine as a whole to be a human discovery (οἱ δὲ ζητήσαντές τε καὶ εὑρόντες ἰητρικήν, 5). Identifying medicine with dietetics, he claims that, since the food, drinks, and very way of life of the healthy do not suit the sick, people, driven by this necessity (ἀνάγκη) and need (χρεία), started to seek medicine and discovered it. It is not the need itself, however, that led to the discovery of medicine. Those who first discovered it (οἱ πρῶτοι εὑρόντες) pursued their inquiries with suitable application of reason to the nature of man (14). To this end, they employed the only true method (ὁδός), which consists in finding the nourishment, the drink, and the mode of life that suits the nature of a sick person (cf. *De arte*, 13). It is the knowledge of all this that makes medicine (3). The same method had been used before, with a view to the nature of the healthy person. Before the nourishment proper for human nature was discovered, people used to live like wild beasts. They ate fruit, grass, and hay, suffered cruelly from it, often got sick and soon died. "For this reason the ancients too seem to me to have sought for nourishment that harmonized with their constitutions, and to have discovered that which we use now." (3) Thus, the author identifies the transition from the savage state to civilization with the discovery

---

44 The translation of the *VM* is throughout by W. Jones.

45 Jouanna. *L'ancienne médicine*, 38f. Second in frequency of using εὑρίσκειν is the treatise *De arte*, 1, 9, 12 (14 times in various forms).

46 καλῶς … ζητήσαντες (14), καλῶς ζητεομένην (12), τὰ εὑρημένα … καλῶς ἔχοντα (2), ὡς καλῶς … ἐξεύρηται (12).

of the appropriate food, which, like the discovery of medicine, came after a long search.

The author's theory of the origin of culture could not have been chronologically the earliest,[47] and such topics as θηριώδης βίος, ἀνάγκη, χρεία, and εὕρεσις were quite current in the fifth-century literature.[48] Still, the views of the Hippocratic on the history and methodology of medicine seem to have had no direct precursors. His treatment of medicine as dietetics, though one-sided, is solidly grounded in theory and practice.[49] This approach in no way prevents him from considering medicine a sphere of *cognitive* activity, which draws it closer to other sciences.

Method. The author assumes his method to be the *only* conceivable way of inquiry into the nature of man and of preventing and curing diseases. All other ways, including the speculations of natural philosophers, are rejected. Medicine starts by applying to the nature of the sick the method that has already been used in the search for food appropriate for the healthy (7). It consists in systematic observations on what food suits the nature of the sick best and even employs experiments with different kinds of food (3.5). From this point of view, an ὁδός, being one of the main characteristics of medicine as τέχνη, appears to be an improved method of research and discovery used outside medicine as well, not just an artifice, like the unverifiable method of natural philosophers. The author is far from a down-to-earth craftsman oriented exclusively toward experience – his empiricism can rather be termed *methodical*. With an awareness of the complex and versatile nature of reality, which does not fit into the schemata of natural philosophers, this empiricism is not averse, however, to using speculative ideas, provided they can be applied to practice.[50] Thus, the notion

---

47 For a discussion on the historical aspects of this theory, see Miller, H.W. *On Ancient Medicine* and the origin of medicine, *TAPA* 80 (1949) 187–202; Herter, H. Die kulturhistorische Theorie der hippokratischen Schrift von der Alten Medizin, *Maia* (1963) 464–483; Jouanna. *L'ancienne médicine*, 34f.; Nickel, D. Bemerkungen zur Methodologie in der hippokratischen Schrift *De prisca medicina*, *Hippokratische Medizin und antike Philosophie*, ed. by R. Wittern, P. Pellegrin, Zurich 1996, 53–61.

48 Aeschylus (*Prom.* 443f., 452, *TrGF* III F 181a), Euripides (*Suppl.* 201f.), Critias (88 B 25), and Isocrates (*Bus.* 25, *Nic.* 5, *Antid.* 254) represented primitive life as 'beastly'; Archelaus believed that the invention of τέχναι separated men from beasts (60 A 4). Aristophanes (*Pl.* 534), and Euripides (fr. 715 Nauck) considered need to be the teacher of wisdom (see Meyer, G. *Laudes inopiae*, Göttingen 1915, 21ff.). On the role of discoveries in medicine, see *De arte* 1, 9, 12; *De victu* III, 69.

49 These views were shared by the authors of the treatises *On Diet*, *On Diet in Acute Diseases*, *On Nutriment*. In *On Diseases* and *On Internal Affections*, diet is suggested as the main therapeutic means.

50 The author objects not to philosophy as such, but only to those of its metaphysical postulates that cannot be applied to medicine (Ducatillon, *op. cit.*, 96f.). On the polemics against philosophers, see also *De nat. hom.* 1; *Hippocrate. La Nature de l'homme*, ed. by J. Jouanna, Berlin 1975, 38f.

of δυνάμεις, which is very important in the treatise and on which health and illness depend, most probably goes back to Alcmaeon (24 B 4). Alcmaeon's fundamental statement that "of things invisible ... only gods have clear knowledge, but men can only judge on evidence" (B 1) also resonates with the ideas of the Hippocratic, who consciously refused to introduce τὰ ἀφανέα τε καὶ τὰ ἀπορεόμενα into medicine, his aim being εἰδέναι τὸ σαφές (1).

Methodical empiricism, while appreciating the art and experience of the physician, does not oppose them to knowledge (1), but, on the contrary, points out the rational aspect of medicine, constantly emphasizing the role of discourse (διάνοια, 5, λογισμοί, 12, 14), method (ὁδός, 2, 4, 8, 15), research, and discovery. The author's defense of medical dietetics has nothing routine about it, for the simple reason that this dietetics itself was, at the time, relatively new. It emerged in the early fifth century, created by the joint efforts of physicians, natural philosophers, and trainers.[51] Maintaining the chief components of the medical method to be exactness and clarity,[52] the author again avoids, here as elsewhere, the extreme opinions of the physicians who tried to find the exact correlation between food, drink, and physical exercise by means of 'mathematical' methods.[53] In searching for the exact measure, the Hippocratic sees its criteria not in weights and numbers, but in the bodily feeling of the patient himself. The task of the physician, according to him, lies in discerning accurately enough to allow only a minor error in either direction (9). Unlike Empedocles, who calculated the proportions in which the four basic elements constitute the human body (31 A 78, B 69, 96–98), he does not cherish vain hopes of achieving mathematical accuracy in medicine.[54]

Discovery of medicine and its history. Fully aware of the hypothetical character of his reconstruction of primitive life (3), the author introduces every new thesis with expressions like 'I presume' (ἔγωγε ἀξιῶ), 'it seems to me' (ἔγωγε δοκέω), 'probably, it looks like' (εἰκός, twice). His ideas, however, though formulated with circumspection, are always novel and never trivial. In his reconstruction, the Hippocratic, like Thucydides, reasons by analogy, observing that "even at the present day such as do not use medical science, foreigners and some Greeks, live as do those in health" (5).[55]

---

51 See Zhmud. *Wissenschaft*, 275f.

52 ἀκρίβεια (9, twice, 12, thrice, 20), e.g. "Many departments of medicine have reached such a pitch of exactness ..." (12); cf. Kurz, D. ΑΚΡΙΒΕΙΑ. *Das Ideal der Exaktheit bei den Griechen bis Aristoteles*, Göppingen 1970, 80ff. τὸ σαφές (1, 20), e.g. "I hold that clear knowledge about natural science can be acquired from medicine and from no other source." (20).

53 See *De victu* I, 2, 8; Delatte, A. Les harmonies dans l'embryologie hippocratique, *Mélanges P. Thomas*, Bruges 1930, 160–171.

54 As Jouanna (*L'ancienne médicine*, 84) pertinently observes, the author of *VM* was a contemporary thinker who had the courage to resist the extreme tendencies of contemporary thought.

55 Ibid., 44f.

The author has no doubt that medicine was found by humans (cf. *De arte* I, 12), who were compelled by need and necessity to undertake the research. He thinks, moreover, that it is the inventors of medicine themselves (οἱ πρῶτοι εὑρόντες) who attributed it to the deity (i.e., Apollo or Asclepius), who was still believed to be its founder. Therefore, the development of the tradition of *prōtoi heuretai* that we traced earlier is logically completed. The first real discoveries crown with glory their inventors; these are followed by the divine *prōtoi heuretai*, to whom human inventions are attributed; later on, historical personages start to figure in the tradition more and more often until, finally, the conclusion is drawn that the inventions of the divine discoverers were, in fact, attributed to them by humans. In this sense, Greek thought of the late fifth century, represented, of course, not solely by the author of *VM*, enters a new stage – that of transition from heurematography to the theory and history of culture and, later, of science. More accurately, this stage is not that of transition, but rather of the divergence of two traditions: heurematography did not disappear, but was reduced to compiling catalogues of inventions without any attempt at their analysis. Its better features were inherited by other genres.

The progress of knowledge. For the author of *VM*, the history of medicine is a history of research and discoveries that multiply our knowledge of human nature and of the causes of man's diseases. His outlook is characterized by an optimistic awareness of the progress of knowledge, which started in the distant past, is moving forward now, and will go on in the future. A progressivism of this kind is a rare case. In the classical period, the notion of progress was, as a rule, retrospective, i.e., founded in the first place on the achievements in knowledge and technology related to the *past* and not to the future. Even if the progress was made in the recent past and thereby bordered on the present, future perspectives were hardly ever considered.[56] But even in cases when an attempt was made to link the past and present with the future, the latter had nothing in common with the fundamentally open and infinite future of the progressivist conceptions of the 19th century.

In the beginning of the treatise, the Hippocratic turns directly from the past to the future:

> Medicine ... has discovered both a principle and a method, through which the discoveries made during a long period are many and excellent, while full discovery will be made (καὶ τὰ λοιπὰ εὑρεθήσεται) if the inquirer be competent, conducts his researches with knowledge of the discoveries already made, and makes them his starting-point (2).

A little later he returns to this subject, this time linking the past with the present and grounding his conclusions on facts:

> Nevertheless the discovery (*sc*. of medicine) was a great one, implying much investigation and art. At any rate even at the present day those who study gym-

[56] Edelstein, *op. cit.*, 98, 145f., 164f. (not clear enough); Thraede. Fortschritt, 162; Meier. 'Fortschritt', 354. On this, see below, 78f.

> nastics and athletic exercises are constantly making some fresh discovery by investigating on the same method (ἔτι γοῦν καὶ νῦν αἰεί τι προσεξευρίσκουσι κατὰ τὴν αὐτὴν ὁδὸν ζητέοντες).

Thus, the correct scientific method found by past generations is not only a guarantee of the present and future progress of medicine, but also of the unity and self-identity of this science for all ages. To put it in modern language, science is method.

The astonishing modernity of this conclusion, to which many contemporary philosophers of science would subscribe without reservation, should not, however, conceal the feature that distinguishes it from the more sober present-day view on the possibility of gaining the final knowledge of things. Unlike modern science, which proceeds from the conviction that knowledge is inexhaustible, the Hippocratic believed that in the future, provided the right direction is taken, medicine can be explored in full. Did he mean the long-distant future, as Herter suggested,[57] or the immediate future? Though the text itself does not answer this question directly, the second option is supported by the excessive optimism of other Hippocratics, who firmly believed that the whole of medicine was *already* discovered. That is, indeed, what the author of *On Places in Man* expressly affirms: ἰητρικὴ δή μοι δοκέει ἀνευρῆσθαι ὅλη ... βέβηκε γὰρ ἰητρικὴ πᾶσα (46). The treatise *On Art* leaves a similar impression: the verb εὑρίσκειν is used here in aorist and imperfect only,[58] and the verb ζητεῖν does not occur at all, as if nothing were left to find! In the introduction, however (1), the author does note the possibility of discovering "what was unknown before", but only in order to "bring to completion what was already accomplished in part" (ἐς τέλος ἐξεργάζεσθαι ὡσαύτως).[59]

Even if the author of *VM* shows more common sense than his colleagues, it remains clear that all these statements rest on very similar notions of the cognitive possibilities of man. Their common feature was the conviction that in the sphere of knowledge the *achievement of the final goal*, τέλος, was a matter of the near future. As a characteristic example of the epistemological optimism of the fourth century, the following lines of the dramatic poet Chaeremon are often quoted: "There is nothing among people that they seek and would not find in due course."[60] His younger contemporary Alexis used a still more aphoristic expression: ἅπαντα τὰ ζητούμενα ἐξευρίσκεται (fr. 31 Kassel–Austin). Aristotle, in his youth at least, believed philosophy was nearing its completion (fr. 53 Rose); Philip of Opus wrote that the Greeks bring to perfection (κάλλιον τοῦτο εἰς τέλος ἀπεργάζονται) all the knowledge they borrow from the 'bar-

---

57 Herter, H. Die Treffkunst des Arztes in hippokratischer und platonischer Sicht, *Sudhoffs Archiv* 47 (1963) 247–290.

58 See *Hippocrate. De L'Art*, ed. by J. Jouanna, Paris 1988, 185.

59 Cf. Isoc. *Paneg.* 10 (see below, 78 n. 143).

60 οὐκ ἔστιν οὐδὲν τῶν ἐν ἀνθρώποις ὅ τι οὐκ ἐν χρόνῳ ζητοῦσιν ἐξευρίσκεται (*TrGF* 71 F 21).

barians' (*Epin.* 987e 1); Eudemus believed that geometry had reached its perfection (fr. 133). This optimism, however, had its drawbacks: the idea of progress was limited by rather narrow bounds. Indeed, most authors of the classical and Hellenistic periods who touched upon this subject regarded progress as either already accomplished or to be completed by the generation to come.[61] The idea that knowledge is inexhaustible is attested in some Roman writers of the first century AD,[62] when scientific progress itself had come to an end in the majority of fields. Later on, the 'horizon of expectations' remains the same as before or gets even narrower. In late Antiquity, perfection in the sciences was associated more with the increasingly distant past than with the future.

To place the ideas of the author of *VM* in the right perspective, however, it is necessary to compare them with the views his contemporaries expressed on τέχναι and their progress. This comparison will make clear, among other issues, the degree to which the ideas of science as knowledge evolving in time and of the scientific method as the most certain path from 'research' to 'discovery' were the personal contribution of the Hippocratic. What does he owe to the common background of his epoch?

3. Archytas and Isocrates

Very interesting comparative material is to be found in two younger contemporaries of the Hippocratic physician, Archytas and Isocrates. Apart from dates of birth[63] and an interest in politics, they had little in common: Archytas is known as a brilliant mathematician and, secondly, as a philosopher, while Isocrates professed a very critical attitude toward exact sciences, as well as toward the claims of philosophy to achieve exact knowledge. Nevertheless, their views on τέχνη and its development are strikingly similar to each other as well as to the notions of the author of *VM* and other Hippocratics.[64] Some of the fragments of Archytas' work show that he used the Sophistic theory of τέχνη,

---

61 Meier. 'Fortschritt', 353; idem. Antikes Äquivalent, 291ff.

62 Edelstein, *op. cit.*, 168f., refers in this connection to Seneca (*NQ* VI,5.31, VII,25. 4–7, VII,30.5–6) and Pliny (*HN* II,15.62); the latter seems to be influenced by Seneca. "Men who have made their discoveries before us are not our masters, but our guides. Truth lies open for all; it has not yet been monopolized. And there is plenty of it left even for posterity to discover." (Sen. *Ep.* 33, 10, cf. 64, 7). It is debatable whether Seneca's ideas were born out the experience of the previous progress in science and technology or derived from Posidonius. For an interesting discussion on this point and an ample bibliography, see Gauly, B. M. *Senecas* Naturales Quaestiones. *Naturphilosophie für die römische Kaiserzeit*, Munich 2004, 159ff.

63 Isocrates was born in 436; Archytas, probably about 435/430.

64 On similarities in the understanding of τέχνη by Isocrates and the author of *VM*, see Wilms, H. *Techne und Paideia bei Xenophon und Isokrates*, Stuttgart 1995.

which was initially oriented toward practical knowledge, in his descriptions of mathematical disciplines as well:

> It seems that arithmetic (λογιστικά) far excels the other arts (τῶν μὲν ἀλλᾶν τεχνῶν) in regard to wisdom (σοφία), and in particular in treating what it wishes more clearly than geometry (γεωμετρικά). And where geometry (γεωμετρία) fails, arithmetic accomplishes proofs ... (47 B 4).

The terminology of this fragment deserves elucidation. Since it deals with demonstrations in which arithmetic surpasses geometry itself, what Archytas means by λογιστική is not practical computation but theoretical arithmetic, i.e., the theory of number based on deduction. Elsewhere he calls arithmetic simply ἀριθμοί, and the four mathematical sciences together μαθήματα (B 1). In the fragment of the treatise Περὶ μαθημάτων, where the social role of arithmetic is discussed, it is called λογισμός (B 3). Is there any difference between λογισμός and λογιστική, and does λογισμός refer, accordingly, to practical or theoretical arithmetic? Leaving this question open for a moment, let us note that Plato often uses λογιστική and ἀριθμητική indifferently in this respect;[65] Aristotle also applies the new term, ἀριθμητική, and the old one, λογισμοί, to one and the same science.[66] The distinction between *practical* logistic and *theoretical* arithmetic is first found in Geminus (Procl. *In Eucl.*, 38.10–12); later it was taken up by the Neoplatonists, who attributed it, naturally, to Plato.[67] None of the passages of Plato usually cited in this context, however, suggests this meaning. This fact was pointed out and explained long ago by J. Klein, who showed that, in Plato, the difference between logistic and arithmetic comes down to the former referring mainly to counting and the latter to computation; both disciplines can be theoretical, as well as practical.[68]

As the material of the fifth and fourth centuries shows, we can hardly expect the names of sciences and their classification to be rigorous and unambiguous. Neither is there any contradiction in the fact that Archytas treats arithmetic and geometry as *mathēmata* (B 1, 3), yet places them elsewhere among τέχναι (B 4). In his time, the *mathēmata*, though constituting, among other τέχναι, a special group, had not yet become model of ἐπιστήμη. Under the influence of the word-formative model of τέχνη, Archytas in one place even changes the traditional term γεωμετρία[69] into γεωμετρικὴ (τέχνη).[70] By "(all) other τέχ-

---

65 *Res.* 525 a 9, *Gorg.* 451 c 2–5, *Tht.* 198 a 5, *Prot.* 357 a 3, *Charm.* 165 e 6, 166 a 5–10.

66 Cf. *Met.* 982 a 26 f., *APo* 88 b 12. λογισμός with reference to theoretical arithmetic, see also Isoc. *Bus.* 23; Xen. *Mem.* IV,7.8; Pl. *Res.* 510 c 3, 522 c 7, 525 d 1; *Pol.* 257 a 7. Arithmetic was often referred to as ἀριθμὸς καὶ λογισμός: Ps.-Epich. (23 B 56); Pl. *Res.* 522 c, *Phdr.* 274 c, *Leg.* 817 e.

67 Olymp. *In Gorg.*, 31.4 f.; *Schol. Gorg.* 450 d–451 a–c; *Schol. Charm.* 165 e.

68 Klein, J. *Greek mathematical thought and the origin of algebra*, Cambridge 1968, 10 ff. (German original: *Q & St* 3.1 [1934] 18–105). See also Burkert. *L & S*, 447 n. 19; Mueller, I. Mathematics and education: Some notes on the Platonic program, *Apeiron* 24 (1991) 88 ff.

69 Hdt. II, 109; Philol. (44 A 7 a); Ar. *Nub.* 202, *Av.* 995.

ναι" he must have meant not only mathematical sciences (he would have used the term μαθήματα in this case), but also other occupations traditionally related to this field. Owing to its σοφία, arithmetic surpassed all these τέχναι, which Archytas considered from the cognitive point of view. In the context of τέχναι (crafts, poetry, music, medicine), σοφία is usually understood as 'skill, craftsmanship, artfulness' and is often associated with 'precision' (ἀκρίβεια).[71] Archytas transfers this quality from the master to τέχνη itself, thus making arithmetic appear more 'artful' and, hence, more 'precise' than all the other τέχναι, including geometry. Arithmetic surpasses the latter in ἐνάργεια, i.e., clearness, evidence, and obviousness, which makes it, in comparison, more demonstrative.[72] This quality of arithmetic is close to what the author of *VM* most of all required of medicine as a τέχνη: clearness and precision in knowledge (εἰδέναι τὸ σαφές, καταμαθεῖν ἀκριβέως).[73] Isocrates, in his own field, held similar ideals: πίστις ἐναργής and ἀπόδειξις σαφής are, for him, the key notions that characterize the conclusiveness of a statement.[74]

Particularly interesting is the fragment of Archytas' work Περὶ μαθημάτων (47 B 3), where he formulates in a concise and aphoristic manner the main notions and ideas of the contemporary theory of τέχνη. The first part of the fragment deals with the scientific method, or, to be more precise, with the cognitive method as such; the second dwells upon the usefulness of arithmetic and on the importance of its discovery for social life and morality. Let us turn to the title of the work first. Initially, μάθημα, a passive derivative from the verb μανθάνω, denoted the result ('what has been learned'), or the subject of study and could refer to different fields of knowledge.[75] It is in this initially large sense, that the title of Protagoras' Περὶ τῶν μαθημάτων (D. L. IX, 55) is to be understood: what is meant here is not mathematics, but various branches of learning.[76] In Archytas, the word μαθήματα acquires terminological character

---

70 Cf. ἀριθμητικὴ τέχνη (Pl. *Ion.* 531e 3, *Gorg.* 451b 1, *Tht.* 198a 5, *Pol.* 258d 4, *Phil.* 55e 1); γεωμετρικὴ τέχνη (*Charm.* 165e 6).

71 τὴν δὲ σοφίαν ἔν τε ταῖς τέχναις τοῖς ἀκριβεστάτοις τὰς τέχνας ἀποδίδομεν (Arist. *EN* 1141a 9).

72 "The analysis of certain classes of problems in geometry, e.g. the construction of irrational lines, can only be completed by means of arithmetical principles." (Knorr, W. R. *The evolution of the Euclidean Elements*, Dordrecht 1979, 311). For Aristotle, too, arithmetic is more exact than geometry (*APo* 87a 34f., *Met.* 982a 26f.). Philolaus, on the contrary, singled out geometry as the 'source' and the 'mother-city' of all mathematical sciences: 44 A 7a; Huffman, C. A. *Philolaus of Croton. Pythagorean and Presocratic*, Cambridge 1993, 193f.

73 Cf. above, 57 n. 52 and σαφῆ διάγνωσις in Archytas (47 B 1).

74 See *Bus.* 37 (cf. *Hel.* 61, *Antid.* 243) and *Antid.* 118, 273. With regard to the gods, εἰδέναι τὸ σαφές is impossible (*Nic.* 26), but here too there is σημεῖον, allowing us to form judgments. Cf. Alcmaeon (24 B 1); *De arte* 12.

75 Snell, *op. cit.*, 76.

76 Burkert. *L & S*, 207 n. 80. μαθήματα was used later with the same meaning: Isoc. *Antid.* 10, 267; Pl. *Lach.* 108c, *Soph.* 224c, *Leg.* 820b. For Hellenistic inscriptions

and designates a particular group of sciences including arithmetic, geometry, astronomy, and harmonics, all of which he regards as akin (B 1 and 4, cf. Pl. *Res.* 530d). Affixing the term μαθήματα to the sciences of the mathematical quadrivium would have been impossible, had it not been preceded, first, by setting them apart as a special group and, second, by turning them into the subjects of learning.

The origin of the quadrivium has been a subject of long discussion. Some scholars date it to the time of the Sophists, others to the time of Plato, and still others relate it directly to Plato.[77] Indeed, the particular importance that he accorded to mathematical sciences is manifest already in the relatively early *Republic*: the ten years dedicated to the study of *mathēmata* were to prepare the future guardians of the ideal state to master the main science, dialectic. Still, Plato, though an expert in mathematical sciences and their advocate, hardly taught them himself.[78] Besides, he never claimed for himself the honor of being the discoverer of the quadrivium. On the contrary, he mentions repeatedly that geometry, arithmetic, astronomy, and harmonics were taught by Hippias of Elis and Theodorus of Cyrene.[79]

There are no reasons to doubt that Hippias, a remarkable polymath for his time, taught all the disciplines of the quadrivium. Since it has become clear, however, that the mathematician Hippias who discovered the curve called quadratrix is not to be identified with the Sophist Hippias of Elis,[80] one can hardly ascribe to the latter any discoveries in mathematics. Meanwhile, the uniting of the four sciences into a special group can only have been effected by someone who directly took up not only mathematics, but (mathematical) astronomy and harmonics as well, since the intrinsic relationship between the latter two sciences is far from evident to a layman. It is not among the Sophists, who picked up and developed the already existing tradition,[81] but among the Pythagorean mathematicians that the origin of the quadrivium is to be sought.[82]

---

relating to school teaching, see Grassberger, L. *Erziehung und Unterricht im klassischen Altertum*, Würzburg 1881, 315, 437.

77 See e.g. Merlan, P. *From Platonism to Neoplatonism*, The Hague 1960, 88; Burkert. *L&S*, 422f.

78 See below, 101f.

79 *Prot.* 316d–e, *Hipp. Mai.* 285b, *Hipp. Min.* 366c, 368e; *Tht.* 145c.

80 Knorr. *AT*, 82f.

81 It is unlikely that any of the Sophists, except for Hippias, taught geometry, let alone arithmetic and harmonics (see above, 45 n. 2). In Aristophanes (*Nub.* 200f.), the entrance into Socrates' 'thinking shop' is flanked by the statues of geometry and astronomy, which does not mean, however, that all the four *mathēmata* were integrated into the Sophists' educational curriculum before the end of the fifth century (so Burkert, *L&S*, 421). In Athens there were other experts in astronomy and geometry: Meton (Ar. *Av.* 997) and Euctemon, Hippocrates of Chios (42 A 2, 5) and Theodorus of Cyrene (43 A 3–5).

82 Heath, T. L. *A history of Greek mathematics*, Vol. 1, Oxford 1921, 10f. The scientific meaning of the word μαθήματα goes back to the Pythagoreans (Snell, *op. cit.*, 77f.).

Soon after Pythagoras added arithmetic and harmonics to Ionian astronomy and geometry,[83] the *mathēmata* began to form a special group. As clearly follows from Philolaus' fragments, he was very familiar with all the four sciences of the Pythagorean quadrivium. If we combine this with a tradition according to which Pythagoras took up all the four *mathēmata* and Hippasus at least three of them (geometry, arithmetic, and harmonics, 18 A 4, 12–15), then we come to the conclusion that as a young man, i.e., before his escape to Thebes (ca. 450 BC), Philolaus was brought up within the framework of the Pythagorean mathematical quadrivium. Distinct traces of this sort of education are also found among his contemporaries, the Pythagorean Theodorus (43 A 2–5) and Democritus, who had Pythagorean teachers (68 A 1, 38).[84] On the other hand, there is no evidence that the Ionian mathematicians Oenopides and Hippocrates of Chios took up arithmetic and harmonics; the very word μαθήματα was hardly ever used in the Ionian dialect.[85]

Let us now return to the text of the fragment, which by all appearances comes from the introductory part of Archytas' work. Like many other works belonging to the genre περὶ τέχνης, the introduction to Περὶ μαθημάτων lays out the major methodological principles of the sciences considered in it and points out their characteristic features, most of which coincide with those of other τέχναι. To judge from the fact that Archytas does not bother to explain and demonstrate them in detail, by the time when Περὶ μαθημάτων was brought out (presumably, at the close of the fifth century), the Sophistic theory of τέχνη was more or less well-known.[86] The first part of the fragment discusses, in succession, different ways of acquiring knowledge:

> δεῖ γὰρ ἢ μαθόντα παρ' ἄλλω ἢ αὐτὸν ἐξευρόντα, ὧν ἀνεπιστάμων ἦσθα, ἐπιστάμονα γενέσθαι. τὸ μὲν ὦν μαθὲν παρ' ἄλλω καὶ ἀλλοτρίαι, τὸ δὲ ἐξευρὲν δι' αὔταυτον καὶ ἰδίαι. ἐξευρεῖν δὲ μὴ ζατοῦντα ἄπορον καὶ σπάνιον, ζατοῦντα δὲ εὔπορον καὶ ῥάιδιον, μὴ ἐπιστάμενον δὲ ζητεῖν ἀδύνατον.

According to Anatolius (an Aristotelian of the third century AD renowned for his mathematical learning), the Pythagoreans gave the name μαθηματική to arithmetic and geometry, which heretofore had not been referred to by a single term (Ps.-Heron. *Def.*, 160.23–162.5).

83 Isocrates presents Pythagoras as a disciple of Egyptian priests, among whose occupations he mentions astronomy, arithmetic, and geometry (*Bus.* 23, 28). See Zhmud. *Wissenschaft*, 183f., 213f., 248f.

84 On Democritus, see Burkert. *L & S*, 421 n. 118.

85 Snell, *op. cit.*, 76.

86 The closeness of Archytas' ideas to the Sophistic theory of τέχνη has not yet been noted, probably because Archytas' philosophy was little studied in the 20th century. Earlier, F. Blass. *Attische Beredsamkeit*, Vol. 1, Berlin 1889, 89, pointed out the similarity of this fragment's style to that of Gorgias. See now Huffman, C. A. Archytas and the Sophists, *Presocratic philosophy: Essays in honour of A. Mourelatos*, ed. by V. Caston, D.W. Graham, Aldershot 2002, 251–270.

> To know what was heretofore unknown, one has either to learn it from another, or to discover himself. What one has learnt, he has learnt from another and with another's assistance, what one has found, he has found himself and by his own means. Discovering without research is difficult and (happens) seldom, by research it is easy and practicable, but without knowing (how) to research it is impossible to research.[87]

Archytas starts by introducing the antithesis of μάθησις–εὕρεσις: knowledge is acquired either by learning or independently.[88] The same pair of notions often occurs in Plato, who uses it to contrast one's creative activity with assimilation of discoveries made by others. True knowledge (τὸ σαφὲς εἰδέναι) can be either learned or discovered by independent research.[89] This is also true for any τέχνη, for instance, the art of training youths. Socrates offers the Sophists, who claim to be masters of this τέχνη, the following choice: they have to prove that they have either discovered it by themselves or learned it from someone else.[90] The currency of the pair μάθησις–εὕρεσις is confirmed by the material of the Hippocratic corpus and Isocrates. The author of *On Diet* believes that the diet he has discovered and considers close to the true one may reflect glory on himself, its discoverer, and be useful to those who learn it (III, 69). Isocrates advises a young man to acquire knowledge both independently and by learning from others: in this way he would learn to find with ease what others had found with difficulty (*Ad Dem.* 18–19; cf. *Antid.* 189, *In Dion.* 4).

The passage from Isocrates' *Panathenaicus* based on the juxtaposition of μάθησις, εὕρεσις and ζήτησις is particularly reminiscent of Archytas' ideas. Talking of the discoverers of the civilization and culture Isocrates remarks that all these things

> are not discovered by any and everyone, but by men who have superior endowments and are both able to learn the most of what has been discovered before their

---

87 In the last period (μὴ ἐπιστάμενον δὲ ζητεῖν ἀδύνατον), it seems most natural to understand ζητεῖν as referring both to μὴ ἐπιστάμενον and to ἀδύνατον: "without knowing how to research – to research is impossible" (cf. ἐπισταμένους λογίζεσθαι in the second part of the fragment). Blass, F. De Archytae Tarentini fragmentis mathematicis, *Mélanges Graux*, Paris 1884, 581–582, preferred the following text: μὴ ἐπιστάμονα δέ, ζητεῖν ἀδύνατον, interpreting it in the sense that he who does not know *what* to seek, cannot seek. As a parallel to it, he cited Plato's words (*Men.* 80e) about the Sophistic theory, according to which one cannot research what one does not know already. Cf. the translation by Diels: "für den freilich, der es nicht versteht, ist das Suchen unmöglich."

88 Cf. already in Aesch. *Prom.*: Prometheus ἐξηῦρον (460), people ἐκμαθήσονται (256).

89 *Phaed.* 85c, 99c 9–d2, see also *Crat.* 439b 7f., *Hipp. Min.* 372c 6–8. In *Meno*, Plato treats the difference between μάθησις and ζήτησις as relative, both being in his view a 'reminiscence' (81d).

90 *Lach.* 186c–187a. Plato, however, does not always understand μάθησις as passive acquisition of knowledge; learning is often merely a stimulus to independent research and discoveries (*Tht.* 150d–151a; *Res.* 455b 7f.).

time and willing more than all others to give their minds to the search for what is new (208–209).[91]

In all these respects the Spartans are, according to him, more backward not only than the Athenians, but even than the 'barbarians', who are both pupils and teachers of many discoveries. Let us note again how crucial these epistemological notions are to the theory and the history of culture: if everything culture consists of is the result of either an independent discovery or learning (borrowing), then every time when faced with two similar things, whether in sciences, or arts, or religious rites, one has to bear in mind that they *both* originate from one first discoverer and teacher, whose knowledge was then disseminated by his imitators and students. This very specific approach of the Greeks to culture, allowing no independent appearance of similar phenomena, led in particular to endless charges of plagiarism and to no less numerous attempts to attribute their own achievements to their neighbors.

Very close to μάθησις–εὕρεσις is another pair of notions – that of μίμησις–εὕρεσις. Isocrates directly contrasts invention with learning and imitation: according to him, sophistic discourses follow one and the same pattern, which is easily enough found, or learned and imitated (οὔθ' εὑρεῖν οὔτε μαθεῖν οὔτε μιμήσασθαι, *Hel.* 11). In many cases μάθησις comes close to μίμησις, though it yields to it in activity and does not have the negative connotations that are often (though not always) associated with imitation. The importance of the pairing of μίμησις and εὕρεσις has been discussed above (1.4); here it interests us as an additional way to contrast borrowing and imitation with independent finding. Let us consider a few examples that demonstrate the currency of this model among Archytas' contemporaries.[92] Speaking of Lycurgus, Isocrates says that the latter did not, in fact, invent the constitution of Sparta, but only imitated ancient Athenian regulations (*Panath.* 153). In the discourse addressed to the young Nicocles, the king of Cyprus, the rhetorician recommends him to improve state regulations and laws. There are two ways to accomplish this task: either by discovering what is best independently, or, if that proves impossible, by imitating what one finds best in others (*Ad Nic.* 17). In Xenophon, Socrates suggests two means of restoring the good morals of past times: either to find out the customs of their ancestors and practice them, or failing that, to imitate those of their contemporaries who have preeminence (*Mem.* III,5.14).

---

[91] Transl. by G. Norlin. Cf. an analogous passage in Plato: a man richly endowed by nature is easy to teach and, after a short period of learning, is apt to discover a great deal more than he has learned (*Res.* 455b 7f.). For comparison between μάθησις, εὕρεσις and ζήτησις, see also *Phaed.* 99c 9–d2; *Men.* 81d; *Res.* 618c; *Crat.* 439b 7f.

[92] As early as Aristophanes we have a writer contrasting his own creative approach to plagiarism and imitation on the part of his rivals: I do not seek to show my pieces twice or thrice and always think up καινὰς ἰδέας, while Eupolis and Hermippus make use of them and imitate (μιμούμενοι) my inventions. Some may like their comedies as well, but the good judgment of those who praise my εὑρήματα will be glorified for ever (*Nub.* 545–562).

Being a teacher as well as a scientist, Archytas must have been well aware of the importance of education as a means of transmitting knowledge.[93] His contrasting of μάθησις and εὕρεσις, however, emphasizes the dependent character of learning, drawing it closer to imitation. The words of Archytas leave no doubt that he, personally, preferred the way of independent research and discovery. This way is then characterized in more detail by means of another pair of notions already familiar to us, that of ζήτησις–εὕρεσις. To make a discovery, conscious research is needed, because one cannot conduct research without knowing how to do it (μὴ ἐπιστάμενον δὲ ζητεῖν ἀδύνατον). What, then, must the one ἐπιστάμενος ζητεῖν know? To all appearances, he must know *what* and *how* to seek – in other words, he must know the object and method of his research. The lack of such knowledge makes any serious *research* – unlike an accidental *find* – a sheer impossibility. This view is close to the notions of the author of *VM*, who claimed that in medicine both principle and method had long been discovered:

> But anyone who, casting aside and rejecting all these means, attempts to conduct research in any other way or after any other fashion, and asserts that he has found out anything, is and has been the victim of deception. His assertion is impossible (ἀδύνατον γάρ, 2).

It follows that for Archytas as well as for the Hippocratic physician, the method, i.e., the art of correct research, is a prerequisite for success in science.[94] Archytas, however, unlike the Hippocratic, did not altogether rule out the chance, small as it might appear, of an accidental discovery: "Discovering without research is difficult and (happens) seldom, by research it is easy and practicable." These words imply one more opposition, that of τύχη–τέχνη (or τύχη–ἐπιστήμη), which is well known from the literature of the fifth century,[95] including the Hippocratic corpus.[96] For the author of *VM*, τύχη, on the one

93 Cf. his praise of his precursors who attained clear knowledge in mathematical sciences (B 1).

94 Describing the efforts of his predecessors to discover the basic principle of the construction of torsion catapults, Philo of Byzantium stresses: "This had to be obtained not by chance or at random, but by a standard method" (ταύτην δ' ἔδει μὴ ἀπὸ τύχης μηδὲ εἰκῇ λαμβάνεσθαι, μεθόδῳ δέ τινι ἑστηκυίᾳ); see below, 282.

95 Gomperz, T. *Die Apologie der Heilkunst*, 2nd ed., Leipzig 1910, 108f.; Snell, *op. cit.*, 85f.; Joos, *op. cit.*, passim; Heinimann, *op. cit.*, 108 n. 18. For the first time, τύχη–τέχνη is found in Euripides (*Alc.* 785) and in the tragic poet Agathon (*TrGF* IV, 39F6, 8). The antithesis of τύχη–ἐπιστήμη is found in the Hippocratic corpus (*De loc. in hom.*, 46) and in the *Alcibiades* by the Socratic Aeschines (fr. 8, 56f. Dittmar). In the latter case, it may go back to Socrates himself; cf. Xen. *Mem.* IV,2.2, *Symp.* VIII, 38f.

96 Villard, L. Les médecins hippocratique face au hasard, *Hippokratische Medizin*, 395–411; Wenskus, O. Die Rolle des Zufalls bei der Gewinnung neuerer Erkenntnisse, ibid., 413–418.

hand, and τέχνη, which is based on knowledge (ἐπιστήμη), on the other, are incompatible:

> Some practitioners are poor, others very excellent; this would not be the case if the art of medicine (as τέχνη) did not exist at all, and had not been the subject of any research and discovery, but all would be equally inexperienced and unlearned therein, and the treatment of the sick would be in all respects haphazard (1).

A similar passage concludes the first part of his work:

> I declare, however, that we ought not to reject the ancient art as non-existent … but much rather, because it has been able by reasoning to rise from deep ignorance to approximately perfect accuracy, I think we ought to admire the discoveries as the work, not of chance, but of inquiry rightly and exactly conducted (12).

Another Hippocratic was not so radical in his repudiation of τύχη: opposing it to τέχνη, he admits, nevertheless, that ruling out luck remains impossible in medicine (*De arte*, 4).[97] The sick person, in particular, could heal himself using the same means that a physician would prescribe (5). But he uses even this case as a proof that medicine is a τέχνη: there is no need to rely on luck as soon as the physician can make the exact diagnosis and prognosis and knows the difference between the correct and the incorrect ways of treatment (6).[98]

To the themes that Archytas and the Hippocratic physicians have in common one should add the epistemological optimism characteristic of that epoch as a whole.[99] Archytas believed making a discovery to be easy and simple, provided one used the right scientific method. The Hippocratics did not reckon a scientific discovery as such among easy accomplishments; they believed, rather, that a thing is easy to use once it has been discovered.[100] Nevertheless, the conviction shared by some of them that in medicine everything either has been discovered already or is going to be discovered in the near future, rested on the same optimistic belief in the possibilities of science and human reason that was typical of Archytas. The progress of science and its approach to perfection are the natural consequence of the scientist's individual progress in the assimilation of knowledge already gained, as well as in the solution of new problems. In many respects this progress depends on the personal endowments of a scientist,[101] but not on them alone. Mathematics, the subject that Archytas studied

---

97 For a similar view close to Archytas' position, see *De loc. in hom.* 46; *De affect.* 45.

98 Jouanna. *De l'art*, 187. For a similar argument, see *Dissoi logoi* 6, 11. Aristotle, as well as Archytas, admitted the possibility of an accidental discovery: in their experiments, it was not wit but chance that made the poets discover how to produce such effects in their plots (*Poet.* 1454a 10f.). See also *Protr.* fr. 11 Ross: συμβαίη μὲν γὰρ ἂν καὶ ἀπὸ τύχης τι ἀγαθόν.

99 See above, 58f.

100 *VM* 2, *De arte* 9. See also: Isoc. *Antid.* 83 (the new is not easy to find) and *Ad Dem.* 18–19 (it is easier to learn what has already been discovered).

101 This point was insisted upon both by Isocrates (*Panath.* 208–209) and Plato (*Res.*

and wrote about, offered better opportunities for progress, both general and individual, than medicine, which deals with a multitude of individual cases and relies on practical experience, not on general rules alone. This is probably what accounts for the disagreement between the mathematician and the doctor on how easily new knowledge is discovered and assimilated. As a sharp observation in one of the Hippocratic treatises shows (*De loc. in hom.* 41), doctors were well aware of the difference between a system of clear and well-defined rules, on the one hand, and medical knowledge as such, on the other:

> Medicine cannot be learned quickly because it is impossible to create any established principle in it (καθεστηκός τι σόφισμα), the way that a person who learns writing according to one system that people teach understands everything; for all who understand writing in the same way do so because the same symbol does not sometimes become opposite, but is always steadfastly the same and not subject to chance. Medicine, on the other hand, does not do the same thing at this moment and the next, and it does opposite things to the same person, and at that things that are self-contradictory.[102]

Then the author gives a number of examples showing how similar means can lead to opposite results and different means to similar results (41–44). As an antithesis to medicine, the Hippocratic cites the generally known rules of writing, but this idea could easily be illustrated by the example of mathematics as well. In *Nicomachean Ethics*, Aristotle, characterizing medicine in practically the same terms,[103] remarks that the difference between scientific knowledge (ἐπιστήμη) and practical reason (φρόνησις) is manifest, in particular, from the following fact (1142a 11–20):

> While young men become geometricians and mathematicians and wise in matters like these, it is thought that a young man of practical wisdom cannot be found. The cause is that such wisdom is concerned not only with universals but with particulars, which become familiar from experience, for it is length of time that gives experience; indeed one might ask this question too, why a boy (παῖς) can become a mathematician, but not a wise man (σοφός) or a natural scientist (φυσικός). Is it because the objects of mathematics exist by abstraction, while the first principles of these other subjects come from experience, and because the young men have no conviction about the latter but merely use the proper language, while the essence of mathematical objects is plain enough to them?[104]

455b 7f.). On the rapid progress in the acquisition of knowledge by those who conversed with Socrates, see Pl. *Tht.* 150d–151a (θαυμαστὸν ὅσον ἐπιδιδόντες … πάλιν ἐπιδιδόασι).

102 Transl. by P. Potter. On the importance of καιρός in medicine, see *De loc. in hom.* 44.

103 It has no fixed principle (οὐδὲν ἑστηκὸς ἔχει, 1104a 4f.), the physicians must τὰ πρὸς τὸν καιρὸν σκοπεῖν (ibid.); what a feverish patient generally benefits from may not prove useful in each particular case (1180b 9).

104 Transl. by J. Barnes. In *EE*, Aristotle illustrates this idea with the example of a famous mathematician: "Hippocrates was a geometer, but in other respects was

An important parallel to the motifs of scientific progress and the ease of learning mathematics is found in *Protrepticus*, written when Aristotle was at the Academy. (Let us note that here Aristotle brings philosophy and mathematics together, rather than opposing them to each other.) Asserting that the acquisition of philosophical knowledge is possible, useful, and (comparatively) easy, Aristotle supports the ease of learning philosophy by the following arguments: those who pursue it get no reward from men to spur them, yet their progress in exact knowledge is more rapid compared with their success in other τέχναι.[105] In the passages parallel to this text, Iamblichus speaks of the rapid progress in both philosophy and mathematics,[106] while Proclus mentions mathematics alone.[107] Considering the passage from the *Nicomachean Ethics* quoted above and several passages in the *Metaphysics* (981b 13–22, 982b 22f.) close in meaning to *Protrepticus*, we can safely surmise that while discussing the (relative) ease of acquiring exact knowledge and the resulting rapid progress of theoretical sciences, Aristotle referred not to philosophy alone, but to mathematics as well.[108]

---

thought silly and foolish, and once on a voyage was robbed of much money by the custom collectors of Byzantium, owing to his silliness, as we are told." (1247a 17f. = 42 A 2, transl. by J. Barnes).

105 τὸ γὰρ μήτε μισθοῦ παρὰ τῶν ἀνθρώπων γινομένου τοῖς φιλοσοφοῦσι, δι' ὃν συντόνως οὕτως ἂν διαπονήσειαν, πολύ τε προεμένους εἰς τὰς ἄλλας τέχνας ὅμως *ἐξ ὀλίγου χρόνου θέοντας παρεληλυθέναι ταῖς ἀκριβείαις*, σημεῖόν μοι δοκεῖ τῆς περὶ τὴν φιλοσοφίαν εἶναι ῥᾳστώνης (Iambl. *Protr.*, 40.19–20 = *Protr.* fr. 5 Ross = B 55 Düring). The rapid progress in mathematics was also mentioned in an early Academic treatise, see below, 87ff.

106 Iambl. *De comm. math. sc.*, 83.6–22 = *Protr.* fr. 8 Ross = C 55:2 Düring: τοσοῦτον δὲ νῦν προεληλύθασιν ἐκ μικρῶν ἀφορμῶν *ἐν ἐλαχίστῳ χρόνῳ ζητοῦντες οἵ τε περὶ τὴν γεωμετρίαν καὶ τοὺς λόγους* καὶ τὰς ἄλλας παιδείας, ὅσον οὐδὲν ἕτερον γένος ἐν οὐδεμιᾷ τῶν τεχνῶν. Cf. μικρὰς ἀφορμάς in Arist. *Cael.* 292a 15.

107 Procl. *In Eucl.*, 28.13–22 = *Protr.* fr. 5 Ross = C 52:2 Düring: τὸ μηδενὸς μισθοῦ προκειμένου τοῖς ζητοῦσιν ὅμως *ἐν ὀλίγῳ χρόνῳ τοσαύτην ἐπίδοσιν τὴν τῶν μαθημάτων θεωρίαν λαβεῖν.*

108 It is revealing that in Isocrates' response to the *Protrepticus* it is mathematics that is in question (see below, 74f.). Another important parallel is the following passage in Plato: lightly esteemed as the studies in solid geometry are by the multitude and hampered by the ignorance of their students as to the true reasons for pursuing them, they nevertheless in the face of all these obstacles force their way ahead by their inherent charm (*Res.* 528b 6f. = *Protr.* C 55:1 Düring). Aristotle shifts the accent a little: at present mathematics and philosophy make more rapid progress than all the other τέχναι, though the studies in the latter are morally and materially stimulated, while those preoccupied with theoretical knowledge are rather hampered than encouraged (see above, n. 105–107).

4. Why is mathematics useful?

Before going on to the second part of Archytas' passage, let us note the obvious similarity of the ideas stated in it, not only to the Sophists' theory of τέχνη and the Hippocratics' notions of their own science, but also to Plato's and Aristotle's views on theoretical knowledge. Archytas' influence on Plato is beyond all doubt,[109] and so is Aristotle's familiarity with Archytas' works.[110] Of course, the classical theory of τέχνη was known to Plato and Aristotle independently of Archytas. His role as intermediary is more likely reflected in the fact that the passage from science as τέχνη to science as ἐπιστήμη took place under the decisive influence of the *mathēmata* in which, in his generation, Archytas was the major expert. Substituting themselves for τέχνη, the *mathēmata* became the standard toward which the Academy and the Lyceum were oriented while they created the new model of science as exact, certain, and irrefutable knowledge, i.e., ἐπιστήμη. From this point of view, Archytas as mathematician was far more important than Archytas as philosopher, the author of *On Mathematical Sciences*. Nevertheless, the influence of this work on the development of the new model of science cannot be ruled out completely.

Unlike the first, methodological part of Archytas' fragment B 3, its second part discusses the main characteristic of τέχνη – its usefulness (χρήσιμον, ὠφέλιμον).

> The invention of calculation (λογισμός) put an end to discord (στάσις) and increased concord (ὁμόνοια). With the invention of calculation greed (πλεονεξία) disappears and equality (ἰσότας) arrives, since it is by means of calculation that we settle our dealings with others. Owing to this the poor receive from the powerful and the rich give to those in need, since both believe that owing to calculation they will have what is fair (τὸ ἶσον). A standard and a barrier to the unjust, it averts those who can calculate (ἐπισταμένους λογίζεσθαι) from injustice, persuading them that they would not be able to stay unexposed when they resort to calculation, and prevents those who cannot calculate from doing injustice by showing through calculation their deceit.

As follows from this solemn praise of arithmetic (which is understood here as the art of calculation), Archytas endeavored to show that *mathēmata* are at least no less useful than other τέχναι. No wonder he relates to the discovery of calculation such important social changes as an increase of concord and an advance toward greater equality. The progress of knowledge leads to social progress, whose main criteria – the absence of inner discords, ὁμόνοια and ἰσότης – are very close to Isocrates' ideals.[111] Moreover, calculation proves ca-

---

[109] See below, 93 n. 58.

[110] See *On Archytas' Philosophy* in three books (D. L. V, 25 No. 92 = fr. 207 Rose) and *Excerpts from Timaeus and the Works of Archytas* (No. 94 = fr. 206 Rose). Archytas is also mentioned in the Aristotelian corpus (*Pol.* 1340b 25; *Rhet.* 1412a 12; *Met.* 1043a 21; *Probl.* 915a 25).

[111] The conditions of attaining εὐδαιμονία are peace and ὁμόνοια, the latter, in its

pable of improving people's moral qualities, keeping them from greed and injustice or, at any rate, exposing these vices. Naive as this view of mathematics may seem, we should not forget that it conforms perfectly with the claims of the Sophists, who asserted that their lessons made young people not only wiser but also better, and with the intellectualism of the ethical doctrines of the time in general. Socrates and Isocrates, Plato and Aristotle shared the conviction that knowledge makes a man and, accordingly, the society in which he lives, better. It was about the kind of knowledge that they were at variance. Closer to Archytas was the position of Plato, who believed, to all appearances, that long and sustained study of mathematics not only strengthens and sharpens a man's intellect (on which Isocrates and Aristotle also agreed), but also leads him to the understanding of what is good and accordingly improves his moral qualities.[112] Otherwise the ten years of studying the *mathēmata* Plato imposed on the future guardians of the ideal polis would have been spent in vain.

It has long been noted that the first and second parts of Archytas' fragment, connected as they are both stylistically and thematically, could hardly have followed each other immediately.[113] What, then, could have filled the lacuna between them? The invention of the *mathēmata* seems to be the most natural theme to bridge the gap between the two parts. It is revealing that the first part deals with the 'methodology' of scientific discovery, while the second begins with the invention of one of the mathematical sciences, the art of calculation (λογισμὸς εὑρεθείς). In other words, the situation described in the second part could only have developed *after* calculation had been discovered and was the result of that discovery. Since the circumstances of the discovery are not mentioned, one can surmise that the passage left out by the excerpter was related to this topic, which seems to be perfectly relevant for a work *On Mathematical Sciences*. From the second part of the fragment, extolling the benefits brought about by the discovery of calculation, it follows that, *before* the discovery, social harmony was not possible. We do not know whether Archytas described the life preceding this discovery as governed by greed and discord, i.e., whether he was developing a theory of the origin of culture. We cannot rule out the possibility that Archytas, like other authors of 'introductions' to various τέχναι, limited himself to a brief digression on the inventors of mathematics.[114]

I have already touched here upon the question whether there is any difference between λογισμός (B 3) and λογιστική (B 4) and whether they referred, respectively, to practical and theoretical arithmetic. The fact that the second

---

turn, being the result of ἰσονομία (*Panath.* 178, *Areop.* 21, 69, *Nic.* 41, 67). See Carpiglione, J. C. Isocrate, sull' idea di progresso, *AAN* 96 (1985) 247–267, esp. 263 f.

112 Burnyeat, M. F. Plato on why mathematics is good for the soul, *Mathematics and necessity. Proc. of British Academy* 103 (2000) 1–81. The same view was held by Nicomachus (*Intr. arith.*, 65.13–16), Ptolemy (*Alm.*, 7.17 f.), and Proclus (*In Eucl.*, 24.4).

113 Blass. De Archytae, 581 f.; *DK* I, 437 n.

114 See above, 51.

part of fragment B 3 deals with the practical art of calculation does not prove that Archytas was consistent in drawing a distinction between λογισμός as the practical part of arithmetic and λογιστική as its theoretical part.[115] Emphasizing the utility of the art of calculation, he might equally as well have used the term λογιστικὴ τέχνη, which, like λογισμός, could denote both practical and theoretical arithmetic. Turning to the traditional subject of the utility of a τέχνη does not imply that the τέχνη itself is regarded as practical. The usefulness of *mathēmata* for society was acknowledged even by Plato,[116] who saw their main value elsewhere. Aristotle, ascribing in his *Protrepticus* an independent value to theoretical knowledge, did not, however, fail to point out that philosophy could well prove useful in practical affairs as well.[117] The fact that mathematics and philosophy have an applied aspect does not undermine their status as theoretical disciplines.

Archytas, of course, could not fail to see the difference between the theoretical and the practical aspects of the *mathēmata*, nor to realize that one and the same term, for example ἀστρολογία, denotes both a mathematical science and applied knowledge used by sailors and farmers. This difference was obvious, at any rate, even for such a layman in mathematics as Xenophon. Speaking of Socrates' attitude toward astronomy, geometry, and arithmetic, he notes that the latter recommended limiting oneself to the practical part of each of these sciences, without going deeply into theory (*Mem.* IV,7.1–8). Hence, one does not need different terms to distinguish between a theoretical discipline and its applied counterpart. Though we do find in Archytas a new term, ἁρμονική, denoting, unlike μουσική, the science of music,[118] μουσική is also repeatedly used in the same treatise for the science of music (B 1–2). Philip of Opus observes that a theoretical science bears the 'ridiculous' name of γεωμετρία (*Epin.* 990d 2), associating it with measuring land; the new term, γεωδαισία, denoting the practical discipline, appears later in Aristotle's *Metaphysics* (997b 26f.).[119]

---

115 It is contradicted by the usage of Isocrates, Plato, and Aristotle, who referred λογισμοί to theoretical arithmetic (see above, 61 n. 65–66).

116 *Res.* 522c, e, 526d, *Leg.* 809c–d.

117 Fr. 46, 51, 54 Düring; *EN* 1172b 5f., 1177a 32–b 4. Elsewhere he stresses again: though theoretical sciences like astronomy or geometry may prove to be useful, their main purpose is knowledge (*EE* 1216b 11f.; cf. Isoc. *Antid.* 262–269, *Panath.* 30–32). For the same in Proclus, see *In Eucl.*, 25.18ff.

118 Archytas' treatise on the theory of music was entitled Ἁρμονικός (B 1), but whether it was he who coined the term remains uncertain. In Aristoxenus (*Harm.* I, 2, 5, 7 etc.), the word ἁρμονικοί denotes a trend in the fifth-century harmonics; according to the papyrus Hibeh 13 (col. I,4) this was a self-definition.

119 Aristotle also pointed out that different disciplines could bear the same name: σχεδὸν δὲ συνώνυμοί εἰσιν ἔνιαι τούτων τῶν ἐπιστημῶν οἷον ἀστρολογία ἥ τε μαθηματικὴ καὶ ἡ ναυτική (*APo* 78b 39). Though he means here a theoretical

Isocrates, speaking about Busiris' invention of philosophy, which included astronomy, geometry, and arithmetic, observes that some praise the utility of these sciences, while others attempt to demonstrate that they are conducive in the highest measure to the attainment of virtue.[120] Though the rhetorician does not state his attitude toward these views directly, his silence seems to indicate that he does not share them.[121] Among those who did hold both these views, one ought, first of all, to cite Archytas, who praised the practical utility of mathematics and its ability to improve moral qualities.[122] It is revealing that *Busiris* also mentions Pythagoras, who borrowed philosophy (including, naturally, the *mathēmata*) from the Egyptians, and his followers, Isocrates' contemporaries, whom he treats with open irony (28–29). If my suggestion is true, it will, first, prove that Archytas wrote about the usefulness of all exact sciences, and not arithmetic only, and second, confirm the dating of his work to the turn of the fifth century. It does not follow, however, that he was the only one to hold these views – they may have been shared, for example, by Hippias, who also used to teach mathematics. As for Plato and his pupils, the dating of *Busiris* excludes the possibility of seeing them among those to whom Isocrates' irony could refer.[123]

The Academics as adherents and advocates of mathematics appear in two of Isocrates' later works. In the *Antidosis* (ca. 353) he characterizes his own position as intermediate between the majority, who regard *mathēmata* as empty talk and hairsplitting with no useful application either to private or to public affairs, and those who praise these sciences, since their words also partake of truth (261–263). Aware of the inconsistency of his position, Isocrates defends it by pointing out that mathematics, unlike other sciences, helps us not in life itself (unless one happens to teach it), but in the very process of learning. Studying it, a young man exercises and sharpens his mind, strengthens his memory and acquires the habit of assiduous work, so that later he learns subjects of greater importance more quickly and easily (263–265).[124] Of course, mathe-

---

science, accounting for the facts, and a descriptive one, establishing them, ναυτικὴ ἀστρολογία refers us to the *Nautical Astronomy*, attributed to Thales (11 A 1 23, A 2, B 1). This work was, of course, of a practical rather than purely descriptive character. μηχανική denoted in Aristotle both a theoretical and a practical science (*APo* 78 b 37, *Mech.* 847 a 18 f.). Anatolius, leaning on Geminus' classification of sciences, referred mechanics to the *mathēmata*, excluding from them τὸ ὁμωνύμως καλούμενον μηχανικόν (Ps.-Heron. *Def.*, 164.9 f.), i.e., a practical τέχνη bearing the same name.

120 ὧν τὰς δυνάμεις οἱ μὲν ὡς πρὸς ἔνια χρησίμας ἐπαινοῦσιν, οἱ δ' ὡς πλεῖστα πρὸς ἀρετὴν συμβαλλομένας ἀποφαίνειν ἐπιχειροῦσιν (*Bus.* 23).

121 Cf. *Antid.* 261–268, *Panath.* 26–28; see further below.

122 Isocrates seems to attribute these views to different people, but since we know that they were typically both held by the same people (Archytas, Plato), οἱ μέν ... οἱ δέ may have been in this case nothing more than a rhetorical figure.

123 Cf. below, 226 f.

124 Cf. Cic. *De orat.* 3, 5, 8; *De fin.* I, 72; Quint. *Inst.* I, 10, 34: mathematics can be used to sharpen and to train the intellect of children.

matics is not yet 'philosophy' (i.e., education indispensable for a citizen),[125] but only a 'gymnastics of the soul' and 'preparation for philosophy', since learning it does not improve the student's skill in discussing affairs and making judgments on them. Isocrates recommends the study of mathematics only to young people and for a short term, so as not to allow their minds to be dried up (266–268).

These ideas are repeated and further elaborated in *Panathenaicus*, written a decade later. Within this period Aristotle had published his *Protrepticus*, containing, in particular, a positive refutation of Isocrates' views on philosophy.[126] Having taken certain passages in *Protrepticus* as false charges and personal prejudice (*Panath*. 25), the elderly orator answers resentfully that in fact he rather approves of the modern program of education (geometry, astronomy, and 'eristic dialogues'), in which, however, young men delight more than they should (26). Then his tone grows harsher. As good as mathematics is for the young, inasmuch as it gives them an occupation and keeps them out of many other harmful things, it is no longer suitable for a grown man. Indeed, even those who have become so thoroughly versed in it as to instruct others fail to use opportunely the knowledge that they possess and, in practical activities, prove less cultivated than their students and even than their servants (27–29).

It is obvious that in the course of the fifty years separating the *Panathenaicus* from the *Busiris*, Isocrates' notions of the exact sciences and their utility remained unchanged. What did change was the target of his criticism. His Busiris realized something that the Pythagoreans and the Academics failed to see: namely, that grown and respectable people should know better than to study mathematics, which is good only for the upbringing of young men.[127] On the whole, however, Isocrates' position lies far from the narrow-minded self-assurance of ignorance. His acknowledgement of the pedagogical importance of mathematical studies is perfectly correct.[128] Of course, he gave preference to his own system of education, but even Plato, who valued mathematical sciences much more highly than Isocrates did, regarded them as but the threshold of dialectic (*Res*. 536d 4f.). Isocrates was also right, of course, in denying mathematics' beneficial influence on morals. In his assertion that certain teachers of mathematics prove extremely foolish in practical affairs, one cannot help seeing a parallel to Aristotle's remark that Hippocrates of Chios was thought foolish in everything except mathematics.[129] Not far from Isocrates' position is

---

125 *Soph*. 21, *Antid*. 50, 270f.; Burk, A. *Die Pädagogik des Isokrates*, Würzburg 1923, 65f.

126 Düring. *Protrepticus*, 20f., 33f.

127 τοὺς μὲν *πρεσβυτέρους* ἐπὶ τὰ μέγιστα τῶν πραγμάτων ἔταξεν, τοὺς δὲ *νεωτέρους* … ἐπ' ἀστρολογίᾳ καὶ λογισμοῖς καὶ γεωμετρίᾳ διατρίβειν ἔπεισεν (*Bus*. 23)

128 Burk, *op. cit.*, 140.

129 See above, 69 n. 104. Isocrates too might have meant Hippocrates, rather than the Academics.

the discourse of the mature Aristotle on the difference between ἐπιστήμη and φρόνησις, which concludes that for a young man mathematics cannot substitute for practical reason based on experience (*EN* 1142a 11–20).

From the point of view of what is primarily required of τέχνη, Isocrates' criticism of mathematics will turn out to touch upon only one point, but that one is decisive: its utility. Isocrates did not deny the possibility of learning mathematics or the existence of specialists in it and a specific purpose to it. Moreover, he was ready to acknowledge it as the ἐπιστήμη that arrives at firm knowledge (*Antid.* 264; *Panath.* 28–30). It is the unattainability of ἐπιστήμη and its resulting uselessness in *human* affairs – the only subject of interest to Isocrates – that made the rhetorician exclude it from the number of important reference points. For Isocrates, the chief measure of human judgments and actions was not ἐπιστήμη, but δόξα.[130] Explaining his ideas of 'wisdom' and 'philosophy', Isocrates writes:

> My view of this question is, as it happens, very simple. For since it is not in the nature of man to attain a science (ἐπιστήμη) by the possession of which we can know positively what we should do or what we should say, in the next resort I hold that man to be wise who is able by his powers of conjecture to arrive generally at the best course (ὡς ἐπὶ τὸ πολύ), and I hold that man to be a philosopher who occupies himself with the studies from which he will most quickly gain that kind of insight (φρόνησις).[131]

This choice characterizes the difference between Isocrates and many of the Presocratics[132] as well as Plato, who were convinced of the attainability of 'knowledge' and hence preferred it to 'opinion'.[133] Among Isocrates' allies in preferring relative certainty and accuracy of knowledge to absolute truth were many of the Sophists and Hippocratic physicians.[134] Later, Aristotle attenuated the contrast between δόξα and ἐπιστήμη by assigning to theoretical science and, in particular, to physics the kind of regularity that he called ὡς ἐπὶ τὸ πολύ.[135]

---

130 Each time that Isocrates compares these notions, he invariably shows his preference for δόξα: *Soph.* 8, *Hel.* 5, *Antid.* 184, 270–271, *Panath.* 30–31. Outside this antithesis, ἐπιστήμη is often used to mean τέχνη (Mikkola, E. *Isokrates. Seine Anschauungen im Lichte seiner Schriften*, Helsinki 1954, 21f., 31f., 65f.). For Isocrates' views on δόξα, see Eucken, C. *Isokrates*, Berlin 1983, 32f., 36f.

131 *Antid.* 271, transl. by G. Norlin; cf. 184.

132 On the antithesis 'knowledge (truth) – opinion' in the Presocratics, see Parmenides (B 1.29–30, 8.51), Empedocles (B 3.13 and 132).

133 See e.g. Sprute, J. *Der Begriff der DOXA in der platonischen Philosophie*, Göttingen 1962.

134 On the role of καιρός in medicine, see above, 69 n. 102–103, in Isocrates: Wilms, *op. cit.*, 288f.

135 See below, 127.

5. From 'progress' to 'perfection'

The example of the notion ὡς ἐπὶ τὸ πολύ shows again how much Aristotle's theory of science owes to the 'technical' literature of the turn of the fifth century. As often happens to be the case, Aristotle applies to ἐπιστήμη the notion used earlier to characterize the cognitive possibilities of τέχνη,[136] giving to this notion a more definite terminological meaning. Very similar is the history of the word ἐπίδοσις, which in the literature of the fourth century usually designates 'progress'; the progress of τέχναι became topical much earlier than the progress of mathematics. The first use of ἐπίδοσις is found in the Hippocratic corpus, where it means (with reference to things and processes) either 'increase, growth', or 'development, aggravation, progress of a disease'.[137] In the meaning of physical growth, increase, and development, ἐπίδοσις continues to be used later, e.g. in the natural-scientific treatises of Aristotle and Theophrastus.[138] Yet already in Isocrates' early speeches, ἐπίδοσις and ἐπιδιδόναι acquire the character of notions that denote *qualitative* development and advancement to a better state.[139] In a similar meaning (that of increase, advance, development), he uses the other verbs from the same semantic group, such as αὐξάνειν, προαγαγεῖν, and προέρχεσθαι.[140] We are going to meet these words later, when discussing the idea of scientific progress.[141]

Very often, the passages in which Isocrates refers to the idea of progress are related to the invention and development of τέχναι. Though this subject is among Isocrates' favorites,[142] his attitude toward first discoverers was ambivalent. On the one hand, he extols Busiris (35) and, still more, Athens (*Paneg.*

---

[136] See in particular Isoc. *Panath.* 30 (ὡς ἐπὶ τὸ πολύ στοχάζεσθαι τοῦ συμφέροντος), *Paneg.* 154, *Ad Nic.* 34, *Areop.* 5, 165, *Antid.* 184, 271 (cited above), *De pac.* 35.

[137] *Acut.* 4, 18: ἐπίδοσις ἐς πλῆθος (growth); *Epid.* II, 1, 6 (bis), VI, 8, 14: ἐπίδοσις (aggravation, 'development' of a disease; *Artic.* 30, 25: ἐπίδοσιν ἔχειν (to grow; cf. 72, 16: ἐπίδοσιν ἐπιδιδόναι, to extend, to yield); *Septim.* 9, 49: ἐπίδοσιν ἔχειν (to develop). The verb ἐπιδιδόναι, from which ἐπίδοσις is derived, is found with the same meaning (to grow, to increase, to strengthen, to develop) in Herodotus (II, 13) and Thucydides (VI,60.2; VII,8.1; VIII,24.4. 83.2); see Edelstein, *op. cit.*, 92 n. 79.

[138] Arist. *GA* 744b 36, *GC* 320b 31, *HA* 560a 20; Theophr. *HP* II,6.3, *CP* V,6.2.

[139] *Paneg.* 10, 103, 189. Sometimes ἐπίδοσις in Isocrates can mean 'development for the worse' as well (*Hel.* 8, *Areop.* 18, *De pace* 127), but in most cases it has a manifestly positive connotation.

[140] αὐξάνειν is often used along with ἐπιδιδόναι as its synonym (*Ad Dem.* 12, *Nic.* 32, *Paneg.* 103). See also προαγαγεῖν (*Paneg.* 37, *Antid.* 185) and προέρχεσθαι (*Antid.* 82; *Ep.* 4, 10).

[141] For ἐπίδοσις in an Academic treatise dealing with progress in mathematics, see 3.1. For ἐπίδοσις, αὐξάνειν, προαγαγεῖν and προέρχεσθαι in Eudemus' *History of Geometry*, see 5.5.

[142] See e.g. *Paneg.* 10, *Nic.* 32, *Antid.* 82, 185 and above, 52f.

38–40) as *prōtoi heuretai* and censures Sparta, which in the field of discoveries is more backward than Greeks and foreigners (*Panath.* 202–209). Isocrates denounces his rivals' proud claims to the novelty of their inventions as futile (*Hel.* 2–3). On the other hand, he finds worthy of admiration not only Homer and the founders of the tragedy (*Ad Nic.* 48), but also those who create new speeches (*Antid.* 81–83), as well as all those who by painstaking thought and endeavor discover some useful things (*Ep.* VIII, 5), though the latter remain, unfortunately, less popular than the winners of athletic competitions (cf. *Paneg.* 1–2, *Antid.* 250). These reproaches betray the wounded ambition of a man who, sharing the common aspiration for priority and creative originality, failed to win acknowledgment for his innovative efforts. It is this resentment that may have made him shift the accent from πρῶτος to εὑρετής, i.e., to the figure of 'innovator' who is, at the same time, a 'perfection seeker', ἐξεργαζόμενος:

> And it is my opinion that the study of oratory as well as the other τέχναι would make the greatest advance if we should admire and honour, not those who make the first beginnings in their crafts, but those who are the most finished craftsmen in each, and not those who seek to speak on subjects on which no one has spoken before, but those who know how to speak as no one else could (*Paneg.* 10).[143]

In the eulogy of Eugoras, Isocrates returns to this idea once more: worthy of praise are not only the heroes of the past, but those of the present as well, though envy prevents people from glorifying the deeds of their contemporaries (5–6). A reasonable man should, however, ignore the envious, particularly because we know that:

> progress is made, not only in τέχναι, but in all other activities, not through the agency of those that are satisfied with things as they are, but through those who correct, and have the courage constantly to change anything that is not as it should be (7).[144]

Unable to claim for himself the status of a discoverer, Isocrates points out that the Greeks owe their high level of cultural development not only to the discoverers who lived in the distant and more recent past, but also to those like himself, who are capable of improving what has already been invented and of bringing it to a state of perfection.

We have had a chance to see how closely the notions of the progress of τέχνη in the past and of its nearing perfection in the present come together.[145] The experience of the man of the 19th century, who could observe steady progress in practically every field in both the past and the present, made him extend this tendency into the future as well. This extrapolation, which still seems natural to us, is not in fact to be taken for granted. The authors of the classical epoch

143 Transl. by G. Norlin. Cf. ἐξεργαζομένους in Isocrates and ἐς τέλος ἐξεργάζεσθαι (*De arte*, 1), above, 59.

144 Transl. by La Rue van Hook.

145 See above, 59f.

reasoned differently: the more striking the progress that had already been made, the more natural it seemed to believe that the efforts of contemporaries, including their own, would soon reach a perfection not, or unlikely, to be surpassed in the future.[146] Generally speaking, the combination of the idea of progress related to the past or the present with a firm belief in perfection to be achieved in the near future is not rare in the history of European thought. Descartes believed that, after the discovery of his principles, humankind was on the verge of mastering nature completely, with only two or three victories left to win. Charles Perrault, an active participant in the 'dispute between the ancients and the moderns', thought of the 18th century as the peak of perfection. The feeling that perfection (a relative one) was an aim that could be achieved used to be widespread in the 18th century as well,[147] combined, as it was in Diderot, Rousseau, and Voltaire, with the notion of a decline likely to follow the peak.

Since the ancient idea of progress is not very different from similar views current in the pre-industrial epoch,[148] one can hardly explain it by the 'aversion to infinity' or the 'predilection for perfect forms' so often attributed to the worldview of the Greeks.[149] It can rather be accounted for by natural limits imposed on the social and cultural experience of those who were the first in the history of humanity to reflect on progress.[150] If the radical shift of views on progress, i.e., the transformation of the idea into ideology, took place at the turn of the 18th century, why expect thinkers of the classical epoch to share the notions typical of Comte or Spencer and, having found none of the kind, deny to them any idea of progress at all?[151] A balanced approach seems to be more productive, one that allows us to find in the views of the classical writers upon this subject something that we share, without blotting out the features that separate theirs from modern notions, on the one hand, or overemphasizing them, on the other.

Let us note for example that Isocrates touches upon social, political, and cultural aspects of 'progress' much more often than upon those related to the growth of knowledge. On the individual level, the ends or results of progress are moral improvement (*Ad Nic.* 29, *Euag.* 81, *Ad Dem.* 12), education (*Antid.* 185, 267), maturity (*Ep.* 4.10), and success in business affairs (*Areop.* 5); on the social level they are wealth (*Nic.* 32, 63, *De pace* 64, *Paneg.* 103), prosperity (*De pace* 20, *Paneg.* 37), the rise and strengthening of the state (*Euag.* 48, *Areop.* 69, *Archid.* 104, *De pace* 140), etc. In this connection, the thesis that the classical epoch understood progress mainly as the development of knowledge

146 On this, see Meier. Antikes Äquivalent, 291 f., 297 f.

147 Koselleck. Fortschritt, 376.

148 A limited notion of progress, with the future playing a negligible role in it, was typical e.g. of the Scottish thinkers of the 18th century (Spadafora, *op. cit.*, 301 f.).

149 See e.g. on the author of *VM*: "Es ist echt griechisch, dass sich ihm das Erkenntnisobjekt nicht ins Unendliche verschiebt." (Herter. Theorie, 172).

150 Meier. Antikes Äquivalent, 303 ff.

151 So den Boer. Progress, 9 f.

and skills needs a certain correction. To be sure, the frequent use of ἐπίδοσις and other notions akin to it does not prove that Isocrates took a 'progressivist' view of things; limited to the past and the present,[152] his 'progress' is of a local, rather than a universal and regular character. Even when he tries to formulate a regularity, this comes down to the popular idea of the Wheel of Fortune (*Areop.* 5, cf. *Archid.* 103f.). It is not for the originality of his ideas, however, that Isocrates remains interesting for us, but for the wide acclaim they enjoyed among his contemporaries. They testify that in the first part of the fourth century the awareness of positive changes in private and social life was strong enough to grow into a concept, dim as it may yet have remained. Considered apart from this social background, philosophers' discussions of the progress of scientific and philosophical knowledge would look more isolated than they actually were.

* * *

Numerous and obvious similarities between the Hippocratics', Archytas', and Isocrates' conceptions of what knowledge is, how it comes into being, is developed, and communicated, a certain kinship between their views on the cognitive abilities of man, as well as the variety in their assessment of the degree of certainty and utility of different kinds of knowledge – all this taken together demonstrates that the philosophy and history of science, which originated in the Academy and the Lyceum, respectively, stemmed from a lasting tradition. Let us now sum up its principal stages and characteristics. In the last third of the fifth century, frequent but isolated mentions of *prōtoi heuretai* are superseded by more systematic attempts to consider culture from the historical point of view – namely, as the *history of discoveries and inventions* in the field of τέχναι and, later, as the history of τέχναι themselves. The genre of heurematography, then in process of formation, was but one of the branches of this tradition, oriented, at that, more toward systematic accounts of τέχναι than toward the history of inventions. More important for the formation of the historical approach to culture are works by Glaucus, Hellanicus, and Hippias, where interest in the development of τέχναι is combined with attempts to show this process in chronological order. The methods of analysis of cultural phenomena worked out in these works were later taken up by Peripatetic historiography, which applied them to new branches of knowledge.

The assessment of man's cognitive abilities, which is optimistic on the whole, now starts to be differentiated according to the subject of knowledge (mathematics, medicine, social sphere) and the degree of exactness attainable in the particular field under discussion. Epistemological optimism leads to the belief that in some fields knowledge can soon be brought to perfection. The

152 With reference to the future (the near future as a rule) only conditional clauses are used: *Nic.* 63, *Euag.* 81, *De pace* 20.

conceptions of progress, cognitive as well as social, are of a limited character and refer to the past and the present rather than to the future. The theory of the origin of culture that is imbued with these conceptions, most likely created by Protagoras and further developed by Democritus, exerts profound influence on a great variety of genres, from history and rhetoric to philosophy and medicine, and helps to establish new notions of the motive forces in civilization.

Still more popular becomes the Sophistic theory of τέχνη, used on the descriptive, as well as on the normative level for the realization of any systematically organized and practically oriented knowledge. It was flexible enough to allow the interpretation of such sciences as arithmetic, geometry, or astronomy, traditionally counted among τέχναι. The formation of the quadrivium of related mathematical sciences, accomplished by the second half of the fifth century, sets the *mathēmata* apart as a special group of τέχναι, which best answer such major criteria of scientific knowledge as exactness and clarity. The further progress of mathematics turns it into an obvious model for the new theory of knowledge, in which practical orientation recedes into the background. Among the prominent features of the old model inherited by the new one is an awareness that science is not just an organized sum total of knowledge, but that it is founded on the true method, which guarantees the correct results. Another important element of science is the search for and the transmission of knowledge, which are explained in terms of the contrasting notions of 'discovery' and 'learning (imitation)'.

Chapter 3

Science in the Platonic Academy

1. Plato as architect of mathematical sciences?

The previous chapter, devoted to the formation of the notions of science that later found their reflection in the Peripatetic historiography of science, focused mainly on three elder contemporaries of Plato: the author of *VM*, Isocrates, and Archytas. This choice of sources was intended, in particular, to emphasize that Plato' role in the development of the new concept of science cannot be adequately defined without a thorough analysis of his predecessors' views. Not every idea to be found in his dialogues belongs to Plato himself. Trivial as it seems, this thesis must be one of the fundamental premises of any research on 'Plato and the exact sciences'.

In the 19th and the early 20th centuries, attention was paid predominantly, although not exclusively, to the questions how great Plato's contribution to specific mathematical research really was and how reliable our sources are that ascribe to him particular discoveries. The general conclusion of these studies was that Plato himself was not an active scientist and that the scientific discoveries and hypotheses attributed to him in the ancient tradition are not really his.[1] There do not seem to have been any later serious attempts to debate this conclusion,[2] and the discussion has been concerned not with Plato as a scientist, but rather with science in the Platonic school. Since the late 19th century, the opinion has been established that even if Plato did not achieve any success in mathematics, he did play a considerable role as an organizer of scientific research and as a methodologist who defined the problems mathematicians and astronomers studied and the methods they used.[3] I quote only one typical opinion:

---

1 See e.g. Blass, C. *De Platone mathematico* (Diss.), Bonn 1861; Allman, G. J. *Greek geometry from Thales to Euclid*, Dublin 1889, 123; Simon, M. *Geschichte der Mathematik im Altertum*, Berlin 1909, 183ff.; Heath. *History* 1, 284ff.

2 See, however, Mugler, C. *Platon et la recherche mathématique de son époque*, Strasbourg 1948; cf. Cherniss, H. Plato as mathematician, *Rev. Met.* 4 (1951) 395–425 (= *Selected Papers*, ed. by L. Tarán, Leiden 1977, 222–252).

3 See e.g. Usener, H. Organisation der wissenschaftlichen Arbeit (1884), *Vorträge und Aufsätze*, Leipzig 1907, 69–102; Wilamowitz-Moellendorff, U. von. *Antigonos von Karystos*, Berlin 1889, 279ff.; Heiberg, I. L. *Geschichte der Mathematik und Naturwissenschaft im Altertum*, Leipzig 1912, 9f.; Shorey, P. Platonism and the unity of science (1927), *Selected Papers*, ed. by L. Tarán, New York 1980, 434ff.; Solmsen, F. Platons Einfluß auf die Bildung der mathematischen Methode (1929),

> Die traditionelle Platosauffassung, wie sie auch von den beteiligten Mathematikern im wesentlichen geteilt wird, besagt: Plato hat natürlich keine mathematische Entdeckungen gemacht; die Überlieferung, die ihm Dodekaeder zuschreibt, ist wegzulegen; aber Plato hat der Mathematik die allgemeinen Direktiven gegeben, die axiomatische Struktur der Elemente, die Beschränkung auf Konstruktionen mit Zirkel und Lineal allein, die analytische Methode sind Platos Werk; die großen Mathematiker seines Kreises, Theätet und Eudoxus, haben die sogenannte Euklidische Mathematik unter seinem Einfluß geschaffen.[4]

Despite the criticism of this position frequently expressed both by philologists and by historians of mathematics,[5] in the last decades it has been developed in many important studies. While differing in approach, these studies share the tendency to present the Academy as a kind of a research institution, where the best mathematicians and astronomers of the time worked under Plato's methodological supervision.[6] This tendency would be of indirect interest to our research, if it did not follow the ancient tradition, which, although represented mainly by the late sources, stemmed from the early Academy itself. The tradition of regarding Plato as an 'architect of *mathēmata*' goes back to a treatise written by one of the Academics and reflects the somewhat paradoxical circumstance that Plato had become the hero of 'historico-scientific' legends even before the historiography of science sprang into being. To detect whether

---

Das Platonbild, ed. by K. Gaiser, Hildesheim 1969, 125–139; Herter, H. *Platons Akademie*, Bonn 1946; Hauser, G. *Geometrie der Griechen von Thales bis Euklid*, Lucerne 1955, 127–138.

4 Toeplitz, O. Mathematik und Antike, *Die Antike* 1 (1925) 201. It is worth pointing out that Toeplitz himself understood the vulnerability of this position.

5 Howald, E. *Die platonische Akademie und die moderne universitas litterarum*, Bern 1921; Frank, E. Die Begründung der mathematischen Wissenschaften durch Eudoxos (1932), *Wissen, Wollen, Glauben*, ed. by L. Edelstein, Zurich 1955, 144f.; Cherniss, H. *Rec.*: Herter, H. *Platons Akademie*, *CQ* 43 (1948) 130–132 (= *Selected Papers*, 217–221); Szabó, Á. Anfänge des Euklidischen Axiomensystems, *AHES* 1 (1960) 99ff. (= *Zur Geschichte der griechischen Mathematik*, ed. by O. Becker, Darmstadt 1965, 450ff.); Fritz, K. von. *Platon, Theaetet und die antike Mathematik* (1932), Darmstadt 1969 (esp. Nachtrag); idem. *Grundprobleme*, 250ff. Neugebauer. *ES*, 152, expressed his opinion very definitively: "I think that it is evident that Plato's role has been widely exaggerated. His own direct contributions to mathematical knowledge were obviously nil ... The often adopted notion that Plato 'directed' research fortunately is not borne out by the facts."

6 Gaiser, K. *Platons ungeschriebene Lehre*, Stuttgart 1963, 293ff.; idem. *Philodems Academica*, Stuttgart 1988; Lasserre, F. *The birth of mathematics in the age of Plato*, London 1964; idem. *Léodamas*, 516f.; Fowler, D. H. *The mathematics of Plato's Academy. A new reconstruction*, Oxford 1987, 342ff.; Hösle, V. *I fondamenti dell' aritmetica e della geometria in Platone*, Milan 1994. I. Mueller is also ready to admit that Plato was a "general mathematical director, posing problems to the mathematicians" (Mathematical method and philosophical truth, *The Cambridge companion to Plato*, ed. by R. Kraut, Cambridge 1992, 175).

there is a grain of historical truth in these legends, we need a critical analysis of ancient evidence of the place of the exact sciences in Plato's Academy.

One of the classic examples of the tradition in question is the story about the famous solution to the Delian problem of the duplication of the cube, preserved by Plutarch, Theon of Smyrna (first part of the second century AD), and several later commentators.[7] It should be noted that for ancient mathematics the Delian problem was not unlike Fermat's theorem for modern mathematics: in the course of the seven hundred years from Hippocrates to Pappus, a great many famous mathematicians tried to find a solution to it.[8] Thus, the tradition connecting it with Plato makes him seem like the originator of one of the central problems in ancient mathematics. The versions of the story found in Plutarch and Theon can be generally summed up in the following way. The people of Delos, tormented by a plague that Apollo had sent upon them, asked Plato to solve the problem of duplicating a cubic altar. The Delphic oracle had posed them this problem – the plague would leave the island if the Delians succeeded in giving it a solution. Plato, having reprimanded the Greeks for their contempt of geometry, commissioned the famous 'Academic mathematicians' Archytas, Eudoxus, and Menaechmus to find a solution.[9] Their approach employed mechanical devices, and Plato rebuked them for ruining the value of geometry by having sunk to the level of crude mechanics.

According to general opinion, the common source of Plutarch and Theon was Eratosthenes' dialogue *Platonicus*.[10] The plot of this dialogue is clearly literary fiction: the problem of the duplication of the cube arose in the mid-fifth century, and was not set for Plato by the Delians. Hippocrates had reduced it to

---

7 Plut. *De E ap. Delph.* 386 E; *De gen. Socr.* 579 B–C; *Quaest. conv.* 718 E–F; *Marc.* 14, 9–11; Theon. *Exp.*, 2.3–12; Eutoc. *In Archim. De sphaer.*, 88.3–96.9; Asclep. Tral. *In Nicom. Intr. arith.*, 61; Philop. *In APo comm.*, 102.12–22; *Anon. Proleg.*, 11. See Riginos, A. E. *Platonica. The anecdotes concerning the life and writings of Plato*, Leiden 1976, 141ff. (no. 99–100); Dörrie, H. *Der Platonismus in der Antike*, Vol. 1. Stuttgart 1987 (Baustein 7.2–5); Geus, K. *Eratosthenes von Kyrene*, Munich 2002, 175ff. Vitruvius wrote on the Delian problem without naming Plato (IX,1. 13–14).

8 Knorr. *TS*, 11ff., offers more than ten solutions. To be sure, unlike Fermat's theorem, this one was already solved in the generation after Hippocrates.

9 Cf. οἱ παρὰ τῷ Πλάτωνι ἐν Ἀκαδημίᾳ γεωμέτραι (Eutoc. *In Archim. De sphaer.*, 90.3).

10 Wolfer, E. P. *Eratosthenes von Kyrene als Mathematiker und Philosoph*, Groningen 1954, 4ff.; Riginos, *op. cit.*, 141; Knorr. *AT*, 17ff., 49ff. Unlike Plutarch, Theon directly referred to this work (*Exp.*, 2.3). A recent work on Eratosthenes attempts to refute – unsuccessfully, it seems to me – the idea that the *Platonicus* was a dialogue (Geus, *op. cit.*, 141–194, esp. 192). Even more difficult is to agree with the author's tendency to regard the *Platonicus* as the *only* source for Eratosthenes' mathematics, denying, e.g., the existence of his work *On Means*, attested by Pappus (*Coll.* VII, 636.24, 672.5, cf. 662.16).

the task of finding two mean proportionals between two given lines,[11] and the brilliant solution to this last problem was first found by Archytas. Eudemus gives a detailed account of it (fr. 141), so that the evidence on Eudoxus' and Menaechmus' solutions that we find in Eratosthenes, as well as his mention of Hippocrates, must go back to the same source.[12] Eudemus, however, does not even mention Plato. To whom does the legend about three great mathematicians of three subsequent generations (Eudoxus was a pupil of Archytas, and Menaechmus a pupil of Eudoxus), all working under Plato's supervision, belong? Was Eratosthenes its author or does it date back to an earlier time?

The answer is made more complicated, since Eratosthenes' letter to King Ptolemy III, preserved by Eutocius (*In Archim. De sphaer.*, 88.3–96.9), gives an entirely different ending to the story. It states that Archytas, Eudoxus, and Menaechmus proposed too *abstract* solutions to the problem and therefore did not deal with the problem in a practical and useful way, with the exception of Menaechmus, though even he met practical criteria only to a very small degree and with difficulty.[13] Knorr, who analyzed this text in great detail, convincingly showed that the letter is not a later forgery (as Wilamowitz thought)[14] and that it belongs to Eratosthenes.[15] Eratosthenes also studied the problem of duplicating the cube, and it is noteworthy that his own solution was mechanical. He manufactured a device for drawing lines, the *mesolabe*, and dedicated a bronze model of it to King Ptolemy, accompanied with a letter and an epigram. Eratosthenes' solution correlates much better with the 'mechanical' ending of the story than with the 'anti-mechanical' one presented by Plutarch, all the more so because the epigram that is widely recognized as authentic also says that Archytas' solution was badly adapted to practice.[16] Hence Knorr concludes that Eratosthenes had two versions: one more historically accurate, in the letter to Ptolemy, and another, more literary version, recorded in the *Platonicus* and carried down to us by Theon and Plutarch.[17] Knorr considers the

---

11 Eratosthenes, by the way, was well aware of this fact (Eutoc. *In Archim. De sphaer.*, 88.18 f.). For more details, see below, 175 f.

12 Cf. Eud. fr. 139–140. See below, 175 f., 207.

13 Eutoc. *In Archim. De sphaer.*, 90.8 f. = 47 A 15.

14 Wilamowitz-Moellendorff, U. von. Ein Weihgeschenk des Eratosthenes (1894), *Kleine Schriften*, Vol. 2, Berlin 1962, 48–70.

15 Knorr. *TS*, 131 ff. Eratosthenes' letter seems to be unknown to Plutarch and Theon.

16 δυσμήχανα ἔργα (Eutoc. *In Archim. De sphaer.*, 96.16 = 47 A 15).

17 Knorr. *AT*, 22. It is much more likely, however, that the 'anti-mechanical' ending of this story belongs to Plutarch, and not to Eratosthenes (Riginos, *op. cit.*, 145). See below, 88 n. 29. Interestingly, Eutocius (*In Archim. De sphaer.*, 56.13–58.14) mentions one more solution to this problem, based on a mechanical device and attributed, strangely enough, to Plato himself! As Knorr (*AT*, 59) points out, we can wonder at the flexibility of the tradition, which ascribed to Plato such a device, on the one hand, while presenting him as a supporter of pure geometry, on the other.

story of the Delian problem to be a legend that arose in the mid-fourth century in the Academy.[18]

A parallel tradition in the history of astronomy depicts Plato as being the first to put forward the principle of 'saving the phenomena' (σῴζειν τὰ φαινόμενα),[19] explaining the apparently irregular movement of heavenly bodies by attributing uniform circular movement to them. Having formulated the problem in this way, Plato posed it to the scientists, who then studied it using their own methods; the first to achieve success was Eudoxus. It is easy to see that the roles in this story are distributed in exactly the same way as in the legend about the Delian problem. Plato's powerful intellect uncovers the essence of the problem and formulates it for the *mathēmatikoi*; they then compete among themselves and in the end come up with an answer. It is revealing that this story occupied a central place in the arguments of those who attempted to present Plato as a forerunner and nearly as one of the founders of European science. Unlike the Delian problem, which despite all its importance cannot be related to the foundations of ancient mathematics, the principle of 'saving the phenomena' is a cornerstone of Greek astronomy,[20] laying the foundations of all astronomical systems from Eudoxus to Ptolemy. If it could be successfully shown that Plato really did have a connection with the formulation of this scientific principle, then this fact alone would be sufficient justification for calling him an 'architect of science'.

If, however, we turn to the only ancient evidence on this story, provided by Simplicius, the bright colors of this picture immediately begin to fade:

> Eudoxus of Cnidus, as Eudemus reports in the second book of his *History of Astronomy* and as Sosigenes repeats on the authority of Eudemus, is said to have been the first of the Greeks to deal with this type of hypotheses. For Plato, Sosigenes says, set this problem for students of astronomy: 'By the assumption of what uniform and ordered motions one can save the apparent motions of the planets?'[21]

Mittelstraß, who analyzed this passage in great detail, came to a compelling conclusion: it is not Eudemus who mentions Plato, but Sosigenes, a Peripatetic commentator of the second part of the second century AD.[22] Actually, with his

---

18 Knorr. *AT*, 22, 24. Wehrli also noted this (Eud. fr. 141, comm. ad loc.). Cf. Geus, *op. cit.*, 176f.

19 'Preserving the phenomena' is admittedly a better translation of σῴζειν τὰ φαινόμενα, but I prefer to preserve the traditional idiom.

20 Lloyd, G. E. R. Saving the appearances, *CQ* 28 (1978) 202–222 (= Lloyd, G. E. R. *Methods and problems in Greek science*, Cambridge 1991, 248ff.).

21 *In Cael. comm.*, 488.18–24 = Eud. fr. 148; cf. below, 273 n. 199.

22 Mittelstraß, J. *Die Rettung der Phänomene*, Berlin 1963, 149ff. See also Krafft, F. Der Mathematikos und der Physikos. Bemerkungen zu der angeblichen Platonischen Aufgabe, die Phänomene zu retten, *BGWT* 5 (1965) 5–24; Knorr, W. R. Plato and Eudoxus on the planetary motions, *JHA* 21 (1990) 313–329. On Sosigenes, see below, 231f.

characteristic pedantry, Simplicius noted that both Eudemus and Sosigenes (who relied on Eudemus) mention Eudoxus, whereas the words concerning Plato belong to Sosigenes only. Thus Simplicius, to whom Eudemus' *History of Astronomy* was still available, could not find in it anything relating to Plato.[23] If Eudemus did in fact mention Plato in the context of posing such an important problem, Sosigenes would surely not be our only source on this. Meanwhile, Theon, Sosigenes' older contemporary, in a special treatise devoted to *mathēmata* in Plato, does not mention this story, though both the principle of 'saving the phenomena' and Eudemus' *History of Astronomy* do appear in this work (*Exp.*, 180.9, 198.14). Plutarch, who lived earlier than Theon, also mentions the principle of 'saving the phenomena' without connecting it with Plato (*De facie* 923 A). Geminus, a still earlier author, ascribed the principle of ordered and circular movement of heavenly bodies not to Plato, but to the Pythagoreans (*Eisag.* I, 19–20). Thus, Sosigenes seems to be the first author to connect Plato with this principle. Sosigenes could have known the legend of the Delian problem, where Plato figures as a methodologist of mathematics, since Plutarch and Theon wrote about it. It is this legend that could have encouraged Sosigenes to ascribe to Plato the most important principle of astronomy. However, Sosigenes could have relied on much earlier sources.

Recent publications of the Herculanum papyrus 1021, which preserved for us the working text of Philodemus' *History of the Academy*,[24] support the suggestion that Plato's image as 'architect of science' goes back to the early Academy. In column Y of the papyrus, which is a quotation from an early and well-informed author, we read the following:

> He says that at this time *mathēmata* were also greatly advanced, with Plato being the architect of this development; he set problems to the mathematicians, who in turn eagerly studied them. In this way, the general theory of proportions (μετρολογία) and research on definitions reached their peak, as Eudoxus and his students completely revised the old theory of Hippocrates of Chios. Especially great progress was made in geometry, as (at that time) the methods of analysis and of

23 Von Fritz. *Grundprobleme*, 179 n. 375, pointed out that the repetition of Sosigenes' name might mean either: 1) that the words about Plato do not belong to Eudemus; or 2) that Simplicius knew about Eudemus' opinion only through Sosigenes, and was not sure exactly where the quotation from Eudemus ends. Since von Fritz did not see any evidence that Eudemus' *History of Astronomy* was available to Simplicius, he was inclined to the second variant. Yet such evidence does exist (7.1), which makes the first suggestion much more plausible. Krafft. Mathematikos, 16, on the other hand, believes that Simplicius knew Eudemus' work only through Sosigenes, but that the latter made it clear that the reference to Plato was his own. Cf. Knorr. Plato and Eudoxus, 319f.

24 The Epicurean Philodemus (mid-first century BC) wrote Σύνταξις τῶν φιλοσόφων in 10 books (D. L. X, 3). His *History of the Academy* might have been the division of this work (Erler, M. Philodem aus Gadara, *Die Philosophie der Antike*, Vol. 4: *Die hellenistische Philosophie*, ed. by H. Flashar, Basel 1994, 297f.).

diorism (τὸ περὶ διορισμοὺς λῆμμα)[25] were discovered. Optics and mechanics also were not (left in neglect) ...[26]

The similarity of this passage[27] to the quotation from Sosigenes,[28] even if it does not allow us to establish a direct connection between the two texts, at least makes it highly probable. Sosigenes' remark about astronomy seems to be a natural development of the main idea of the papyrus passage, where all *mathēmata* including mechanics and optics are mentioned.[29] It is hardly possible to tell whether Sosigenes learned about this idea from the work used by Philodemus or from a different source. But what inspired Sosigenes to present Plato as the methodologist of astronomy is not that important.[30] It is much more important that the treatise quoted by Philodemus obviously precedes Eratosthenes and could have been among the sources of his *Platonicus*.

Since the papyrus omits the name of the author of this passage, several theories have been proposed about his identity. Lasserre suggested that the passage comes from Περὶ Πλάτωνος by Philip of Opus.[31] Burkert and later Dorandi supported his opinion,[32] while Gaiser ascribed column Y to the Peripatetic Dicaearchus.[33] Without going into the details of the papyrological problems, one has to admit that the first hypothesis is much more plausible than Gaiser's suggestion.[34] Whoever the author of this passage is, it is obvious that he belonged

---

25 The methods of diorism allows one "to determine when a problem under investigation is capable of solution and when it is not" (Procl. *In Eucl.*, 66.22f.).

26 Gaiser. *Academica*, 152f.; Dorandi. *Filodemo*, 126f.

27 ἀρχιτεκτονοῦντος μὲν καὶ *προβλήματα διδόντος τοῦ Πλάτωνος, ζητούντων δὲ μετὰ σπουδῆς* αὐτὰ τῶν μαθηματικῶν.

28 πρῶτος Εὔδοξος ἅψασθαι λέγεται τῶν τοιούτων ὑποθέσεων, *Πλάτωνος*, ὥς φησι Σωσιγένης, *πρόβλημα τοῦτο ποιησαμένου τοῖς περὶ ταῦτα ἐσπουδακόσι.*

29 The mention of mechanics and optics in this passage makes the criticism of the mechanical methods, which Plutarch (*Quaest. conv.* 718 E–F; *Marc.* 14.9–11) ascribed to Plato, even more unreliable. In Aristotle's *Second Analytics* mechanics and optics figure as theoretical sciences, ἐπιστῆμαι (75b 16, 76a 24, 77b 2, 78b 37). Aristotle himself wrote works on optics and mechanics (D.L. V, 26 No. 114, fr. 380 Rose). Several books on optics are ascribed to Philip of Opus (Lasserre. *Léodamas*, 20 T 1).

30 Cf. below, 289f.

31 Lasserre. *Léodamas*, 20 F 15 a, 611 f. Lasserre considered the Academic Hermodorus to have been the intermediary between Philip and Philodemus. See below, 89 n. 37.

32 Burkert, W. Philodems Arbeitstext zur Geschichte der Akademie, *ZPE* 97 (1993) 87–94; idem. *Platon in Nahaufnahme. Ein Buch aus Herculaneum*, Stuttgart 1993, 26f.; Dorandi, T. La tradizione papirologica di Dicearco, *Dicaearchus of Messana*, ed. by W.W. Fortenbaugh, E. Schütrumpf, New Brunswick 2001, 347f. Cf. Dorandi. *Filodemo*, 207f.

33 Gaiser. *Academica*, 76f., 97f., 342ff.

34 The quotation from Dicaearchus takes up the column I and the beginning of the column II of the papyrus, whereas the column Y is on its reverse side. Thus, it is an addition made by Philodemus after he had already finished his work with Dicaearchus' text.

to the Academy, since nobody but a member of the Academy could ascribe to Plato so important a role in the development of mathematics.[35] It is revealing, for example, that the fragments of Eudemus' *History of Geometry* and *History of Astronomy* do not mention Plato's name at all. The few places it does occur are later interpolations, as in fr. 148 of the *History of Astronomy* that we have already considered, or in the well-known *Catalogue of geometers* (fr. 133), an excerpt from the *History of Geometry* found in Proclus.

In this case, we have at our disposal a fragment from an Academic treatise discussing, among the other themes, the flourishing state of mathematics due to Plato. Many ideas of this passage can be found in Eudemus' works on the history of science.[36] But even if this is a fragment of the history of mathematics (which we have good reasons to doubt), it is a history *sub specie Academiae* – its main protagonist is Plato. So we will put aside for a while the analysis of the historical evidence contained in this fragment and try to find out its source as well as the grounds substantiating its main thesis – that of Plato as 'architect of science'. To verify this thesis, we will consider the *Catalogue of geometers*, whose picture of the development of mathematics in Plato's time resembles that of the papyrus passage but goes into much greater detail.

2. The *Catalogue of geometers* about mathematicians of Plato's time

First, it should be noted that Lasserre failed to give any convincing arguments either that Philip was the author of the papyrus passage or that it was taken from his book *On Plato*, of which not a word has survived. Philip was not the only one among the first generation of the Academics who took an interest in both Plato's biography and his mathematics. Hermodorus of Syracuse, another student of Plato, may well have been its author: he also wrote a book titled Περὶ Πλάτωνος.[37] Philodemus mentions this book (col. VI), so it is likely to have

---

35 It would be quite unnatural if Dicaearchus, a partisan of βίος πρακτικός, and having never been seriously interested in theoretical mathematics had so enthusiastically praised Plato's leading role in its progress. Dicaearchus' preserved fragments reveal his very critical opinion of Plato (fr. 42, 43, 44, 71), which is supplemented and strengthened by the new evidence in Philodemus: Plato "more than all people elevated philosophy and ruined it at the same time" (col. I, 10f.).

36 More details about this see below, 114f.

37 Fr. 4–5, 7–8 Isnardi Parente = *FGrHist* 1008 T 3, F 1–2. Though Lasserre also related this passage back to Hermodorus' book, he believed that the latter had drawn it from Philip (*Léodamas*, 20 F 15 a, 220, 433 f., 611 f.). Lasserre's unconfirmed view of Hermodorus as an intermediary between Philip and Philodemus was based *inter alia* on a wrong interpretation of the columns III–IV, in which an unknown author retells Philip's oral story about Plato's last night. Despite Lasserre's opinion, the author of this story was not Hermodorus, but Neanthes of Cyzicus, whose name is also at-

been available to him. Hermodorus' work Περὶ μαθημάτων seems to attest to his interest in methodology and, possibly, in the history of mathematics as well.[38] Another possible author of this passage is Speusippus, the author of Πλάτωνος περίδειπνον whom Philodemus refers to,[39] as well as of several treatises on mathematics.[40] And, finally, we have to consider Xenocrates, who, apart from many works on mathematics,[41] wrote the book entitled Περὶ τοῦ Πλάτωνος βίου.[42] Since Philodemus apparently had no problems in using the works of Academics and their contemporaries, it is difficult to decide from whom our passage derives. The text itself contains nothing that could allow us to identify one of the aforementioned Academics.

One of the main reasons adduced by Lasserre in favor of Philip is that the ending phrase of the *Catalogue*, in which Philip is mentioned, looks like an illustration of the papyrus passage.[43] Indeed, according to the *Catalogue*, Philip was precisely one of those 'Academic mathematicians' who studied under Plato's methodological direction:

> Philip of Mende, a pupil whom Plato had encouraged to study mathematics also carried on his investigations according to Plato's instructions and set himself to study all the problems that he thought would contribute to Plato's philosophy.[44]

It is also essential that the passage from Philodemus closely matches the description of Plato given in the *Catalogue*:

> Plato greatly advanced mathematics in general and geometry in particular because of his zeal for these studies. It is well known that his writings are thickly sprinkled with mathematical terms and that he everywhere tries to arouse admiration for mathematics among students of philosophy.[45]

These words used to be regarded as a later insertion by either Proclus or one of his Neoplatonic predecessors,[46] but now it is possible to connect them with the

---

tested in column II and in the marginal note after column V (Gaiser. *Academica*, 180; Dorandi. *Filodemo*, 222; Burkert. Arbeitstext, 91).

38 D. L. I, 2 and 8 = fr. 6 Isnardi Parente.

39 Col. VI and Pap. Herc. 164, fr. 12. See Gaiser. *Academica*, 185, 441f.; Dorandi. *Filodemo*, 178. This work might be identical to Πλάτωνος ἐγκώμιον (D. L. IV, 5 = fr. 1 Tarán = *FGrHist* 1009 T 2–3, F 1–3). See Tarán, L. *Speusippus of Athens*, Leiden 1981, 231 n. 15.

40 Μαθηματικός (D. L. IV, 5); Περὶ τῶν Πυθαγορικῶν ἀριθμῶν (fr. 28 Tarán).

41 Περὶ τὰ μαθήματα in six books, Περὶ ἀστρολογίας in six books, Περὶ γεωμετρῶν in five books, Περὶ γεωμετρίας in two books, Περὶ διαστημάτων, Περὶ ἀριθμῶν, Ἀριθμῶν θεωρία (D. L. IV, 13–14 = fr. 2 Isnardi Parente). To these we can add Περὶ ἐπιστήμης and Περὶ ἐπιστημοσύνης (ibid.).

42 Fr. 264–266 Isnardi Parente = *FGrHist* 1010 F 1a–c.

43 Lasserre, F. Le Barbare, le Grec et la science selon Philippe d'Oponte, *Mus. Helv.* 40 (1983) 169–177; idem. *Léodamas*, 611f.

44 Procl. *In Eucl.*, 67.23f. = Eud. fr. 133, transl. by G. Morrow.

45 Ibid., 66.8f. = Eud. fr. 133, transl. by G. Morrow.

46 See e.g. van der Waerden. *EW*, 91.

papyrus passage. This connection seems all the more likely because, further on, the *Catalogue* mentions Eudoxus as the one who "extended the number of theorems relating to the section by applying to them the method of analysis, which originated with Plato" (ibid., 67.5f.), as well as another geometer, Leon, who discovered the method of diorism. Although the similarity of the two texts is not that striking, it prompts us to take seriously the version according to which the author of the *Catalogue* used the material of the same treatise that column Y goes back to. The weak point of Lasserre's argumentation is not the similarity of the *Catalogue* and Philodemus' passage, but the attribution of both texts to Philip. Coincidences between them could be explained without calling into question the traditional view, which traces the *Catalogue* back to Eudemus' *History of Geometry.*

The problem of the *Catalogue*'s authorship will be considered in detail (5.3). Meanwhile it should be pointed out that Proclus received the *Catalogue* through intermediary sources. The main source was Porphyry, who, in turn, could use both Eudemus' *History of Geometry* and the books of the Academics, especially when the subject was Plato and his students. The passages about Plato and Philip could scarcely belong to Eudemus, but they might have been inserted into the *Catalogue* by a Neoplatonic redactor. Indeed, in the context of the *Catalogue*, which lists particular achievements in mathematics, Philip's characteristics look rather odd: his foremost merit in mathematics is that he studied problems that he believed to be connected with Platonic philosophy! Such an assessment can hardly come from Eudemus. It is more natural, rather, to expect it from Philip's Academic colleagues or from their Neoplatonic followers.[47] Evaluating the little that is known about Philip's scientific work, one should admit that in the field of *mathēmata* Philip scarcely had any other achievements that could do honor to the Academy. [48]

Plato occupies a central place in the second half of the *Catalogue*, and such a perspective, of course, brings it closer to the papyrus passage. But even then, it mentions only one mathematician as a pupil of Plato and says nothing about the posing of the problems. What is said about his contribution to the development of mathematics is supported with a reference to his dialogues, but not to his role as an 'architect of science'. The author (or redactor) of the *Catalogue* uses more subtle means to express what is directly said in Philodemus: all the mathematicians of Plato's time worked under his methodological supervision. This effect is achieved mainly by situating all these mathematicians in the text between Plato and Philip, the latter being described as a devoted student working in accordance with Plato's instructions. Because of this circular arrangement, Plato's figure overshadows those of his contemporaries. This impression is reinforced by the constant emphasis on the close relationship between Plato and

---

47 The similarity of characteristics the *Catalogue* attributes to Philip and Euclid points to a Neoplatonic redactor rather than to an Academic author. See below, 182.

48 See below, 102 f.

the mathematicians: some 'lived at the time of Plato', others 'communicated with him' or 'were his students', and still others 'were friends of his students', etc. Since for our present analysis it is not very important whether this perspective derives from the early or from the later Platonists, I would propose the following approach to the second part of the *Catalogue*. If, in spite of its clear bias, it does not specifically mention that someone was a pupil of Plato or that he worked at the Academy, this fact was unknown in the late fourth century BC. It seems very unlikely that Porphyry or Proclus would have omitted such a fact, had they found it in an Academic or Peripatetic source.

The first three mathematicians of Plato's time mentioned here are Leodamas of Thasos, Archytas, and Theaetetus. Nothing is said about their connection with the Academy or about their personal relationships with Plato. Since the chronology in this part of the *Catalogue* is very accurate, one can suggest that Leodamas was the oldest of the three, or at least that he was not younger than Archytas. It is with him that Lasserre begins his collection of sources concerning the 'Academic mathematicians', although there is absolutely no evidence that Leodamas worked at the Academy.[49] The only thing linking him with Plato is Favorinus' (second century AD) statement, repeated with some hesitation by Proclus, that Plato taught him a method of analysis[50] and the pseudo-Platonic 11th letter addressed to a certain Leodamas. But then why should not Archytas be included in this collection as well? After all, there is much more evidence concerning him: we have Plato's authentic 7th letter that mentions the help he gave to Plato, Eratosthenes' evidence, and the fact that Archytas (but not Leodamas!) occurs in several lists of Plato's students.[51] Certainly, Archytas, unlike Leodamas, was too independent a figure to be easily turned into an Academic mathematician (moreover, he was known as a Pythagorean). But this is not easy to do with Leodamas himself, either. Even if Leodamas was the same age as Archytas (born ca. 435/430),[52] then at the time the *Meno* was written (ca. 385/380) – the first dialogue where Plato shows an interest in mathematics and gives, in particular, a description of the method ἐξ ὑποθέσεως (86e–87c),

---

49 Fritz, K. von. Leodamas, *RE Suppl.* 7 (1940) 371–372; Tarán, L. Proclus on the Old Academy, *Proclus. – Lecteur et interprète des Anciens*, ed. by J. Pépin, H. D. Saffrey, Paris 1987, 273. Lasserre himself admitted this (*Léodamas*, 24, 445).

50 D. L. III, 24 = Favor. fr. 25 Mensching. Cf. "Plato, it is said, taught this method to Leodamas, who also is reported to have made many discoveries in geometry by means of it." (Procl. *In Eucl.*, 211.18f.). Whereas Favorinus calls Plato the discoverer of analysis (πρῶτος εἰσηγήσατο), Proclus only says that he passed it (παραδέδωκε) to Leodamas. See Mensching, E. *Favorinus von Arelate*, Berlin 1963, 103f. On probable reasons for confusions in Favorinus and Proclus, see Heath. *History* 1, 291; Cherniss. Plato as mathematician, 418f.

51 So in Philodemus (col. VI) and in Theon of Smyrna, whose list is preserved in Arabic translation (Gaiser. *Academica*, 439f., 444).

52 Mathieu, B. Archytas de Tarent pythagoricien et ami de Platon, *BAGB* (1987) 239–255; Zhmud. *Wissenschaft*, 73.

which is one of the forms of analysis, Leodamas must have been 45–55 years old. If he was at least five years older than Archytas (which the order in which the names in the *Catalogue* are arranged seems to suggest) then, accordingly, he must have been 50–60. Is this not too late to study analysis, even under such a teacher as Plato?[53]

The improbability of such an apprenticeship is increased by the following. 1) One should hardly attach any importance to Favorinus' statement about analysis, which is repeated by Proclus, since elsewhere, referring to an anonymous source (which might have been Eudemus), Proclus attributes the discovery of the method of reduction (ἀπαγωγή), i.e., one of the earliest forms of analysis, to Hippocrates of Chios.[54] 2) Plato himself, describing the method ἐξ ὑποθέσεως, says that it was already in use by geometers (ὥσπερ οἱ γεωμέτραι *πολλάκις* σκοποῦνται, *Men.* 86e); the method he describes is identical to the method of reduction, which Hippocrates used in trying to solve the problem of the duplication the cube.[55] 3) To study analysis on the basis of the *Meno* (or the whole of Plato's works) would not only be embarrassing for the not very young Leodamas, but utterly impossible: despite endless interpretations of this passage, a clear understanding of what Plato had in mind has not been achieved to this day.[56]

Despite the mention of Archytas in the Academic legend about the duplication of the cube, there is no information whatsoever about whether he ever went to Athens.[57] Sources talk about his friendship with Plato, who visited him several times in Magna Graecia. But he was never Plato's pupil, rather vice versa: Plato learned a lot from him. Archytas' influence on Plato has been repeatedly noted,[58] but no one has yet succeeded in tracing the opposite influence.

---

53 Relying on the *Catalogue*, Mensching, *op. cit.*, 104f., suggested that Leodamas was born about 435/30 and considered Favorinus' statement "more than implausible". Lasserre eventually comes to the conclusion that it was Leodamas who influenced Plato, rather than vice versa (*Léodamas*, 457f.), but it is scarcely possible to verify this assertion either.

54 *In Eucl.*, 212.24–213.11; see below, 175, 203.

55 Knorr. *AT*, 71f.

56 The old literature is given in Heiberg, I. L. Jahresberichte, *Philologus* 43 (1884) 469f. (about ten interpretations). See also Bluck, R. S. *Plato's Meno*, Cambridge 1964, 322f., 441ff.; Klein, J. *A commentary on Plato's Meno*, Chapel Hill 1965, 205ff.; Thomas, J. E. *Musings on the Meno. A new translation with commentary*, The Hague 1980, 165f.; Lasserre. *Léodamas*, 451f.; Knorr. *AT*, 71f. – On Plato's intended ambiguity in mathematical matters, see especially Lloyd, G. E. R. The Meno and the mysteries of mathematics, *Phronesis* 37 (1992) 166–183.

57 Lasserre. *Léodamas*, 434; Tarán. Proclus, 273; Gaiser. *Academica*, 448.

58 Krafft. *Mechanik*, 143ff.; Mathieu, *op. cit.*, 251f.; Lloyd, G. E. R. Plato and Archytas in the Seventh letter, *Phronesis* 35 (1990) 159–173. If the 7th letter emphasizes Plato's independence from Archytas, it only means that Plato unwillingly acknowledged this dependency. This tendency coincides with the scarcity of his mentioning the Pythagoreans in the dialogues and with his total silence about Archytas.

Where it is possible to find comparable material, the position of Archytas has mostly differed from or been directly opposed to Plato's.[59]

According to the *Catalogue,* Theaetetus was of Leodamas' and Archytas' generation, so there was not much difference in age between him and Plato. Theaetetus does not occur in any list of the Academics; Plato himself describes him as a student of Theodorus of Cyrene (*Tht.* 145c). The biography of Theaetetus[60] remains extremely confused. Eusebius places his acme in 438/5, which, if we take it to be the date of his birth, would explain his synchronization with Leodamas and Archytas, as well as his study with Theodorus. In the *Suda* we find two Theaetetuses, one a student of Socrates who lived at the time of the Peloponnesian War, and the other a student of Plato. E. Sachs' suggestion that his dates were 415/412–369 relied mainly on the fact that in the *Theaetetus*, whose dramatic date is 399, he is depicted as an adolescent.[61] But she failed to explain either the confusion in Eusebius or the appearance of the two entries in the *Suda.*[62] Recently H. Thesleff proposed returning to the old date of Theaetetus' death, i.e., about 390, without changing the date for his birth, about 415.[63] Yet this revision would make much more sense if we preferred the chronology of the *Catalogue*, which implies that Theaetetus belonged to the generation of Archytas and Leodamas, to all the other versions. This would perfectly match the revised chronology of Eusebius, with date of birth instead of acme. In this case we should date Theaetetus ca. 438/5–ca. 390. His main achievements in mathematics were the theory of the regular solids and the general theory of irrationals. Both of these theories point to his Pythagorean predecessors (among them Hippasus)[64] and teachers (Theodorus), which makes the influence of Plato entirely redundant. If one relies on Theaetetus' traditional chronology (ca. 415–369), he *might have been* one of the older associates of Plato working at the Academy. However, the absence of any evidence of his activity there on the one hand, and his studies with Theodorus on the other, make this suggestion very unlikely.[65]

Nothing is known about Neoclides, who follows Theaetetus in the *Catalogue,* and he is not mentioned anywhere else. His student Leon is named as the

---

59 47 A 23–25. Cf. 47 B 1 and *Res.* 531c, 47 B 3 and *Res.* 525c–d; see below, 105. For some points of agreement see above, 74 n. 122 and below, 110.

60 Lasserre. *Léodamas*, 3 T 1–3.

61 Sachs, E. *De Theaeteto Atheniensi mathematico*, Berlin 1914, 13ff. It is known, however, that Plato sometimes changed the age of his personages depending on the dramatic situation in the dialogue.

62 Lasserre. *Léodamas*, 461.

63 Thesleff, H. Theodoros and Theaetetus, *Arctos* 24 (1991) 147–159.

64 Waterhaus, W.C. The discovery of the regular solids, *AHES* 9 (1972) 212ff.; Neuenschwander, E. Die stereometrischen Bücher der Elemente Euklids, *AHES* 14 (1974) 104; Zhmud. *Wissenschaft*, 171f. According to Eudemus, the Pythagoreans constructed the first three regular solids and Theaetetus the last two; see below, 171 n. 19.

65 Tarán. Proclus, 273; Lasserre. *Léodamas*, 463.

author of *Elements* and the discovery of the method of diorism is attributed to him, which, as we remember, has a parallel in the papyrus passage, although Leon himself is not mentioned there. Unless we want to regard Plato as responsible for this discovery – and there is not the slightest reason to do so[66] – then no connection between him and Leon can be definitely established.[67]

Eudoxus is a key figure for understanding the nature of the real relationship between the Academy and mathematicians of the time, because in this case it is possible to make comparisons with an independent tradition. In the *Catalogue*, Eudoxus is carefully named as ἑταῖρος τῶν περὶ Πλάτωνα γενόμενος, and nothing is said here about his being at the Academy, so Lasserre rightly does not include him in his list of 'Academic mathematicians'.[68] The reasons for this are more than sufficient. Let us turn first to Eudoxus' chronology. His traditional dates (408–355), which still appear in some works, relied first on his acme as given by Apollodorus, i.e., 103 Ol. (368/5), and second on Diogenes Laertius (VIII, 90), who says that Eudoxus lived to the age of 53. Apollodorus connects the acme with the most important event in Eudoxus' life, the discovery of curved lines (καμπύλαι γραμμαί), and this unmistakably indicates his source: Eratosthenes' *Platonicus*.[69] The dramatic date of the dialogue was probably 368/7 – an attempt to synchronize Archytas, Plato, Eudoxus, and Menaechmus.

Eudoxus' traditional dating has been criticized for a long time. Susemihl suggested 390–337, Gisinger 395–342; both of them relied on the fact that Eudoxus mentioned the death of Plato (fr. 342) and could not therefore have died

66 Actually, the method of diorism was used even before Leon (Heath. *History* 1, 319f.; Lasserre, *Léodamas*, 516f.).

67 Tarán. Proclus, 273f. Though Tannery thought it impossible to make any reliable identification of the mathematician Leon, he gave names of two 'platoniciens' with the same name. One of them was a sophist from Byzantium and possibly the author of the pseudo-Platonic dialogue *Alcyon*, the other was from Heraclea and took part in the assassination of the tyrant Clearchus, a former student of Plato (Tannery, P. *La géométrie grecque*, Paris 1887, 130). In Lasserre (*Léodamas*, 513f.), the mathematician Leon becomes the author of the *Alcyon*, which serves as the main evidence that he belonged to the Academy. All this is absolutely unsubstantiated. 1) Leon of Byzantium has nothing in common with the alleged author of the *Alcyon*, which was written in the Hellenistic period. 2) The name of Clearchus' assassin was Leonides, which was later corrupted into Leon. 3) These two contemporaries of Plato are 'platoniciens' only in the sense that they have the same (or *almost* the same) name as the alleged author of the *Alcyon*. 4) None of these three persons can be identified as the mathematician Leon.

68 Lasserre. *Eudoxos*, 141. Cf. Krämer, H. J. Die Ältere Akademie, *Die Philosophie der Antike*, Vol. 3: *Ältere Akademie, Aristoteles, Peripatos*, ed. by H. Flashar, 2nd ed., Basel 2004, 56f.

69 In Eratosthenes' letter to King Ptolemy Eudoxus finds the solution to the Delian problem διὰ τῶν καμπύλων γραμμῶν (Eutoc. *In Archim. De sphaer.*, 90.7 = 47 A 15).

before 347.[70] Von Fritz proposed his 'minimal' dates as 400–347,[71] but in a special article about Eudoxus' chronology, Santilliana reasonably returned to 390–337.[72] Lasserre accepts the latter dates and gives a detailed argument in support of them in his edition of Eudoxus' fragments.[73] Since then, no one has seriously tried to defend the old chronology, though it has been tacitly used even after Lasserre's edition.[74]

Eudoxus' teacher in geometry was Archytas,[75] and it is not by chance that Diogenes Laertius finishes his Pythagorean book (VIII) with a biography of Eudoxus. He visited Athens twice (VIII, 86–88). The first time, when he was 23, i.e., in the year 367, he went there for two months. He attended the Sophists' lectures and possibly visited the Academy, but nothing is said about his acquaintance with Plato, since the latter was in Sicily at that time.[76] The second time, he was already a grown man and came to Athens "bringing with him a great number of pupils: according to some, this was for the purpose of annoying Plato who had originally passed him over".[77] According to Santilliana and Lasserre, Eudoxus probably spent a few years in Athens, from about 350 to about 349, and then returned to his homeland in Cnidus, where he died in 337. It seems that one may relate Eudoxus' participation in Academic discussions on the relationship between Forms and things and on what is the highest Good to his second visit to Athens. His answers to both problems were so un-Platonic in

70 Susemihl, F. Die Lebenszeit des Eudoxos von Knidos, *RhM* 53 (1898) 626ff.; Gisinger, F. *Die Erdbeschreibung des Eudoxos von Knidos*, Leipzig 1923, 5.

71 Fritz, K. von. Die Lebenszeit des Eudoxos von Knidos, *Philologus* 39 (1930) 478–481.

72 Santillana, G. de. Eudoxus and Plato. A study in chronology, *Isis* 32 (1940) 248–282.

73 Lasserre. *Eudoxos*, 137ff. See also Waschkies, H.-J. *Von Eudoxos zu Aristoteles*, Amsterdam 1977, 34ff.; Trampedach, K. *Platon, die Akademie und die zeitgenössische Politik*, Stuttgart 1994, 57ff.

74 P. Merlan's (*Studies in Epicurus and Aristotle*, Wiesbaden 1960, 98ff.) alternative chronology for Eudoxus (395–342) depends on the highly unlikely supposition that, at the age of 27, he came to Athens with a group of his students and, at 28, during Plato's absence, became a scholarch at the Academy.

75 D. L. VIII, 86, with reference to Callimachus, who was a bio-bibliographer and a librarian at the Museum in Alexandria.

76 It is to this visit that the well-known statement from the late biography of Aristotle refers: Ἀριστοτέλης φ<οιτᾷ Πλάτωνι ἐπὶ Εὐδ>όξου (*Vita Marciana* 10). These words used to be taken as evidence that during Plato's absence Eudoxus played the role of scholarch. The impossibility of this reconstruction has been repeatedly shown (Waschkies, *op. cit.*, 41f.; Krämer, *op. cit.*, 56f.; Trampedach, *op. cit.*, 59). The point of this statement is probably that Aristotle, joining the Academy in 367, met Eudoxus there (Lasserre. *Eudoxos*, T 6a–b). This fully correlates with the chronologies of Santilliana and Lasserre. Cf. Waschkies, *op. cit.*, 41f.

77 The tradition about the personal hostility between Plato and Eudoxus has hardly any historical basis. In any case, the one time Eudoxus mentions Plato, it is with great respect (fr. 342 Lasserre).

character[78] that it is totally inconceivable that he should have served his apprenticeship at the Academy.

There seems to have been still less of Plato's influence in Eudoxus' famous work *On Velocities*, in which he put forward his system of homocentric spheres. The impulse to create this system comes not from Plato's metaphysics, but from professional astronomy, in which, in the mid-fourth century, the problem of anomalous movements of the planets became very important.[79] The fact that both Plato and Eudoxus were adherents of the principle of uniform circular movement shows the common Pythagorean source of their astronomical ideas,[80] which was most likely Archytas. Although Archytas is practically unknown as an astronomer,[81] there are strong grounds for suggesting that it was exactly his mathematical and mechanical research that led Eudoxus to discover the *hyppopede* – the curve that is created by the rotation of several interconnected spheres and describes the visible looped motion of the planets.

Archytas' research in mechanics was a mirror image of his mathematical research. On the one hand, he introduced movement into geometry, while on the other he applied geometry to mechanical movement (D.L. VIII, 83). According to one source, Archytas claimed that natural movement (ἡ φυσικὴ κίνησις) "produces circles", according to another, the unequal (τὸ ἄνισον) and the uneven (τὸ ἀνώμαλον) are the causes of movement (47 A 23–23a). It is exactly on this principle that the Aristotelian treatise *Mechanical Problems* is based, which, as Krafft suggested, derives its main features from Archytas' mechanics.[82] *Mechanical Problems* reduced all mechanisms described (the lever, the windlass, the pulley, the winch, etc.) to the principle of unequal concentric circles. Further, it established that the linear speeds of two concentric circles moving with equal angular speed are different and gave a mathematical analysis of this movement.[83]

---

78 Arist. *Met.* A 9, M 5; *EN* I, 12; X, 2. See Krämer, *op. cit.*, 57f., 64f.

79 Knorr. Plato and Eudoxus, 323. See below, 7.6.

80 See below, 271 f.

81 Partly this is because he was not listed among the physicists whose views were considered in Theophrastus' doxography of astronomy (see below, 132). On the astronomical aspect of Archytas' work, see Zhmud. *Wissenschaft*, 219ff.

82 The authorship of the *Mechanical Problems* had been contested for a long time; after F. Krafft's study (*Dynamische und statische Betrachtungsweise in der antiken Mechanik*, Wiesbaden 1970, 3f., 13ff., 149ff.), many scholars are inclined to regard Aristotle as its author (Schneider, *op. cit.*, 227, 234; Oser-Grote C. Physikalische Theorien in der antiken Mechanik, *Antike Naturwissenschaft und ihre Rezeption*, Vol. 7 [1997] 25 n. 2; Schürmann, *op. cit.*, 48ff.). Even if Aristotle did not write the treatise, it certainly belongs to the early Peripatos.

83 Whoever the author of this treatise was, it is obvious 1) that it treats mechanical movement geometrically, which is untypical for the Peripatetics, and 2) that such a principle could have been introduced into mechanics only by a gifted mathematician. I do not know of such a person among the early Peripatetics. Since Aristotle in his earlier works (see above, 47 n. 11) related mechanics to the theoretical sciences,

Eudoxus' treatise *On Velocities* developed Archytas' research,[84] conceiving every planet as fixed to a rotating sphere, whose axis, in turn, is linked with another sphere, etc. The curve resulting from the rotation of these spheres can be regarded as the intersection of the inner sphere with the cylinder. This construction is very similar to the one that helped Archytas solve the problem of duplicating the cube. Here the necessary curve is made by the intersection of the three rotating bodies – the cone, the torus, and the half-cylinder (47 A 14).[85] Thus, the Pythagorean tradition contained all the mathematical elements necessary for the development of Eudoxus' theory. Eudoxus' book *On Velocities* was most likely written during the last period of his activity, when he was living in Cnidus,[86] and it is only reasonable to suppose that Plato knew nothing about it. Theoretically, he *might have* learned about the basics of Eudoxus' astronomical system in 350 when the *Timaeus* was already written and the *Laws* had not yet been finished. However, no one has succeeded in finding in the *Laws* convincing evidence of his knowledge of the system of homocentric spheres, so Eudoxus' influence on Plato remains as unproved[87] as Plato's influence on Eudoxus.

Let us return to the point where Diogenes Laertius talks about Eudoxus' second visit to Athens from Cyzicus, where he had his own school with a large

---

it must have been existed already by the mid-fourth century. Thus, Archytas who is known to work on mechanics (47 A 10a; Athen. Mechan. *De machinis*, 5.1; D. L. VIII, 83), is the best possible candidate for its founder. See also Cambiano, G. Archimede meccanico e la meccanica di Archita, *Elenchos* 19 (1998) 291–324, and below, 129 n. 45.

84 Krafft. *Mechanik*, 145 f.; Neugebauer. *HAMA*, 678. Not accidentally, Archytas' definition of astronomy begins with περὶ τᾶς τῶν ἄστρων *ταχυτᾶτος* (cf. Pl. *Phaed.* 98a; *Gorg.* 451c), and he attributes to his Pythagorean predecessors a 'clear knowledge' of this subject (47 B 1).

85 Heath. *History* 1, 333 f.; Knorr. *AT*, 54 f. Riddell, R. C. Eudoxian mathematics and Eudoxian spheres, *AHES* 20 (1979) 1–19.

86 Lasserre. *Eudoxos*, 142, 193.

87 Ibid., 181 f.; Tarán, L. *Academica: Plato, Philip of Opus and the Pseudo-Platonic Epinomis*, Philadelphia 1975, 107. Mittelstraß, *op. cit.*, 133 ff., although a keen adherent of the idea of such an influence (he relied on the old chronology for Eudoxus), nevertheless admitted that Plato did not change his former astronomical system, as proposed in the *Republic* and *Timaeus,* and that only by some occasional hints in the *Laws* can we conclude that Plato was acquainted with Eudoxus' theory. The fact remains, however, that the most important elements of Eudoxus' theory are missing from the *Laws*, primarily the idea that all planets are attached to spheres by which they rotate. Is it possible to be under the influence of Eudoxus' theory and not mention a sphere at all? Besides, there are no traces of Eudoxus' theory even in the *Epinomis* Philip wrote after Plato's death (Tarán. *Academica*, 110; Knorr. Plato and Eudoxus, 323). Cf. Gregory, A. Eudoxus, Callippus and the astronomy of the *Timaeus*, *Ancient approaches to Plato's Timaeus*, ed. by R.W. Sharples, A. Sheppard, London 2003, 5–28.

number of pupils (VIII, 87). I believe it was this group of Eudoxus' students that formed the main body of 'Academic mathematicians' of the younger generation.[88] After Eudoxus, the *Catalogue* mentions six mathematicians (after whom follows Philip): Amyclas of Heraclea, Menaechmus, Dinostratus, Theudius of Magnesia, and Athenaeus of Cyzicus. They are said to have spent their time together in the Academy and collaborated in their research. This group is followed by Hermotimus of Colophon, mentioned independently. The *Catalogue* names Menaechmus and his brother Dinostratus as Eudoxus' students, and to them two mathematicians from Cyzicus must be added: Athenaeus and Helicon (who was mentioned by Plutarch)[89] and perhaps Hermotimus who "continued the work that had been done by Eudoxus and Theaetetus" (*In Eucl.*, 67.20f.). Theudius' origin (whichever of the two Magnesia's he was born in) may also indicate that he studied with Eudoxus in Cyzicus and traveled with him to Athens. Although the last possibility remains mere conjecture, it is revealing that almost the entire group of Eudoxus' young contemporaries came from Asia Minor.[90] From this group, only Amyclas is *directly* named as 'one of Plato's followers' (*In Eucl.*, 67.9). Meanwhile Amyclas, an acquaintance of Plato, is presented by Aristoxenus as a Pythagorean![91] In any case, we do not know anything about Amyclas' mathematical research.

---

88 See already Allman, *op. cit.*, 178.

89 *De gen. Socr.* 573 C; *Dion.* 19,4 = Lasserre. *Léodamas*, 16 T 2–3; cf. pseudo-Platonic 13th letter (360b–c). Two other students of Eudoxus, Callippus and Polemarchus, were also from Cyzicus (Simpl. *In Cael. comm.*, 493.5, 504.20, 505.21). Another native of Cyzicus, Timolaus, is mentioned in two lists of Plato's students (D. L. III, 46; Dorandi. *Filodemo*, 135, col. VI), but it is not known whether he was a mathematician and a disciple of Eudoxus.

90 The origin of Menaechmus and Dinostratus is unknown. The identification of the mathematician Menaechmus with a certain Menaechmus of Alopecae or Proconnesus who is mentioned in the *Suda* is unconvincing. See Schmidt, M. Die Fragmente des Mathematikers Menaechmus, *Philologus* 42 (1884) 72–81. The *Suda* says: φιλόσοφος Πλατονικός. ἔγραψε φιλόσοφα καὶ εἰς τὰς Πλάτωνος Πολιτείας βιβλία γ' (Lasserre. *Léodamas*, 12 T 2). If these referred to a contemporary of Plato, he would be called Plato's student and not just 'a Platonic philosopher'. Meanwhile, Dercyllides calls Menaechmus and his fellow student Callippus mathematicians, separating them from philosophers (Theon. *Exp.*, 201.25f., cf. Procl. *In Eucl.*, 254.4). Furthermore, in the *Suda* it is not said that the philosopher Menaechmus was concerned with mathematics. When this Menaechmus was alive is not clear, but it is well known that in the fourth century there were no *special* commentaries to the Platonic dialogues. Proclus names Xenocrates' student Crantor as the first interpreter of Plato (*In Tim.*, 76.1–2). See Tarán. Proclus, 270f.; Dörrie, *op. cit.*, 328f.; Krämer, *op. cit.*, 122f.

91 Aristoxenus tells us that Plato wanted to collect all Democritus' books and burn them, but the Pythagoreans Amyclas and Cleinias persuaded him not to do this, explaining that too many people had copies of them (fr. 131 = *DK* 54 A 2). Cleinias, unlike Amyclas, occurs in a catalogue of Pythagoreans, compiled by Aristoxenus (*DK*, 446.28; Zhmud. *Wissenschaft*, 67f.). *Pace* Lasserre (*Léodamas* 7 T 6), who con-

It seems quite likely that after Eudoxus' return to Cnidus (before 348, according to Lasserre), his students remained for some time in Athens and worked at the Academy. How long they stayed there is unknown, as is the nature of their relationship with Plato, who was nearly eighty. In the earliest known list of his students, preserved by Philodemus,[92] the names of five of these mathematicians are missing; only Amyclas is named (strictly speaking, Amyntas of Heraclea), and he is the only one of this group who appears in Diogenes Laertius' list of the Academics (III, 46).[93] The other five are not on any list of early Academics and practically nothing is known about their connection with the Academy.[94] This could mean either that their stay at the Academy was very short and did not leave any traces outside the *Catalogue*, or that they worked there only after Plato's death, or that the *Catalogue*'s information about their activity at the Academy is not reliable. Whichever of these explanations we favor, none of them supports the Academic legend of Plato as the architect of mathematical science.

3. Mathematics at the Academy

The *Catalogue* names three predecessors of Euclid who had written *Elements*: Hippocrates, Leon, and Theudius. The first of these is well known and the last two are not mentioned at all outside the *Catalogue*. However, that is not the question. Whoever followed the tradition of writing *Elements*, it is obvious that its originator was Hippocrates. It is very likely that there were attempts to systematize geometrical knowledge before Hippocrates,[95] but his achievement was more successful and served as an example to later generations. Is there anything especially significant in the fact that all the authors of *Elements* were

---

sidered Aristoxenus' evidence doubtful, the latter did not say that Amyclas was born in Magna Graecia or that he was an opponent of Plato; therefore I do not see any problems in identifying Amyclas of Heraclea with the hero of this anecdote.

92 Gaiser. *Academica*, 110ff., 443ff.

93 In the case of Amyclas (Amyntas), we cannot be sure, probable as it seems, whether the same person is meant in all our sources. Philodemus mentions Amyntas of Heraclea, Proclus Amyclas of Heraclea, Diogenes Ἄμυκλος (and not Ἀμύκλας as in all other sources) of Heraclea, Aelianus Plato's student Amyclas (*VH* III, 19), and finally, Aristoxenus the Pythagorean Amyclas, Plato's acquaintance. See Amyclas, Amyclos, Amyntas, *DPhA* 1 (1994) 174f.

94 Menaechmus appears in Eratosthenes as one of the 'Academic mathematicians', so his description in the *Catalogue*, Μέναιχμος ἀκροατὴς ὢν Εὐδόξου καὶ Πλάτωνι δὲ συγγεγονώς (*In Eucl.*, 67.10), very likely goes back to the *Platonicus*. On Menaechmus' alleged discussion with Speusippus (Procl. *In Eucl.*, 77.7–79.2 = Lasserre. *Léodamas*, 12 F 4–5), see Bowen, A. C. Menaechmus versus the Platonists: Two theories of science in the early Academy, *AncPhil* 3 (1983) 13–29; cf. Tarán. Proclus, 237 n. 36f. and below, 5.4.

95 On the Pythagorean compendium, see below, 195f.

contemporaries of Plato – one older than him and the two others younger? The 'Platocentric' view of ancient philosophy is honored because of its antiquity and because of the number of celebrities who shared it; however, for a long time now, the majority of experts has not shared this view, and it has brought nothing except misunderstanding to the history of Greek science. What is behind it except the natural desire to see genius in everything? Primarily, the obvious fact that from the pre-Euclidean period not a single mathematical writing is preserved, whereas the *Corpus Platonicum* was handed down through the generations in its entirety. Certainly, Plato knew and valued mathematics and often used mathematical examples in his reasoning.[96] But was this love mutual? To judge from the *Elements* of Euclid, whom Proclus or his source enlisted as a Platonist, this was not the case.[97] One can only guess about the contents and nature of the books of Euclid's predecessors, but it is more reasonable to base these guesses on the natural tendency of all the sciences of that time to systematize accumulated knowledge rather than on Plato's demand for the axiomatization of geometry[98] or on his more prosaic demand for textbooks for the Academy.

What is the basis of the current general opinion that geometry and possibly other mathematical sciences were *taught* at the Academy? There is no reliable historical evidence of this,[99] and we actually know very little about what exactly was taught at the Academy. Most reconstructions rely on the Platonic dialogues, and in particular on book VII of the *Republic*, where a solid program of mathematical education is put forward (for those between the ages of 20 and 30). Nevertheless, an expert like Krämer notes that we have no knowledge of a stable program of education at the Academy like the one described in the *Republic* and in the *Laws*. "Anyway, the educational curriculum outlined in the *Republic* VII and the *Laws* XII cannot be directly transferred to the reality of the Academy."[100]

---

96 The mathematical passages from the dialogues are collected in Brumbaugh, R. S. *Plato's mathematical imagination*, Bloomington 1954; Frajese, A. *Platone e la matematica nel mondo antico*, Rome 1963.

97 Knorr, W. R. On the early history of axiomatics: A reply on some criticism, *Theory change, ancient axiomatics and Galileo methodology*, ed. by J. Hintikka et al., Dordrecht, 1981, 194 ff.; idem. What Euclid meant: On the use of evidence in studying ancient mathematics, *Science and philosophy in classical Greece*, ed. by A. C. Bowen et al., New York, 1991, 141 ff.; Mueller, I. On the notion of a mathematical starting point in Plato, Aristotle, and Euclid, ibid., 59–97.

98 In practice, this demand meant the construction of the 'philosophical base' for mathematical definitions: Taylor, C. C.W. Plato and the mathematicians, *PhilosQ* 17 (1968) 193–203.

99 The famous inscription ἀγεωμέτρητος μηδεὶς εἰσίτω is a late literary fiction: Saffrey, H. D. ΑΓΕΩΜΕΤΡΗΤΟΣ ΜΗΔΕΙΣ ΕΙΣΙΤΩ: Une inscription légendaire, *REG* 81 (1968) 67–87.

100 Krämer, *op. cit.*, 5.

To judge from Plato's dialogues, the mathematical element in his work increases toward the end of his life. Might we conclude that during the last decade of his life mathematics was especially intensively taught at the Academy? Almost none of the younger Academics show any special interest in geometry.[101] As for the older Academics, at the time of Plato's death, Speusippus, Xenocrates, Heraclides, Philip and Aristotle were long past the students' age. They were more suited to teaching than to learning mathematics. Their writings imply that they received some mathematical education, but did this take place at the Academy? It is difficult to imagine Plato himself teaching mathematics, but if he did not, then who did, and what kind of mathematics?[102] Cherniss developed his idea about the teaching of mathematics at the Academy only because it was necessary to support his thesis that Platonic metaphysics was *not* taught there.[103] What, then, was taught at the Academy if not mathematics? The easiest answer to this is dialectic; the most honest answer is that we do not know.

Whether Plato adopted his educational program from the Pythagoreans or the Sophists is not so important;[104] the significant point is that his predecessors realized it in practical teaching and produced generations of brilliant mathematicians such as Theodorus, Hippocrates, Archytas, Theaetetus, and Eudoxus and his pupils. In Plato, we come across this program only in the dialogues and even there only as a preparation for the study of dialectic (*Res.* 531d), which was for him far more important than any other science. He handed this attitude down to his students: despite all their fertility in the field of the philosophy of mathematics,[105] none of them left any mark in the exact sciences. To judge, for instance, from the large fragment from Speusippus' *On Pythagorean Numbers* (fr. 28 Tarán), the material he was interested in was very far from the real problems of contemporary mathematics and his approach could in no way be described as professional.[106] Speusippus, Xenocrates, and Hermodorus are no exceptions. Strictly speaking, none of Plato's immediate students achieved anything remarkable in mathematics. If we look at the sciences as a whole, then it is only Aristotle who achieves any real success; and significantly, that success was primarily in biology, i.e., in an area that was *not* studied at the Academy.

Philip was known as an astronomer; the *Suda* attributes to him a number of mathematical and astronomical writings, which we know practically only by

---

101 For lists of the Academics, see Lasserre. *Léodamas*, I T 2–9; Gaiser. *Academica*, 181f.; Dorandi. *Filodemo*, 135. The only exception is Amyclas, who is discussed above.

102 Although Eudemus was the major authority on the exact sciences, Dicaearchus on geography, and Aristoxenus on musical theory, no one has yet come to the conclusion that Aristotle himself taught these sciences at the Lyceum.

103 Cherniss, H. *The riddle of the early Academy*, Berkeley 1945, 60ff.

104 On the Pythagorean origin of the quadrivium, see above, 63.

105 See above, 89f. n. 37–41.

106 See Zhmud. Philolaus, 263ff. Speusippus, in particular, believed that, in a sense, an equilateral triangle has only one angle (fr. 28 Tarán).

their titles.[107] It is hardly possible to prove whether Philip really was the author of all these books; Neugebauer expressed serious doubts about the authenticity of the majority of astronomical treatises.[108] Only in some cases have Tarán and Lasserre succeeded in linking the flimsy surviving evidence with titles known only from the *Suda*.[109] Paradoxically, most of the astronomical material connected with Philip's name relates to the so-called parapegma, i.e., to observational astronomy and meteorology, which his teacher, Plato, held in very low opinion (*Res.* 529a–530c) and could hardly have encouraged Philip to study them. What is significant, however, is that we do not know about any discoveries Philip personally made in astronomy.[110] Specifically, in the *Epinomis,* written by Philip, there are no astronomical ideas that cannot be found in the *Timaeus* or the *Republic*,[111] and there is nothing astronomically original at all. In short, if Philip really was converted by Plato to study the exact sciences and worked under the latter's guidance, then the results of this work seem rather poor.

Tradition connects two interesting astronomical hypotheses with another Academic, Heraclides Ponticus, who later collaborated with Aristotle (fr. 104–110). One of these hypotheses – that Venus and Mercury rotate around the sun, which in turn rotates around the earth – is based on an incorrect interpretation of the sources, as Evans and Neugebauer showed.[112] The other hypothesis – that the Earth rotates on its own axis – has nothing in common with Platonic astronomy.[113] In all probability, Heraclides borrowed it from the Pythagorean Ec-

---

107 IV, 733. 24–34 Adler = 20 T 1 Lasserre. Mathematics: Ἀριθμητικά, Μεσότητες, Περὶ πολυγώνων ἀριθμῶν; astronomy: Περὶ πλανητῶν, Περὶ μεγέθους ἡλίου καὶ σελήνης καὶ γῆς α′, Περὶ ἐκλείψεως σελήνης, Περὶ τῆς ἀποστάσεως ἡλίου καὶ σελήνης; meteorology: Περὶ ἀστραπῶν, Περὶ ἀνέμων; optics: ᾿Οπτικῶν β′, ᾿Ενοπτ<ρ>ικῶν β′.

108 Neugebauer. *HAMA*, 574.

109 Tarán. *Academica*, 115ff., 135f.; Lasserre. *Léodamas*, 596f. The identification of the book on lunar eclipses and the meteorological writings seems relatively safe.

110 In the mid-fourth century, a "demonstration of the sphericity of the moon" (Tarán. *Academica*, 136) cannot be regarded as a discovery. Even in the field of parapegmata, Philip was not original (Neugebauer. *HAMA*, 740 n. 12).

111 Tarán. *Academica*, 98–114.

112 Evans, P. The astronomy of Heraclides Ponticus, *CQ* 20 (1970) 102–111; Neugebauer, O. On the alleged heliocentric theory of Venus by Heraclides Ponticus, *AJP* 93 (1972) 600–601. Gottschalk's arguments in favor of Heraclides' epicyclical model do not seem convincing (Gottschalk, H. B. *Heraclides of Pontus*, Oxford 1980, 69ff.). Our main source, Chalcidius, was by no means an expert in astronomy (*In Tim.*, 176 = fr. 109), and the fact that he attributes the same epicyclical model to Plato makes his evidence about Heraclides especially suspicious.

113 Heraclides (fr. 106) interpreted a controversial passage in *Timaeus* (40b) in this sense; it turned out to be a point of great debate among the Academics (Arist. *Cael.* 293b 30f.). It is interesting that Proclus, seeing such a divergence between Plato and Heraclides, refused to consider him a student of Plato (Tarán. Proclus, 263f.).

phantus (51 A 1, 5), who continued the line of Philolaus.[114] According to Diogenes Laertius, Heraclides studied with the Pythagoreans and wrote a special book about them (V, 86); his ideas have a whole series of other similarities with the Pythagorean astronomy (fr. 104, 113).

Is it really necessary to recall that the Academy never produced even one significant mathematician or astronomer? It does seem necessary, especially when one takes into account the exaggerated significance usually attributed to the program of mathematical education described in the Platonic dialogues. The *Republic, Theaetetus*, and *Laws* probably persuaded not a few talented youths to take up mathematics, but having begun the study of it, they inevitably had to comply with the demands worked out by the *mathēmatikoi*. If they still considered Plato more worthwhile than mathematical truth, then they developed a mathematical theology in the spirit of Anatolius or Iamblichus, or compiled a commentary to the mathematical passages in the Platonic dialogues, or in the best case, wrote a philosophical commentary on Euclid, as Proclus did.[115]

4. Plato on science and scientific directorship

It is evident that tracing back all the stories about Plato as an organizer of science (the duplication of the cube, the 'saving the phenomena', the discovery of analysis and general progress in mathematics) to their Academic sources does not prove their reliability. That these stories are not supported by sources outside the Academy, especially Peripatetic, is not decisive in itself: one can always object that, if the Academics exaggerated the role of their teacher, the Peripatetic attitude toward Plato was anything but objective, as well.[116] However, neither the independent evidence on the mathematicians of the fourth century, nor the writings of the Academics themselves – unlike the Academic *legends* – actually support the idea of the exact sciences flourishing under Plato's directorship. The source of these legends, therefore, was not the real relation-

---

114 According to Philolaus, the earth rotates round the Central Fire in 24 hours; Ecphantus transformed his idea into that of the earth's rotation around its own axis. *Pace* Krämer (*op. cit.*, 75 f.), who tries to prove Heraclides' priority, there is no evidence of Plato's influence on Ecphantus (νοῦς in A 1 is clearly from Anaxagoras), nor are there any reasons to date the latter in the last third of the fourth century: by that time no Pythagoreans remained.

115 Interestingly, Proclus himself studied mathematics with the mathematician Hero of Alexandria, not in the Academy of Athens (Marin. *Vit. Procl.* 9). – For a different view of the relations between Platonism and mathematics, see Burkert, W. Konstruktion und Seinsstruktur: Praxis und Platonismus in der griechischen Mathematik, *ABrWG* 34 (1982) 125–141.

116 Aristoxenus gathered all the gossip about Plato (fr. 61–68, 131), Dicaearchus wrote that Plato raised and then destroyed philosophy (Dorandi. *Filodemo*, 125, col. I), Eudemus sometimes preferred Archytas to Plato (fr. 60, but see fr. 31), and Aristotle himself was known for his inordinate criticism for his teacher.

ship between Plato and contemporary mathematicians, but his dialogues. It is here that we should look for, and can find, the basis for the idea of Plato as an architect of the sciences, which the Academics then developed further.

So far I have neglected the issue of the extent to which the Academics' efforts to emphasize Plato's role in establishing the methodology of the exact sciences reflected his own position. Plato often criticized the scientific methodology of his contemporaries, especially in books VI–VII of the *Republic*, where he outlines a program of education for future guardians of the ideal polis. Let us compare, for example, Archytas' description of the numerous acoustic observations and some simple experiments (47 B 1) with Plato's remark that the true science of harmonics must be independent of all this, measuring mathematical and not audible consonances, which is exactly what the Pythagoreans fail to realize (531 c). While Archytas sings the praises of the social and even moral consequences of practical arithmetic,[117] Plato insists that arithmetic should be pursued mainly for the sake of pure knowledge (525 c–d). The geometers derive their propositions from several premises, which they consider self-evident and do not further explain (510 c–e); solid geometry is in a very undeveloped state (528 b–c). For Plato, true astronomy is concerned not with the movement of the visible heavenly bodies, but with ideal kinematics of mathematical heavens (529 a–530 c).[118]

These well-known passages were discussed and interpreted many times, now in support of Plato's anti-empiricism and of his hostility toward the real sciences of that time, and now as an example of his foresight of future mathematical astronomy.[119] I do not think it possible to add anything significantly new to what has already been said on this subject. If, however, one tries to concentrate on what is uncontroversial, or at least to avoid the extreme points of view, then it should be said that the position of external and competent critic was only natural for Plato, as were his efforts to put the results and methods of the exact sciences to the use of his favorite science – dialectic. It is also obvious

---

117 47 B 3. See above, 71 f.

118 Admittedly, the emphasis in the *Timaeus* is rather different.

119 See e.g. Taylor, *op. cit.*; Cornford, F. M. Mathematics and dialectic in the *Republic* VI–VII (1932), *Studies in Plato's metaphysics*, ed. by R. E. Allen, London 1965, 61–95; Hare, R. M. Plato and the mathematicians, *New essays on Plato and Aristotle*, ed. by R. Brumbaugh, London 1965, 21–38; Barker, A. Σύμφωνοι ἀριθμοί: A note on *Republic* 531 c 1–4, *CPh* 73 (1978) 337–342; *Science and the sciences in Plato*, ed. by J. P. Anton, New York 1980; Mourelatos, A. P. D. Astronomy and kinematics in Plato's project of rationalist explanation, *SHPS* 12 (1981) 1–32; Annas, J. *An introduction to Plato's Republic*, Oxford 1981, 272 ff.; Gaiser, K. Platons Zusammenschau der mathematischen Wissenschaften, *A & A* 32 (1986) 89–124; Robins, I. Mathematics and the conversion of the mind, *Republic* vii 522 c 1–531 e 3, *AncPhil* 15 (1995) 359–391; Gregory, A. Astronomy and observation in Plato's *Republic*, *SHPS* 27 (1996) 451–471; Kouremenos, T. Solid geometry, astronomy and constructions in Plato's *Republic*, *Philologus* 148 (2004) 34–49.

that the disciplines that constituted the quadrivium variously suited Plato's goal: arithmetic and geometry to a greater extent and astronomy and harmonics to a lesser, for they were connected with mathematical interpretation of natural phenomena, which, according to Plato, could not be a subject for scientific study. The controversy begins when, based on Plato's often rather vague remarks, we try to understand what stands behind his criticism: is he proposing an alternative program for developing the exact sciences, anticipating the work of Euclid and Ptolemy, or is he simply worried about how to adapt the exact sciences to his own pedagogical purposes, how to make them a true preliminary for dialectic. I personally prefer the second answer,[120] but I am ready to admit that these passages *could* be interpreted as valuable methodological instructions on how to develop the exact sciences. I think they were understood exactly in this way in the Academy.

The first indication here is the term πρόβλημα, which we came across in the quotations from Philodemus and Sosigenes: Plato sets the problems to the specialists.[121] This is the approach insistently put forward in the *Republic*. When discussing astronomy, Socrates proposes: *προβλήμασιν* ἄρα … χρώμενοι ὥσπερ γεωμετρίαν οὕτω καὶ ἀστρονομίαν μέτιμεν (530b 6). He returns to this when discussing harmonics, reprimanding the Pythagoreans: ζητοῦσιν, ἀλλ' οὐκ εἰς *προβλήματα* ἀνίασιν, ἐπισκοπεῖν τίνες ξύμφωνοι ἀριθμοὶ καὶ τίνες οὔ (531c 3). Whatever Plato meant by these appeals,[122] the appeals themselves, urging the necessity to study the *real* problems of a *true* science, have to remain in the memories of the readers of the *Republic*.[123] The resemblances become even greater if one compares Plato's reprimands for the contempt of geometry, known from the legend about the Delian problem,[124] with Socrates' description of the situation in solid geometry (528b–c). His definition of solid geometry, ἔστι δέ που τοῦτο περὶ τὴν τῶν *κύβων αὔξην* καὶ τὸ βάθους μετέχον, contains, as was noted long ago, a clear reference to the problem of the duplication of the cube.[125] Glaucon agrees with this definition and remarks that this field has not yet been properly investigated. Socrates

---

120 See Lloyd, G. E. R. Plato on mathematics and nature, myth and science, *Methods and problems*, 333–351; Hetherington, N. S. Plato and Eudoxus: instrumentalists, realists, or prisoners of themata?, *SHPS* 27 (1996) 278.

121 Plutarch (*Marc.* 14.9–11) also mentions the 'problems', but here the term has a special mathematical meaning; Philodemus' passage and Sosigenes use it in a wider sense.

122 "It seems … that for Plato to proceed in geometry, astronomy, and harmonics by means of the problems meant to formulate the questions and to find the cause or explanation of certain phenomena in an abstract way." (Tarán. Proclus, 237 n. 36).

123 One of these readers might have been Sosigenes; he was the first to step up from geometry to astronomy (see above, 86f.). Knorr. Plato and Eudoxus, 324f.

124 Plut. *De. gen. Socr.* 579 B–C; *Quaest. conv.* 718 E–F; Theon. *Exp.*, 2.8–12.

125 *The Republic of Plato*, ed. by J. Adam, Vol. 2, Cambridge 1902, 122; Robins, *op. cit.*, 370.

gives two reasons for this situation: first, the state does not support these studies and, being very complex, they develop rather slowly and, second, "the investigators need a director, without whom they will hardly discover anything" (ἐπιστάτου τε δέονται οἱ ζητοῦντες, ἄνευ οὗ οὐκ ἂν εὕροιεν). It is hard to find a clearer expression of the need for philosophical or even state-philosophical patronage of science. The passage from Philodemus (ἀρχιτεκτονοῦντος μὲν καὶ προβλήματα διδόντος τοῦ Πλάτωνος, ζητούντων δὲ μετὰ σπουδῆς αὐτὰ τῶν μαθηματικῶν) thus becomes an immediate reflection of Plato's words.

The suggestion that Plato designed the role of this ἐπιστάτης for himself[126] acquires substance from the following words of Socrates:

> It is not easy to find such a director, and then if he could be found, as things are now, investigators in this field would be too arrogant (μεγαλοφρονούμενοι) to submit to his guidance. But if the state as a whole join in superintending these studies and honour them, these specialists would accept advice and continuous and strenuous studies would bring out the true nature of the studied subject.[127]

As long as this is not so, mathematicians are prompted exclusively by their intellectual interest in solid geometry and cannot even account for the practical use of their research.[128] Interpretations taking this 'director' as some famous mathematician of that time, e.g. Archytas or Eudoxus,[129] seem naive and to impute to Plato an unlikely generosity. Obviously what is meant here is not a specialist but a dialectical philosopher, one who would be obeyed only in the ideal state and only with the support of this state. Hippocrates, Archytas, and Eudoxus did not need such support and they definitively would react to the dialectician's advice with a μεγαλοφροσύνη, so characteristic of all specialists. Earlier in the *Euthydemus* (290c), Plato did not yet lay claim to setting problems to the scientists, but only to a true interpretation of scientific achievements. Mathematicians and astronomers themselves do not know how to use their discoveries (cf. *Res.* 528c 5), so they have to hand them over to the dialecticians to use properly. This concerns, at least, those mathematicians "who are not utter blockheads" (μὴ παντάπασιν ἀνόητοι). How then were Archytas and Eudoxus supposed to respond to such advice?

One more line leading to the *Republic* is the reference in the *Catalogue* to a certain section, which originates from Plato: Eudoxus augmented τὰ περὶ τὴν

---

126 *Plato's Republic*, ed. by P. Shorey, Vol. 2, Cambridge, Mass. 1935, 177; Cornford, *op. cit.*, 78.

127 528b 8–c 4, transl. by P. Shorey.

128 528c 5f. It is interesting that Aristotle, who writes in the *Protrepticus* about the rapid progress of mathematics and philosophy in comparison with all other τέχναι, explains it by the inner attractiveness of these sciences, rather than by measures of encouragement on the part of the state (see above, 70 n. 105–107). Evidently he saw no need of such measures.

129 See Adam, *op. cit.*, 123f.; Heath. *History* 1, 12f.

τομὴν ἀρχὴν λαβόντα παρὰ Πλάτωνος (*In Eucl.*, 67.6). The only place where Plato mentions the geometrical section is the well-known passage about the division of a line into extreme and mean ratio (golden section): the ratio between the segments of this line symbolizes the relationship between the material world and the world of Forms (*Res.* 509d–e). Meanwhile, the golden section was already known to the Pythagoreans,[130] so only someone who was absolutely sure that everything Plato says about mathematics derives from him could have regarded him as the author of this discovery.

Can we conclude that book VII of the *Republic*, in which Plato gives valuable instructions on how to develop mathematical sciences in order to make them most useful for dialectic, and similar passages from other dialogues were necessary and sufficient conditions for the creation of the Academic legend of Plato as the architect of science? If we take into account the previous analysis, which shows an absence of any firm historic evidence that he really did play this role, such a conclusion seems to me very compelling. The legend about his Apollonian ancestry, mentioned by Speusippus (fr. 1 Tarán), serves here as an excellent parallel, since it was also born out of an interpretation of a Platonic dialogue, in this case the *Phaedo*.[131] The tendency to reconstruct or rather to construct a biography relying on the author's writings was widespread in Antiquity. If the image of Plato giving instructions to the mathematicians and astronomers originated in the image of Socrates, the hero of the dialogues, such a transformation would have been well justified in the eyes of the Platonists, since it corresponded to the basic intention of their teacher: to see further and to penetrate deeper than any of those whose knowledge he used.

5. The theory and history of science in the Academy

The role of Plato and his school in the formation of mathematics and astronomy proves quite negligible, as far as we can trace it. However, in history and especially in the theory of science, the situation looks different. Plato was one of the many who, disappointed with the heurematographic tradition, sought different explanations for cultural achievements and innovations.[132] To a considerable extent, the break with heurematography can be explained by its totally ignoring the cultural and social context out of which the invention of arts and sciences could hardly be conceived. The total or partial rejection of inherited myths, attested as early as in Herodotus, led either to the reconstruction of the primeval past of humankind in which such figures as Asclepius, Orpheus, and Palamedes did not yet exist, or to the history of the oldest civilization – Egypt (2.1). Democritus, Protagoras, the author of *VM*, and others turn

---

[130] Heath. *History* 1, 324f.; Lasserre. *Eudoxos*, 176f.

[131] Riginos, *op. cit.*, 9ff., 30f.

[132] On Plato's attitude toward the traditional Greek *prōtoi heuretai*, see below, 225f.

to the primeval past, while Herodotus and Isocrates develop the Egyptian version. Plato used both approaches in different dialogues. Hardly original in either of them, he must nevertheless have exerted considerable influence on Aristotle and, through him, on the history of culture and science that was born in the Lyceum.

Aristotle develops Plato's theory of recurrent catastrophes,[133] in which acquired knowledge is lost to be regained anew.[134] Both versions, the primeval as well as the Egyptian, dwelled upon in many of Aristotle's works, were taken up and further developed by his disciples. It is revealing that the traditional *prōtos heuretēs* discredited by Plato never makes an appearance in Eudemus' history of science, in Theophrastus' and Meno's doxography, or in Dicaearchus' cultural history. That does not mean, however, that it ceased to exist. Rather, it was pushed aside, joining the 'unserious' genre of heurematography (indulged in, however, by the Peripatetics) and popular history, where it lingered until the end of Antiquity.

At the same time, Plato never attempts to give a schematic historically oriented outline of the development of knowledge from its earliest forms, determined by necessity, up to the perfect ones that lead to wisdom, such as we find, e.g., in the *Epinomis*.[135] Plato seems to have been only peripherally interested in the history of knowledge; he does not as much as mention the historical development of mathematics. Gaiser's attempt to prove that Plato divided the development of knowledge into clearly defined periods is unsatisfactory.[136] As a rule, we find only schemes borrowed from Protagoras and Democritus, where political τέχναι (or arts) follow the necessary ones. As a result, Plato's variant of *Kulturentstehungslehre* proved even less historical than that of many of his predecessors: the invention of arts and sciences is usually represented as the gift of gods, the story of the invention taking the form of a myth.[137]

Turning to Plato's *theory* of science, let us stipulate that it interests us only insofar as it influenced the formation of the historiography of science, providing it with indispensable theoretical tools. The first steps in the development of the theory of exact sciences were made by the Pythagorean school, which was most closely connected with mathematics. The notion that knowledge is impossible without number is found in Philolaus.[138] This notion laid the basis for

---

133 Pl. *Tim.* 22c 1f., 23a 5f., *Crit.* 109d–110a, *Leg.* 677a–681e; Festugière, A.-J. *La révélation d'Hermès Trismégiste*, Vol. 2, Paris 1949, 99f.

134 See below, 212 n. 225.

135 See below, 112f.

136 Gaiser. *Platons ungeschriebene Lehre*, 223ff. On the specific character of Plato's attitude toward history, see Weil, R. *L' "archéologie" de Platon*, Paris 1959, esp. 18f., 42f.

137 *Menex.* 238b, *Phileb.* 16c, *Polit.* 274e. So the invention of writing and exact sciences (geometry, astronomy, arithmetic) is ascribed to the Egyptian god Thoth (*Phdr.* 274c–d; *Phileb.* 18b–d); Plato, however, does not insist on it. See below, 224ff.

138 "And indeed all the things that are known have number, for it is not possible that any-

the classification of τέχναι in *Philebus* and was embraced with enthusiasm by the author of *Epinomis*.[139] Archytas considered *mathēmata*, arithmetic in particular, to be the most exact of the τέχναι and insisted on its wholesome effect upon virtue (47 B 3–4). Plato developed both ideas,[140] along with Archytas' theory of four related sciences (47 B 1).

It is to be emphasized that mathematics, which was not part of Socrates' legacy, entered the sphere of Plato's philosophical interests through his contacts with the Pythagoreans, first of all with Archytas and Theodorus. Plato, however, treats mathematics in a way substantially different from the Pythagoreans. Archytas considered *mathēmata* within the framework of the Sophistic theory of τέχνη, which served to account for every systematically organized and practically oriented kind of knowledge (2.3). This orientation is not to be interpreted as purely utilitarian. The practical utility of *mathēmata* seemed still another argument in favor of their being made part of the τέχναι. In the further differentiation of τέχναι into sciences, arts, and crafts, it is the problem of utility, however, that comes to the fore. While Archytas emphasized the utility of mathematics, and Socrates and Isocrates tried to refute or downplay it, Plato offers a radically different solution to the problem. The necessary and the useful (crafts) hold the lowest grade in his hierarchy of activities; μουσική, based not solely on knowledge but on inspiration as well, is differentiated from the sphere of the τέχναι, while *mathēmata* and ἐπιστῆμαι do not serve any end but knowledge itself (*Res*. 525c–d). Without denying the applied value of scientific knowledge, Plato derives his model of science from mathematics, its least utilitarian and most thoroughly theoretical branch. His own science, dialectic, which aims at the knowledge of Forms, was to surpass mathematics in both purity and exactness, being still further removed from the corporeal world.

The particular attention the Academy paid to mathematics played an important role in the new approach to science. The exact and irrefutable character of mathematical knowledge, the transparency of the criteria of mathematical certainty, the absence of disagreement on essential points, so typical of other sciences – all these factors concurred to make mathematics an attractive model for the development of a conception of theoretical knowledge. In this respect, mathematics proved unrivaled by any other τέχνη; in time, all other models sank into the background.[141] At the same time, it would be wrong to take the attractiveness of mathematics for granted: the Sophists and Socrates, e.g., did not

---

thing whatsoever be understood or known without this." (44 B 4, transl. by C. Huffman).

139 *Phileb*. 55d 5–8, 55e–56c, *Leg*. 747b 1f., *Epin*. 977d 7f. On Philolaus' epistemology and its influence on Plato, see Huffman. *Philolaus*, 172ff.

140 Burnyeat. Plato. See above, 72 n. 112.

141 Note, however, the special role of medicine in Aristotle's model of τέχνη, especially in his ethico-political treatises (Fiedler, W. *Analogiemodelle bei Aristoteles*, Amsterdam 1978, 180ff.). For the Stoic view on τέχνη as a model for philosophy, see below, 287.

notice it. The elementary character of fifth-century mathematics can hardly explain this: in the Hellenistic period, the exact sciences could boast of greater achievements, which did not prevent the major philosophical schools of the period from criticizing or ignoring them (8.1). Plato himself criticized the mathematical sciences; his criticism, however, showed interest and was sympathetic, and aimed at the common task of achieving a more exact knowledge. His suggestion was to bring astronomy and harmonics closer to geometry and arithmetic by 'removing' the physical foundations that prevented them from becoming mathematical sciences proper. This is what underlies Plato's critical attitude toward these two sciences in the *Republic* (530b, 531c) and his suggestions to reform astronomy and harmonics by making them follow methods accepted in geometry.[142]

For Speusippus and Xenocrates, τὰ μαθηματικά proved still more important than for Plato, since in their ontology mathematical numbers, magnitudes, and bodies took the place of the Platonic Forms, which they rejected.[143] Freed from Plato's interest in τέχναι, which he inherited from Socrates, the Academics vigorously develop the philosophy of mathematics.[144] "Mathematics has come to be identical with philosophy for modern thinkers, though they say that it should be studied for the sake of other things." (*Met.* 992a 31) This comment by Aristotle on his former colleagues from the Academy leaves no doubt about his critical attitude toward the place they accorded to mathematics. In fact, even while theorizing about mathematical objects, the Academics did it for the sake of philosophy, and not for mathematics proper. Aristotle's range of interests was richer and more varied. Still in the Academy, he defends and advocates in the *Protrepticus* the ideal of *vita contemplativa*, while developing in his logical works, especially in the *Second Analytics*, the methodology of scientific research based chiefly on those means of acquiring new knowledge that were elaborated in mathematics. In spite of the obvious influence of mathematics on Aristotelian logic and the wealth of mathematical examples and analogies that we find in his writings,[145] only a few small works of his enormous heritage are devoted to exact sciences as such.[146] Aristotle does not seem to be

---

142 Plato emphasizes that harmonics' application of numbers to real physical phenomena is the same as what is done in astronomy (*Res.* 531c 1). Hence, his criticism of both sciences is identical: both should "rise to the consideration of general problems" following the example of geometry.

143 Krämer, *op. cit.*, 28f.

144 See above, 90 n. 40–41. Speusippus' Τεχνῶν ἔλεγχος (D. L. IV, 5) must have been devoted to rhetorical treatises (Tarán. *Speusippus*, 195).

145 On mathematical analogies and examples in Aristotle, see Fiedler, *op. cit.*, 47ff., 64ff.

146 See the titles in the list of Aristotle's works (D. L. V, 24–26): Ἀστρονομικόν (No. 112), Ὀπτικόν (No. 113), Μηχανικόν (No. 122). Moraux, P. *Les listes anciennes des ouvrages d'Aristote*, Louvain 1951, 111f. The only one which has sur-

interested in mathematics as a researcher, which makes his stay at the Academy a still more important factor in the formation of his views on science.

The historical development of mathematics, unlike its philosophical implications, was of comparatively little concern for the Academics. Though we have no grounds for asserting that this subject preoccupied them *only* in connection with Plato, the passage quoted by Philodemus is one of very few pieces of evidence we can rely on. In spite of a wealth of treatises devoted to exact sciences, the extant fragments of Speusippus and Xenocrates do not allow us to consider them predecessors of Eudemus in the historiography of science. The extract from Speusippus' work *On Pythagorean Numbers* does contain some material on Pythagorean arithmetic. It is mixed, however, with Speusippus' own arithmological speculations, which have little to do with either mathematics or history.[147] The only evidence to be found in Xenocrates is the mention of Pythagoras' discovery of harmonic intervals (fr. 87 Isnardi Parente). Heraclides is known to have written a historical treatise *On the Pythagoreans*. But the extent to which it concerned scientific discoveries is not clear.[148]

A much more interesting source is the *Epinomis*. It contains a few ideas also found in Aristotle, Eudemus, and Aristoxenus. According to the scheme suggested by Philip, the first to appear were τέχναι and ἐπιστῆμαι necessary for human life; these were followed, in succession, by τέχναι that serve pleasures and those used for defense (medicine, maritime and martial arts, etc.). The last to appear was ἐπιστήμη, which gives people the knowledge of number and leads them to wisdom.[149] The author identifies this wisdom with astronomy. The knowledge of number is the most valuable and important knowledge, the lack of it making the exercise of reason impossible.[150] But rather than discovering number by themselves, people received it as a gift from a deity, which Philip identifies with the visible cosmos (976e 3f.). This deity taught and still teaches people to distinguish the numbers one and two through the alternation of day and night, the numbers three to fifteen through the phases of the moon, etc. (978b 7f.). The inhabitants of Egypt and Syria were the first to discover the

---

vived, Μηχανικόν (= *Mechanical Problems*), shows the influence of Archytas (Krafft. *Mechanik*, 149f.).

147 Tarán. *Speusippus*, 257ff.; Zhmud. Philolaus, 261ff.

148 D. L. V, 88 = fr. 22, 41–42. Wehrli traces to his book only the two mentions of the prohibition on beans. Some of Heraclides' astronomical ideas obviously have Pythagorean origin (see above, 103); we do not know, however, in what particular book they appeared.

149 974d 3–977b 8. The chronological sequence of the two last stages is less pronounced than that of the first two. Philip, like Plato, combines systematic classification with historical periodization (Gaiser. *Platons ungeschriebene Lehre*, 223f., 245). For detailed analysis, see Tarán. *Academica*, 69ff.

150 Number plays an important role in the τέχναι necessary for survival, which are good only insofar as they possess number (977d 7). See Pl. *Phileb.* 55d 5–8, 55e–56c, *Leg.* 747b 1f.

planets and given them divine names, since the sky in those countries is clear and propitious for astronomical observations (986e 9f.).[151] Having inherited astronomy from the 'barbarians', the Greeks are going to turn it into real wisdom owing to their ability to bring to perfection everything they borrowed, as well as to their climate, which favors the practice of virtue (987d 3f.).

We know many of the elements of this theory from the earlier literature on the origins of culture (2.1). They include the division of the τέχναι into the necessary and the pleasurable (Democritus); the appearance of sciences after the main material needs have been satisfied (Isocrates); the borrowing of knowledge from the barbarians, Egyptians in particular (Herodotus, Isocrates); the divine origin of the τέχναι (Ps.-Epicharmus); the connection of knowledge with number (Philolaus); the bringing of sciences to perfection (Hippocratic treatise *On Art*, Isocrates); the favorable influence their climate exercised on the character of the Greeks (the Hippocratic corpus). All these elements are found, in one form or the other, in Plato's dialogues,[152] which further confirms the continuity of these ideas with earlier literature on the τέχναι.

The fact that these subjects were discussed in the Academy is also supported by the similarity between Philip's scheme and the aforementioned Aristotelian theory, which distinguished three stages in the development of knowledge: 1) the necessary τέχναι; 2) arts, music in particular; 3) sciences and philosophy directed toward pure knowledge.[153] This theory, familiar to us from the *Metaphysics* (981b 13–30), goes back to *Protrepticus* and *On Philosophy*, two early works written while Aristotle was still in the Academy.[154] It means that, against Tarán, we cannot rule out Aristotle's possible influence on Philip.[155] Both of them discussed the development of crafts, arts, and sciences according to the degree to which they participate in σοφία. Unlike Philip, who denied wisdom to all kinds of knowledge except the science of number, Aristotle believed that while in olden times wisdom had been accorded even to the inventors of useful τέχναι, later it was only granted to the inventors of arts and, finally, the notion of σοφία came to be associated with theoretical science, ἐπιστήμη.[156] Interest-

---

151 Cf. Pl. *Tim.* 24b7–c3, *Leg.* 747b–e.

152 Tarán. *Academica*, 69ff.

153 See above, 52 n. 34.

154 Spoerri, *op. cit.*, 54 n. 19. Book I of Aristotle's *On Philosophy* combined the theory of the origin of culture (as well as its fall as a result of catastrophes) with the history of philosophy, which ends with Plato (Wilpert, P. Die aristotelische Schrift "Über die Philosophie", *Autour d'Aristote*, Louvain 1955, 99–118; Effe, B. *Studien zur Kosmologie und Theologie der Aristotelischen Schrift "Über die Philosophie"*, Munich 1970, 62ff.).

155 Tarán. *Academica*, 140ff. Gaiser, on the contrary, does not deny the connection of the *Epinomis* with the *Protrepticus*, but believes that the influence of Plato can account for their similarities (*Platons ungeschriebene Lehre*, 244f.).

156 Note the parallelism with a fragment of Archytas (47 B 4) comparing different τέχναι from the point of view of σοφία they participate in (see above, 61f.). The 'wi-

ingly, the passage from the *Metaphysics* ends with the invention of geometry in Egypt, and Philip's survey of sciences with the invention of astronomy in Egypt and Syria (whether this subject was touched upon in the *Protrepticus*, remains unknown). On the whole, Aristotle's theory looks more logical and less schematic; besides, it is closely linked with his theory of science. Considering that the *Epinomis* was written later than the *Protrepticus* and *On Philosophy*, Philip could have drawn the historical part of his scheme largely from Aristotle.

Unlike the theories of Aristotle and Philip, the Academic treatise on Plato places the exact sciences in a much more topical context. It deals with contemporary mathematical sciences, considers actual, rather then reconstructed, discoveries, and mentions such real historical figures as Hippocrates of Chios, Plato, and Eudoxus. Here we meet a number of elements familiar to us, in this form or another, from Aristotle and Eudemus, such as the rapid progress made recently in the exact sciences (which already include optics and mechanics), further, the rapid progress in geometry, which found expression in the substitution of new theories for old ones as well as in the emergence of such new methods as analysis and diorism, and, finally, the completion of the general theory of proportions (τὰ περὶ μετρολογίαν ἦλθεν ἐπὶ κορυφήν).

The notion of the rapid progress of different branches of knowledge and their approaching perfection dates back to the late fifth – early fourth centuries.[157] Was the author of the Academic treatise the first to apply it to mathematics? Though we do not know when the treatise was written, we can safely date it after Plato's death.[158] Meanwhile, in the *Protrepticus*, written still in the 350s, Aristotle notes rapid progress in mathematics – without, of course, mentioning Plato.[159] According to Philip, the Greeks will bring to perfection the astronomical knowledge they borrowed in the Orient (*Epin.* 987d–e). Thus we have to admit that the progress and perfection of exact sciences and the development of new methods were discussed in the Academy, though they did not become a subject of special studies, as they were later in the Lyceum.

That Eudemus could have known the treatise quoted by Philodemus and shared some of its ideas is hard to test, ignorant as we are of both its author and

---

sest', i.e., the most exact among them appeared to be arithmetic, surpassing in this respect geometry as well as all other τέχναι. Aristotle also considered arithmetic more exact than geometry (*APo* 87a 34f.; *Met.* 982a 26f.). The comparison made by Archytas is systematic rather than historical. But considering Aristotle's interest in his philosophy (see above, 71 n. 110), it might well have drawn the latter's attention.

157 See above, 58f., 70f., 77f.

158 This agrees with the mention of Eudoxus, as well as of optics and mechanics (see above, 47 n. 11).

159 ἐν ὀλίγῳ χρόνῳ τοσαύτην ἐπίδοσιν τὴν τῶν μαθημάτων θεωρίαν λαβεῖν (fr. 5 Ross = fr. C 52:2 Düring); τοσοῦτον δὲ νῦν προεληλύθασιν ἐκ μικρῶν ἀφορμῶν ἐν ἐλαχίστῳ χρόνῳ ζητοῦντες οἵ τε περὶ τὴν γεωμετρίαν καὶ τοὺς λόγους καὶ τὰς ἄλλας παιδείας, ὅσον οὐδὲν ἕτερον γένος ἐν οὐδεμιᾷ τῶν τεχνῶν (fr. 8 Ross = fr. C 55:2 Düring). See above, 70 n. 105–107.

the time of its appearance. Eudemus' works on the history of science most probably date from the period between the foundation of the Lyceum and the death of Aristotle, i.e., between 335/4 and 322/1 (5.1). Speusippus, the oldest of the possible authors of the book on Plato, died in 338, while Xenocrates, who was ten years older than Philip, lived until 314; Hermodorus' chronology and the date of Philip's death remain unknown. Hence, we have no conclusive evidence that the Academic work was written before Eudemus' *History of Geometry.* Even if we accept this dating for purposes of argument, the place of this work in the historiography of science presents a problem.

In spite of the similarities between the papyrus passage and the part of Eudemus' *Catalogue* devoted to mathematics in the time of Plato,[160] the differences between them – in general approach as well as in details – remain obvious.[161] The main difference is that the *Catalogue* focuses on the history of geometry, whereas the treatise quoted by Philodemus is primarily concerned with Plato. The scope of the second part of the *Catalogue*, let alone the first, is much broader than that of the passage from Philodemus, which ends in polemics against some disciples of Plato who made the 'fruits of knowledge' serve their own ends. The end of column Y (after the words 'optics and mechanics') is seriously damaged, and Gaiser's reconstruction is only tentative.[162] Yet nothing of what remains legible points to a work on the history of science as a possible source of this quotation. The text quoted by Philodemus is clearly related to Plato, while the passage concerning progress in mathematics seems to be a digression,[163] intended to show that his influence extended to this science as well. Particular facts mentioned in the text, such as the development of the theory of proportions, of analysis and of diorism, are indeed found in Eudemus. But Eudemus' knowledge of it, however, must have been first-hand rather than borrowed from a book on Plato.

To judge from its title, Hermodorus' work Περὶ μαθημάτων was broader in scope than the treatise quoted by Philodemus. Unfortunately, we know very little about its subject matter. It remains unclear whether it treated mathemat-

160 See above, 87f.

161 Eudemus does not mention the term μετρολογία, which is now considered to have referred to the general theory of proportions (for other variants, see Dorandi. *Filodemo*, 209). His words concerning Eudoxus being the first to have increased the number of so-called general theorems must have referred, however, to the general theory of proportions (see below, 206f.). Eudemus ascribes the discovery of diorism to Leon, who is not mentioned in the papyrus passage. Associating analysis with Eudoxus, the *Catalogue* does not, however, call him its discoverer. Mentioning Plato as the author of the method of analysis, Favorinus (D. L. III, 24 = fr. 25 Mensching) might well have been developing the ideas of the Academic work.

162 Gaiser. *Academica*, 153f.

163 This is also indicated by the beginning of the quotation: [κατε]νενόητο δὲ φησί, *καὶ* τῶν μαθημάτων ἐπίδοσις πολλὴ κατ' ἐκεῖνον τὸν χρόνον. Mathematics is not mentioned before this or after the words about mechanics and optics.

ical sciences or, in the traditional sense, various branches of learning in general. The fact that the two short references to this work concern Zoroaster and the Persian Magi as his successors,[164] rather than mathematics, seems to favor the second alternative. Diogenes Laertius, on the other hand, mentions Hermodorus in the context of a discussion of the origin of philosophy, quoting Aristotle on the Magi being older than the Egyptians, and Eudemus on the Magi's belief in the immortality of the soul.[165] The quotation from Aristotle goes back to his dialogue *On Philosophy,* which traces the Oriental origins of philosophy and its further development by the Greeks, while Eudemus' fragment must be taken from his *History of Theology*.[166] To guess by analogy – and we do not seem to have any alternative – Hermodorus' work *could have been* related to the history of knowledge, mathematical in particular, starting from its origin in the Orient. All of this, however, remains highly conjectural.

Summing up our survey of the testimonies of the Platonists' works related to the history of science, we can conclude that, beyond doubt, they did discuss the development of the exact sciences, though the evidence for this is very poor. Which means, in turn, that in the Academy this subject did not become a topic of special studies that later authors could use, as was the case with Eudemus' works. For example, if Xenocrates' numerous works on exact sciences[167] contained any historical data on the development of geometry or astronomy, traces of it would have survived in Greek literature. Their absence seems to indicate that the Platonists' preoccupation with exact sciences was not motivated by historical interests.[168] Though many ideas that later entered the historico-philosophical conception of Eudemus are occasionally to be found scattered through the works of Plato and his students, it is only in the Lyceum that history, the exact sciences, and philosophy were united in such a way as to produce a new subject, the history of science.

---

[164] D. L. I, 2 and 8 = fr. 6 Isnardi Parente.

[165] I, 8–9 = Arist. fr. 6 Rose = fr. 23 Gigon; Eud. fr. 89.

[166] See below, 130 n. 51, 131 n. 53.

[167] See above, 90 n. 41.

[168] An excellent parallel is Speusippus' work *On Similar Things* (D. L. IV, 5), where similarities and differences in the vegetable and the animal realms provided material for a purely logical classification on the model of Plato's diaeresis. See Tarán. *Speusippus*, 64f. and F 6–27.

Chapter 4

The historiographical project of the Lyceum

1. Greek science in the late fourth century BC

Among the trends in Greek thought that we have already considered as sources and/or precursors of the Peripatetic historiography of science, two main groups of ideas can be discerned. Most ideas of the first, historical group – such as heurematography, the early historiography of poetry and music, the theories on the origin of culture by Presocratics, Sophists, and Hippocratic physicians, the rudiments of doxography – date from the pre-Platonic period. To the second, theoretical group belong the Sophistic theory of τέχνη and the Platonic notions of τέχνη and ἐπιστήμη, which came to be integrated into the Aristotelian theory of science. Let us now examine another factor that predetermined to a large extent the forms in which the historiographical project of the Lyceum was realized, namely, the concrete configuration of sciences that took shape in the late fourth century and the related ideas of the scientists regarding the nature of science (cf. 2.3).

The more rapid development of the exact sciences in comparison with the natural ones doubtless played a decisive role in the fact that mathematics became a model science for Plato and Aristotle. By that time it had grown into an axiomatico-deductive system that guaranteed the truth of final conclusions deduced from indemonstrable and self-evident principles. Science, understood in this way, determined parameters for the history of science as well. Since the distinctive features of Greek geometry were the setting of problems in general form and their deductive proof, Eudemus' *History of Geometry* started with Thales, the first Greek mathematician in whose work both of these qualities are clearly apparent. Even at present, the history of science remains, indeed, the history of those results whose significance is acknowledged by the contemporary scientific community. In this sense, it depends directly on the expert knowledge of scientists, in accordance with which the sorting out and the assessment of the historical evidence normally takes place. This does not mean that the past is rewritten each time science takes a step forward. This is impeded first and foremost by the cumulative character of scientific development, which allows the integration of old notions and long-acknowledged facts into new theories. Nevertheless, any analysis of the science of the past cannot help relying on its present condition as the specialists understand it. There is no reason to believe that in the earliest period of the history of science the situation was substantially different in this respect. To be sure, the first histories of geometry, arithmetic, and astronomy were written by a Peripatetic philosopher, not a mathematician. Yet his idea of the exact sciences is almost wholly derived from the professional milieu of his time.[1] In other words, the scientific disciplines contemporary to Eudemus were not a mere *subject* of the history of science – in a sense, they shaped that genre in itself. The situation was similar in physics. For Theophrastus, the major expert in this area was Aristotle, so that Peripatetic doxography interpreted the theories of the Presocratics from the point of view and in terms of Aristotelian physics.

Unfortunately, the sources of the classical period contain much more information about philosophical theories of science than about the views of science held by mathematicians, astronomers, or natural scientists. Apart from medical treatises, perhaps, these views were left outside the framework of scientific writings. We should not, however, jump to conclusions and mistake the scarcity of our sources for the lack of any general idea of science among Greek scientists; and even less should we presume that philosophical theories of science reflected a generally accepted attitude toward science within the scientific community. Even the little we know about Archytas indicates that his idea of mathematics was substantially different from that of Plato.[2] On the contrary, neither Euclid's *Elements*, nor the program of 'saving the phenomena', formulated by Eudoxus, show any traces of Aristotle's or Plato's definite influence.[3] Let us take an example from a later epoch. Ptolemy, being versed in philosophy, held an eclectic view of science (as nearly every scientist does). From the theories familiar to him, he used to choose those more in keeping with his own scientific views. Turning in the preface to *Almagest* to the division of theoretical knowledge into theology (metaphysics), physics, and mathematics, a division traditional since Aristotle, Ptolemy notes:

> The first two divisions of theoretical philosophy should rather be called guesswork than knowledge: theology because of its completely invisible and ungraspable nature, physics because of the unstable and unclear nature of the matter... only mathematics can provide sure and unshakable knowledge to its devotees, provided one approaches it rigorously. For its kind of proof proceeds by indisputable methods, namely arithmetic and geometry.[4]

This evaluation contradicts the views of Aristotle himself, who believed theology to be the highest and the most valuable kind of knowledge, with physics coming second.[5] Ptolemy may have been influenced by Plato's ideas that physical reality cannot be fully known, but he shared them only insofar as they

---

1 See below, 168f. Aristotle and Eudemus obviously followed the opinion of the professionals in considering unscientific the attempts of Antiphon and Bryson to square the circle (see below, 178 n. 50).

2 See above, 94 n. 59, 110.

3 See above, 101 n. 97, and below, 271ff.

4 *Alm.*, 6.11–21, transl. by G. Toomer.

5 *Met.* 1026b 24f., 1064b 1–4. Still, mathematics is the most exact of sciences (*Cael.* 306a 27).

agreed with his own conviction that the main science is mathematics, and not Platonic dialectic or Aristotelian metaphysics.

The expert knowledge the Peripatetics relied on is accessible to us mostly in its objectivised form, i.e., in the form of scientific treatises contemporary to them. What was Greek science at the end of the fourth and the beginning of the third century like? The earliest of the surviving works in mathematics, astronomy, harmonics, optics, and mechanics are not numerous: Autolycus of Pitane's *On the Moving Sphere* and *On Risings and Settings*, Euclid's *Elements*, *Sectio canonis*, and *Phaenomena*. Adding to them Euclid's *Optics* and the Aristotelian *Mechanical Problems*,[6] we get an almost exhaustive list of the surviving texts that allow us to judge the achievements of the exact sciences in Greece in the first three centuries of their development.[7] The early Greek natural sciences are represented by a larger number of texts, yet their distribution in time is similar to that in the domain of exact sciences. At our disposal are Aristotle's treatises on biology and physics, as well as Theophrastus' research on botany, mineralogy, and other natural sciences. Theophrastus is also the author of the first doxographical compendium on the problems of natural philosophy, *Physikōn doxai*. Finally, early Greek medicine reached us in the form of the numerous treatises of the Hippocratic corpus, dated mostly to the late fifth and the fourth centuries, and its selective doxography was compiled by Meno, a colleague of Eudemus and Theophrastus in the Lyceum.

The bulk of the information on the first three centuries of Greek science dates to the end of the fourth century. No wonder we know this period much better than earlier ones. An analysis of scientific texts of this time shows that, in the majority of cases, we are dealing with scientific disciplines that took shape after a long period of formation. The level achieved by this time in different fields was disproportionate, of course: whereas mathematics and mathematical astronomy fully satisfied the major scientific criteria, the natural sciences, let alone medicine, were still far from this standard. Still, Greek science of this period can be regarded as formed, at least in the sense that its conceptual foundations had already been laid down, its basic methods worked out, and the priority of its problems set. In the following centuries, each of the sciences went its own individual way, some of them branching out into completely new directions, such as Archimedes' statics and hydrostatics, or spherical trigonometry, or Diophantus' 'algebra'. It is revealing, however, that not a single new science appeared in Antiquity after the fourth century.[8] This means that the

---

6 See above, 97 n. 82–83.

7 Euclid also wrote two other small treatises, *Data* and *On Division (of Figures)*. His authorship of *Catoptrics* is debatable. Aristoxenus' *Elements of Harmonics* is not related to the exact sciences.

8 Astronomy and geometry appeared in the sixth century, arithmetic and harmonics in the late sixth to early fifth century, optics and mechanics in the early fourth century. Descriptive geography stems from the sixth century; the application of mathematical

foundations laid down by the end of the classical period remained unchanged on the whole.

This conclusion applies not only to the disciplines that later underwent little changes, such as Aristotle's zoology or his theory of motion,[9] but also to those that were far from stagnating. Euclid's mathematics, though developed by Archimedes, Apollonius, and a dozen other less eminent scientists, remained until the end of Antiquity, with very few exceptions, the same Euclidean mathematics. Mathematical harmonics and optics, improved to a certain extent by Ptolemy, still do not reveal any remarkable progress. The greatest changes seemed to occur in astronomy, in the first place due to the Babylonian observational and numerical data on planetary motions, which became accessible to the Greeks from the second century BC. This gave astronomy a new impetus and permitted it to achieve a hitherto unattainable accuracy by passing from qualitative to quantitative models.[10] All the same, there is no reason to speak of a radical transformation of Greek astronomy under the influence of Babylonian data: its conceptual basis remained mostly unchanged, even in Ptolemy's epoch. The main aim of astronomy, as formulated by Eudoxus, was to create a kinematic theory of the motion of heavenly bodies, which would explain their visibly irregular motion in the firmament through the postulation of uniform circular motion. Developed by Autolycus and Euclid, the method of exposing an astronomical theory as a system of deductive arguments from initial principles eventually became a standard for any serious astronomical treatise. Among the basic axioms and definitions figuring in the preface to Euclid's *Phaenomena*, we find practically the same fundamental notions as in the preface to Ptolemy's *Almagest*.[11]

Conventional as analogies ever remain, there is no doubt that, after the fourth century BC, nothing happening in Greek science could be compared to the 16th- to 18th-century transformation of astronomy connected, in particular, with the transition to the heliocentric model, the invention of the telescope, and

---

methods to it is connected with Dicaearchus (fr. 104–115; Keyser, P. The geographical work of Dikaiarchos, *Dicaearchus of Messana*, 353–372).

9 Though some aspects of Aristotle's dynamics were modified by Strato of Lampsacus, a decisive step away from it was made only by Ioannes Philoponus (sixth century AD).

10 Also important in this respect was Apollonius' and Hipparchus' research on spherical trigonometry, completed later by Menelaus (ca. 100 AD). See Björnbo, A. A. *Studien über Menelaos' Sphärik*, Leipzig 1902, 124ff., 133f. Sidoli, N. Hipparchus and the ancient metrical methods on the sphere, *JHA* 35 (2004) 71–84, makes Hipparchus' crucial contribution to spherical trigonometry more feasible.

11 Ptolemy proceeds from the general assumptions that were formulated by the end of the fourth century BC: 1) the skies have a spherical shape and rotate as a sphere; 2) the earth has a spherical shape; 3) it is situated in the center of cosmos; 4) in terms of its dimensions and its distance from the stellar sphere, the earth relates to the latter as a point to a sphere; 5) the earth does not take part in any motion (I, 2).

the creation of Newtonian dynamics. Nor was there anything comparable to the development of algebra and analytic geometry, or later to the revolution in biology, starting with the discovery of the cell and the emergence of evolutionary theory. Accordingly, we have every reason to assert that, on the whole, by the end of the fourth century, the *foundations* of Greek science were complete. It is very likely that Aristotle and his students had reached a similar conclusion. The growing awareness that the formation of science and philosophy had been completed was probably one of the incentives for the research on the history of knowledge undertaken in the Lyceum, which embraced the entire period from the beginnings of mathematics, astronomy, theology, natural philosophy and medicine until the second third of the fourth century. In fact, ideas had been expressed even earlier that a particular field of knowledge or skill, e.g. medicine or rhetoric, had already reached its perfection.[12] Generalizing these ideas into a theory that can be called teleological progressivism (5.5), Aristotle applied it to the whole of Greek culture, which in many (though by no means all) of its aspects had by his time, in fact, reached the level that later proved to be unsurpassed. In philosophy, Aristotle and his disciples regarded his system as the consummation of the entire tradition from Thales until Plato, in a sense as the consummation of philosophy as such. It is not by chance that physical doxography ended with Plato: for Theophrastus, Aristotle's physics was no longer δόξα, but ἐπιστήμη. Eudemus thought contemporary mathematics (and probably astronomy, too) had reached its perfection. Similar motives can be seen in the attempts of Euclid, his younger contemporary, to summarize in 'concluding' writings the most indisputable results of previous investigations in geometry, arithmetic, astronomy, harmonics, and optics.

The leading position of mathematical sciences was reflected also in the Greek classification of sciences. It originated with the Pythagoreans, who set apart geometry, arithmetic, astronomy, and harmonics as a specific group of *mathēmata,* which was eventually extended to include many new spheres of knowledge (2.3). This classification was based on the idea, formulated by Archytas (47 B 1), of the close relationship between all sciences that use mathematical methods. Aristotle included in the *mathēmata* not only astronomy and harmonics, but also optics and mechanics, although he regarded them as 'more physical' than pure mathematics (*Phys.* 193b 22f.). According to a Hellenistic classification preserved by Geminus, *mathēmata* included geometry, arithmetic, canonics (harmonics), astronomy, logistics, geodesy, optics, and mechanics.[13] Thus Greek 'mathematics', expanding at the expense of the applied

---

[12] See above, 2.2, 2.4 and 3.5. It is revealing that, when speaking of the rapid progress and soon completion of philosophy, Aristotle (fr. 53 Rose) rejected analogous pretensions of his predecessors who claimed that, due to their talents, philosophy had already reached perfection. The difference between them lies, therefore, not in the character of their pretensions, but in their validity.

[13] Gemin. ap. Procl. *In Eucl.*, 38,4–42,8. For the same classification, see Ps.-Heron. *Def.*, 164.9–18.

sciences, embraced a number of key disciplines that in the modern period have been transferred to the domain of physics, whereas ancient 'physics' retained only natural philosophy and the life sciences. Eudemus' history of science accordingly covered only *mathēmata*, while Theophrastus' and Meno's doxographical works treated physics and medicine. Likewise, Eudemus' *History of Astronomy* included only problems related to mathematical astronomy, whereas physical astronomy constituted a special division of the *Physikōn doxai*. The latter clearly indicates that the Peripatetics viewed the configuration of the sciences formed by the middle of the fourth century in the light of the Aristotelian classification, which we, therefore, now ought to consider.

2. Aristotelian theory of science and the Peripatetic historiographical project

Strictly speaking, what we find in Aristotle does not constitute a unified and elaborated system.[14] Rather, it represents attempts to classify various kinds of knowledge undertaken at different times and from different points of view. His separate observations about the interrelations between particular disciplines and about their epistemological status quite often contradict each other. This is largely explained by Aristotle's well-known tendency, when examining a question, to draft a classification or to give an occasional definition that, due to its *ad hoc* character, does not always agree with classifications and definitions he gives on different occasions in other writings. We will attempt, nonetheless, to give a general outline of Aristotle's classification of knowledge in its correlation with the forms in which the historiographical project was realized, without aiming to reconcile *more scholastico* all the small- and large-scale contradictions.

In its most general form, Aristotle's classification of the sciences is presented in book E of the *Metaphysics*, which may go back to his lost work Περὶ ἐπιστημῶν.[15] All sciences (ἐπιστῆμαι) and all mental activities (διάνοιαι) are divided into three kinds: πρακτική, ποιητική, and θεωρητική. Πρακτικὴ ἐπιστήμη includes practically-oriented sciences that regulate nonproductive human activity (πρᾶξις), like ethics and politics; ποιητική includes productive sciences, or arts (τέχναι); and θεωρητική embraces theoretical sciences.[16] Ar-

14 On this class of problems, see McKirahan, R. Aristotle's subordinate sciences, *BJHS* 11 (1978) 197–220; Owens, J. The Aristotelian conception of the sciences, *Collected papers*, ed. by J. R. Catan, Albany 1981, 23–34; Taylor, C. C.W. Aristotle's epistemology, *A companion to ancient thought. Epistemology*, ed. by S. Everson, Cambridge 1990, 116–165.

15 D. L. V, 26; Moraux, *Listes*, 46. The same classification is repeated in *Metaphysics* K.

16 *Met.* 1025b–1026a, 1063b 36–1064b 6. Along with the tripartite, Aristotle often uses the bipartite division, in which practical and productive sciences are brought to-

istotle produced this classification comparatively early; it was already mentioned in his *Topics*.[17] Further, θεωρητικαὶ ἐπιστῆμαι are divided into μαθηματική, φυσική and θεολογική.[18]

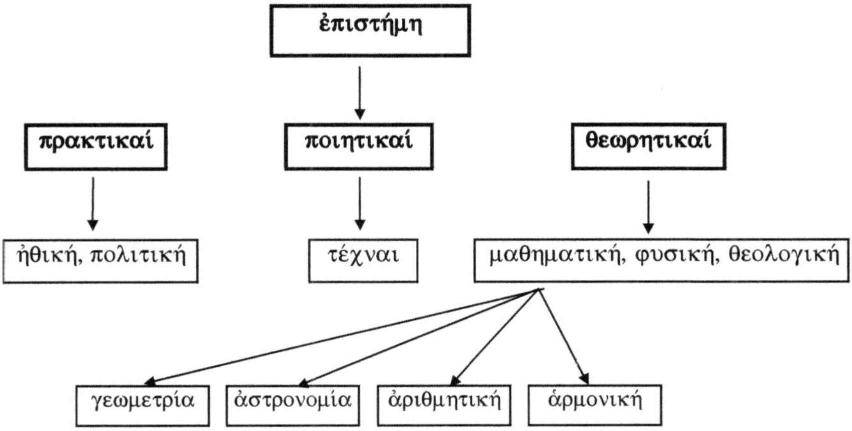

The latter subdivision occurs only in the *Metaphysics* E and K and comes, according to all evidence, from a later date. Stemming from the tripartite division of being into Forms and mathematical and corporeal objects (τὰ εἴδη, τὰ μαθηματικά and τὰ αἰσθητά), which Aristotle attributes to Plato,[19] it has, nevertheless, an obviously different character. 1) The object of metaphysics that Aristotle several times calls θεολογικὴ ἐπιστήμη (φιλοσοφία) was not Forms, but the fundamental principles and causes, or being *qua* being, or separate, unmoved and eternal entities.[20] 2) In contrast to the Platonists' views, Ar-

gether in one category of ποιητικαὶ ἐπιστῆμαι (*Met.* 982a 1, b 9–12, 1075a 1–3; *EE* 1216b 10–19, 1221b 5–7, etc.).

17 *Top.* 145a 14–18, 157a 10. It agrees with the passages from *EN*, where we find the division into τέχνη, μέθοδος, πρᾶξις (1094a 1), πράξεις, τέχναι, ἐπιστῆμαι (1094a 6–7), πράττειν, ποιεῖν, θεωρία (1178b 20–21). In *EN* 1140a 2–3, Aristotle notes that he considered this matter in exoteric writings as well, probably those of the Academic period. Cf. classification of τέχναι in Pl. *Phileb.* 55c.

18 *Met.* 1026a 6–19, 1064b 1–3; cf. *Phys.* 193b 22–36. Aristotle related logical sciences to the field of propaedeutics, which precedes the study of science itself (*Met.* 1005b 2–5).

19 *Met.* 987b 14–16, 28–29, 1028b 19–21, 1059b 6–8. Merlan. *From Platonism to Neoplatonism*, 59f.

20 In different parts of the *Metaphysics*, Aristotle gives this science various names (σοφία, πρώτη φιλοσοφία, θεολογική) and provides various definitions for its subject matter: *Aristotle's Metaphysics*, ed. by W. D. Ross, Oxford 1924, lxxvii; Flashar, H. Aristoteles, *Die Philosophie der Antike*, Vol. 3, 333ff.

istotle came to deny that mathematical objects belong to substances and can exist apart from sensible things. He did not deny, however, that they are existing things (ὄντα), so that his classification of science retained *mathēmata* as an independent branch of knowledge. 3) Physics was not a priority in the Academy; Plato, though discussing, by way of exception, many of its problems in his *Timaeus,* never accorded to physics the status of a theoretical science. Aristotle rehabilitates the Presocratic περὶ φύσεως ἱστορία, turning it into a theoretical science of the objects of sense perceptions (τὰ αἰσθητά).

Let us point out how this classification differs from the Sophistic theory of τέχνη and the contemporary conception of science. Aristotle applies his fundamental notion, ἐπιστήμη, to subjects now covered by theology and metaphysics, natural philosophy and science, arts and handicrafts. In modern languages, the only term embracing all these kinds of activities is 'culture'. Aristotle, however, does not invent a new term that would *equally* correspond to all the fields he endeavors to classify, but extends the notion of ἐπιστήμη, signifying *one* of them, to all the rest. The opening words of the *Metaphysics*, "All men by nature desire to know", best illustrate that his classification emphasizes the cognitive element even in those fields we hardly associate with cognition at all.[21] Aristotle considers all these 'sciences' from a purely intellectual point of view as containing a certain kind of knowledge (ἐπιστήμη) used by people to different ends: cognition, action, production.[22] When applied to the cognitive sphere itself, ἐπιστήμη, from 'knowledge', formerly the cognitive aspect of τέχνη, or 'skill' equivalent to τέχνη, turns into a branch of theoretical science. In this sense, its principal goal is not practical utility, but cognition as such, with *mathēmata* as its privileged model. Being already a part of the theoretical sciences, *mathēmata* presented no serious problems for Aristotle, while, as regards physics, he still had to prove that it really belongs to this kind of sciences; as for the subject matter and tasks of theology, he had to formulate them himself. At the same time, outside theoretical sciences, the notion of ἐπιστήμη does not so much oppose τέχνη as include it or serve as its synonym.[23] Aristotle's productive sciences correspond to what before (as well as after) him were referred to as τέχνη, e.g. medicine; the same is also true of many practical sciences – rhetoric, politics, etc.

Enriched with many new features, ἐπιστήμη at the same time inherited three of its four major characteristics from the old model of τέχνη: it can be acquired by learning; it has a particular aim; there are specialists able to achieve

---

21 On the intellectualism of the Sophistic theory of τέχνη see above, 46f.

22 The latter two spheres of human activity could be called ἐπιστήμη, since they relied on correct knowledge, and not on routine skills or blind luck.

23 On Aristotelian usage, see Walzer, R. *Magna Moralia und aristotelische Ethik*, Berlin 1929, 37ff.; Vogel, C. de. Quelques remarques à propos du Premier chapitre de l'Ethique de Nicomaque, *Autour d'Aristote*, ed. by A. Mansion, Louvain 1955, 315f.; Fiedler, *op. cit.*, 169ff.

this aim.[24] As we have already noted, the principal difference between the new and the old model is that while τέχνη was aimed ultimately at the practical application of knowledge, (θεωρητική) ἐπιστήμη aspired to pure knowledge. This opposition goes back to Plato, yet he was far from keeping to it systematically on the terminological level. In his dialogues, τέχνη and ἐπιστήμη quite often appear interchangeable,[25] and even the notion of *mathēmata* is not originally, or exclusively, limited to the sphere of mathematics.[26] In Aristotle, the distinction between τέχνη and ἐπιστήμη, found occasionally in his early treatises (*APo* 89b 7f., 100a 8), was theoretically grounded only in the *Nicomachean Ethics* (1139b 14ff.; cf. *Met.* 981b 26f.), which is one of his latest works. Like Plato, he could write of μαθηματικαὶ τέχναι,[27] the more so because in certain contexts his theory allowed not only opposing τέχνη and ἐπιστήμη, but also bringing them together. Τέχνη, e.g., is similar to ἐπιστήμη in being the knowledge of the general (τὰ καθόλου), not the particular, as in the case of ἐμπειρία.[28] The point, therefore, is not only that Aristotle proved unable to draw a clear terminological distinction between the notions of τέχνη and ἐπιστήμη, but also that his theory substantiated their kinship in a new way. One has to note that in Antiquity the consistent distinction between τέχναι and ἐπιστῆμαι was never made, to say nothing of their differentiation into sciences, arts, and crafts.[29] After the fourth century BC, exact sciences could again be called τέχναι (sometimes σεμναί, ἐγκύκλιοι or λογικαὶ τέχναι), while grammar and rhetoric were identified as μαθήματα.[30]

How did the division of cognitive space suggested by Aristotle affect the Lyceum's historiographical project? Students of Aristotle's thought have long noted that he distributed the historiography of different branches of knowledge among his disciples,[31] presumably taking their professional interests into account. So far, however, many important aspects of this project have not been

---

24 See above, 46ff.

25 Schaerer, *op. cit.*; Cambiano G. I rapporti tra episteme e techne nel pensiero platonico, *Scienza e technica nelle letterature classiche*, Genoa 1980, 43–61. Similar terminological uncertainty is characteristic of the *Epinomis* as well. Xenocrates devoted special works to τέχνη and ἐπιστήμη (D. L. IV, 13).

26 In the *Laws*, μαθήματα, as a rule, are included in mathematics (817e, 822d, 846d, 967e, 968e), but still there are exceptions (810b). In the earlier dialogues, μάθημα could mean the study of τοῦ καλοῦ (*Symp.* 211c), one of the τέχναι (*Men.* 90e), a doctrine (*Prot.* 313c), a field of knowledge (*Tht.* 206b, *Tim.* 87b), a tactic (*Lach.* 182b), etc.

27 *Met.* 981b 23. In *APr* 46a 19–22 ἀστρολογικὴ τέχνη is synonymous to ἀστρολογικὴ ἐπιστήμη.

28 *APo* 100a 6–9; *Met.* 981a 16; *Rhet.* 1356b 29; *EN* 1138b 2.

29 In the modern period, this also happened rather late; see above, 21f.

30 For examples, see Fuchs, *op. cit.*, 373, 376f.

31 Leo, F. *Die griechisch-römische Biographie nach ihrer litterarischen Form*, Leipzig 1901, 99f.; Jaeger, W. *Aristotle: Fundamentals of the history of his development* (1923), 2nd ed., Oxford 1948, 334ff.

sufficiently clarified. 1) The project as a whole was devoted to the theoretical kind of sciences. 2) Individual parts of the project correspond to the division of the theoretical sciences into mathematics, physics, and theology. 3) These three sciences were distributed in such a way that Eudemus received the first and the third kind, the histories of mathematics and theology, while Theophrastus dealt with the second, physical doxography (Meno's medical doxography belongs to this kind as well).[32] 4) The material of each science was, in turn, arranged in a way designed to avoid duplication as far as possible, e.g. between physics and medicine, physics and theology, mathematical and physical astronomy, etc.

All this demonstrates that the historiographical project of the Lyceum was carefully worked out. Both in its general plan and in the concrete forms and methods of the Peripatetics' work, including their choice of problems, the selection and arrangement of the material etc., we see the traces of Aristotle's influence. This influence can hardly be accounted for solely by his students' excellent knowledge of his writings and theories; rather, it is backed by the presence of the head of the school himself – the real, not legendary, architect of sciences. The degree of Aristotle's involvement in the project is hard to determine for certain, but equally hard is to believe that he did not participate in any form in his students' coordinated efforts to cover the three areas in the history of knowledge that he related to the theoretical sciences. At the least, he must have prompted his students to undertake a research in the history of knowledge.

Proceeding from the assumption that Eudemus' history of the exact sciences and history of theology, Theophrastus' physical doxography, and Meno's medical doxography were parts of a common project, rather than works written at different times and with different aims, we can better understand the interconnections, similarities, and distinctions between them. In contrast, the idea that Meno's *Medical Collection* is not a part of a common project and may even prove to be Aristotle's, and not Meno's, work,[33] appears unconvincing. Still less grounded seem to me the attempts to isolate Theophrastus' doxography

---

32 See below, 127.

33 See e.g. *Aristotelis opera*, Vol. 3: *Librorum deperditorum fragmenta*, ed. by O. Gigon, Berlin 1987, 511ff. (Gigon listed this work among Aristotelian fragments); Manetti, D. 'Aristotle' and the role of doxography in the Anonymus Londiniensis, *AHM*, 98f, 129. The papyrus' author, who made excerpts from Meno's book, believed it to have been written by Aristotle. It is more natural, however, to ascribe a book of an obscure Peripatetic to the founder of the Lyceum than vice versa. The very obscurity of Meno, who is only known as Aristotle's student, makes the invention of his authorship highly improbable. According to Galen, 'Ιατρικὴ συναγωγή, ascribed to Aristotle, is generally agreed to have been written by his disciple Meno, and that is why some call it Μενώνεια (*In Hipp. De nat. hom. com.* I, 25–26 = fr. 375 Rose). The title Μενώνεια occurs in Plutarch (*Quaest. conv.* 733 C), who, nevertheless, attributes this work to Aristotle. See Zhmud, L. Menon, *Die Philosophie der Antike*, Vol. 3, 564f.

from other branches of the project (in particular, from the history of science) and to consider it in the context of a 'dialectical practice' of the Lyceum.[34]

Aristotle regarded the theoretical sciences as the most important ones and as the worthiest occupation for a free man. It is not by chance, therefore, that mathematics, physics, and theology became the subject of a special historiographical project. To judge from the amount of material and, accordingly, the effort put into it, physical doxography was certainly the central part of the project. Theophrastus' compendium consisted of 16 (or 18) books (D. L. V, 46, 48), whereas none of Eudemus' histories of the exact sciences exceeded four books. Such correlation can be largely explained by the amount of material in each science and by the methods of its selection. At the same time, it reflects Aristotle's own interests as a physicist, interests shared by Theophrastus and Eudemus as well. The subject matter of the *Physikōn doxai* largely coincided with that of most, though by no means all, Presocratic writings – if, of course, we keep in mind what Aristotle himself understood by περὶ φύσεως ἱστορία. It included fundamental principles, notions, and categories of physics (matter, causes, space, time, void, etc.), as well as its separate branches: astronomy, meteorology, psychology, physiology, and embryology. (This approach, naturally, left many of Heraclitus', Parmenides', and Zeno's ideas outside the framework of doxography.) As for the composition of this treatise, the physicists' *doxai* were set forth in accordance with two main principles: systematic (thematic) and chronological. The first allowed the opinions of different philosophers to be put together in books and chapters devoted to individual topics and problems; the second, to place them in chapters according to their historical sequence. For example, the first chapter of the *Physikōn doxai*, which deals with ἀρχαί of philosophers, starts with Thales, the first physicist according to Aristotle (*Met.* 983b 20), and ends with Plato, whose dialogue *Timaeus* was the main source of his physical doctrines for the Peripatetics.

Following Plato, Aristotle believed that the subject of theoretical science is phenomena and processes of a general and regular character (τὰ καθόλου). Yet he modified this postulate, extending the province of theoretical science to what occurs 'as a general rule' (ὡς ἐπὶ τὸ πολύ).[35] This modification not only strengthened the theoretical status of physics, but allowed Aristotle to consider the theoretical part of medicine, related to physics, along with physical theories, without changing medicine's status as a productive science aiming to attain health (*EE* 1216b 18). Aristotle notes several times that, in regard to the causes (αἰτίαι) of health and disease, physicists and doctors have a common task, so that the former often complete their works with medical topics, where-

---

34 See below, 134.

35 *APr* 43b 30–38, *APo.* 96a 8–19; *Met.* 1027a 20–24, 1064b 32–36, 1065a 1–6. See De Ste. Croix, G. E. M. Aristotle on history and poetry (1975), *Essays on Aristotle's Poetics*, ed. by A. O. Rorty, Princeton 1992, 23–32; Mignucci, A. ὡς ἐπὶ τὸ πολύ et nécessaire dans la conception aristotélicienne de la science, *Aristotle on science. The Posterior Analytics*, ed. by E. Berti, Padua 1981, 173–204.

as the latter start from physical principles.[36] Meno's medical doxography was built on the same principles as Theophrastus' physical doxography, but was limited to a much narrower set of problems, namely to the causes of diseases.[37] This stressed its proximity to the investigation of the physical ἀρχαί and αἰτίαι and fully agreed with Aristotle's views on this subject. Meno, following Theophrastus, divided doctors into two groups on the basis of similarities in their theories. Within each group, they seem to be placed in a more or less chronological order.[38] Meno started with the most ancient doctors whose writings were available at the time, Euryphon and Herodicus of Cnidus,[39] and evidently finished with Plato.[40] Earlier Crotonian doctors, like Calliphon and Democedes, either did not leave writings that corresponded to the topic of Meno's doxography, or like Alcmaeon were considered natural philosophers (φυσικοί) and correspondingly figured in Theophrastus' work.[41]

Particularly interesting is the way Meno's etiology of diseases was fitted into the narrow space between ἰατρικὴ τέχνη, on the one hand, and physical doxography, on the other. Meno completely ignores everything that concerns the application of general regularities to particular cases or methods of treatment.[42] Theophrastus, by contrast, pays particular attention to physiology and psychology, whereas his work lacks a special section on the causes of diseases.[43] The presence in Meno's book of three natural philosophers – Hippon, Philolaus, and Plato – who suggested their explanations for the causes of dis-

---

36 *De sens.* 436a 17–b2; *De resp.* 480b 21–31. On medicine as ἐπιστήμη that investigates the causes of health, see also *Met.* 1026a 1f., 1064a 1f. In the passage describing how the 'production' of health takes place (1032b 2–22), the process is divided into two stages. In the first, intellectual stage (νόησις), the doctor compares the patient's condition with his knowledge about health (ἡ δὲ ὑγίεια ὁ ἐν τῇ ψυχῇ λόγος καὶ ἡ ἐπιστήμη) and selects appropriate methods of treatment. In the second, productive stage, these methods are applied in accordance with the aim.

37 ἀρχαὶ (τῶν νόσων) 4.28, 18.31, 18.47; αἰτίαι (τῶν νόσων) 4.41, 5.35, 9.40, 33.8 Diels.

38 See below, 164. Cf. Manetti, *op. cit.*, 102.

39 Grensemann, H. *Knidische Medizin*, T. 1, Berlin 1975, 197ff., suggests that Euryphon was born before 500 BC, but this seems to be too early (Zhmud. *Wissenschaft*, 243 n. 67).

40 In the papyrus text, the last person mentioned is Plato's contemporary Philistion of Locri, but according to a very plausible reconstruction by Manetti (*op. cit.*, 118f.), Plato came after him.

41 Although Alcmaeon did practice medicine (Zhmud. *Wissenschaft*, 239f.), the content of his book corresponds generally to the Presocratic περὶ φύσεως ἱστορία, which rightly placed him among the physicists. On the other hand, his omission can be explained by the highly lacunose state of the papyrus.

42 This restriction is obviously artificial and was not characteristic of the later medical doxography. See Eijk, Ph. van der. The Anonymus Parisinus on "the ancients", *AHM*, 302f.

43 Aët. V, 29 on the causes of fever was added later; see below, 295f.

eases, can be regarded as another example of similarity between physical and medical theories. At the same time, this example shows that physicists could encroach on the territory of medicine more easily than doctors could on the field of physics. Although the doxographical work of Aëtius, which goes back to Theophrastus, did include opinions of some early physicians, like Hippocrates and Polybus, it is after Theophrastus that these must have been added.[44]

Turning to the mathematical division of theoretical sciences, let us notice that not all the sciences regarded in the fourth century as *mathēmata* became subject matter for historical studies. In the case of optics and mechanics, the reason is obvious: both disciplines were too young to have a history of their own.[45] It is harder to say why the Lyceum, leaving a number of essays on the history of music, never produced a history of mathematical harmonics. This omission might be explained by the fact that Aristoxenus, the main expert on music in the Lyceum, was highly critical of the mathematical harmonics the Pythagoreans developed.[46] Theophrastus also criticized the Pythagorean theory of music, opposing to it a purely qualitative approach (fr. 716 FHSG). Aristotle, by contrast, treated mathematical harmonics with due respect, never casting doubt on its scientific status.[47] The disagreement between the experts on harmonics could have made the history of this discipline problematic.[48]

As for the characteristic features of the mathematical division of the project, for the time being I only note that, in contrast to doxography, Eudemus' material was not organized systematically. Strictly following the chronology, the *History of Geometry* and the *History of Astronomy* start with Thales, who was re-

---

44 Diels. *Dox.*, 232; Runia, D. The *Placita* ascribed to doctors in Aëtius' doxography on physics, *AHM*, 189–250, 248f. Cf. Mansfeld, J. Doxography and dialectic: *The Sitz im Leben* of the 'Placita', *ANRW* II 36.4 (1990) 3058f.

45 The founder of mechanics and optics must have been Archytas: Krafft. *Mechanik*, 3f, 144ff.; Schneider, *op. cit.*, 227; Schürmann, *op. cit.*, 33, 48ff.; Cambiano. Archimede meccanico; Burnyeat, M. Archytas and optics, *Science in context* 18 (2005) 35–53. Aristotle and his contemporaries mention both disciplines (see above, 47 n. 11), Aristotle devoted to them two special treatises, Μηχανικόν and the lost 'Οπτικόν (fr. 380 Rose). The list of Philip's works includes two writings on optics (Lasserre. *Léodamas*, 20 T 1). Euclid's *Optics* must have been based on these works.

46 In the *Elements of Harmonics*, he stated that the Pythagoreans "used arguments quite extraneous to the subject, dismissing perception as inaccurate and inventing theoretical explanations, and saying that it is in ratios of numbers and relative speeds that the high and the low come about" (I, 41.17f., transl. by A. Barker). See Bélis, A. *Aristoxène de Tarente et Aristote: Le traité d'harmonique*, Paris 1986; Barker, A. Aristoxenus' harmonics and Aristotle's theory of science, *Science and philosophy*, 188–226.

47 See e.g. *APo* 75b16, 76a10, a24, 78b38, 79a1; *Top.* 107a16; *Met.* 997b21. McKirahan, *op. cit.*, 220; Barker. Aristoxenus' harmonics, 190f. On the whole, Pseudo-Aristotelian *Problems* follows the Pythagorean viewpoint (Barker. *GMW* II, 85ff.).

48 The only fragment from Eudemus' *History of Arithmetic* (fr. 142) is related rather to mathematical harmonics than to arithmetic. See below 5.2, 6.1.

garded as the progenitor of both sciences, and finish with Eudemus' own contemporaries, the students of Eudoxus. In the *History of Geometry,* and probably in the *History of Astronomy* as well, Eudemus mentions the Oriental predecessors of these sciences. The *History of Arithmetic* must have been written on the same principles, though the only preserved fragment does not give any conclusive evidence on the matter.

Aristotle's 'theology', expounded in book Λ of the *Metaphysics*, culminates in the doctrine of the divine Unmoved Mover that sets in motion the whole system of celestial bodies. Aristotle regarded the subject of this science to be the first principles of the divine (τὸ θεῖον), eternal, motionless, immutable, and separable from matter (*Met.* 1026a 15f.).[49] At first sight, this definition of theology made writing its history problematic. Nevertheless, Aristotle does not fail to find predecessors of this science, as well. His tendency to regard his theories as the development of earlier ones and his readiness to use even traditional wisdom to the degree that it did not contradict his own views made him appeal even to incipient, imperfect forms of theology, which he found in the early mythical cosmogonies and theogonies. Usually he called their authors θεολόγοι.[50] The incomplete and even wrong answers the theologians offered to questions of the principles of the divine became the subject of Eudemus' *History of Theology.*[51] This book was a chronologically organized outline of the specific principles of the divine introduced by Greek and 'barbarian' theologians. First came Orpheus, who introduced Night as the first principle; he was followed by Homer, Hesiod, Acusilaus, Epimenides, and Pherecydes.[52] The principles of the Babylonians, Persian Magi, Sydonians, and Egyptians were treated separately.

---

49 Elders, L. *Aristotle's Theology. A commentary on the Book Λ of the Metaphysics*, Assen 1972.

50 *Met.* 983b 29, 1000a 9, 1071b 27, 1091a 34; Palmer, J. Aristotle on the ancient theologians, *Apeiron* 33 (2000) 181–205. Palmer relates Xenophanes to the theologians, but in Theophrastus he figures among the physicists.

51 The only fragment of this work (fr. 150), preserved by Damascius, does not contain a title, but Usener, *op. cit.*, 64, rightly related it to Τῶν περὶ τὸ θεῖον ἱστορίας α′–ς′, listed among Theophrastus' works (251 No. 2 FHSG). Wehrli, who entitled it "Geschichte der Theologie?" nevertheless argued against this identification (Eud. fr. 150, comm.). Meanwhile, we have, on the one hand, a title in Theophrastus' list, which does not agree with any of his known fragments, and, on the other, Eudemus' fragment that perfectly matches this title (cf. Damascius' reference: κατὰ τὴν Εὐδήμου ἱστορίαν, p. 70.6 Wehrli). See also Betegh, G. On Eudemus fr. 150 (Wehrli), *Eudemus of Rhodes,* ed. by I. Bodnár, W. W. Fortenbaugh, New Brunswick 2002, 337–357; Zhmud, L. Eudemos aus Rhodos, *Die Philosophie der Antike*, Vol. 3, 558–564.

52 The chronological sequence of theologians was broken in one case: Acusilaus (ca. 500 BC) is mentioned after Hesiod, whom he seems to have closely followed in his *Genealogiai* (9 A 4 *DK*), but before Epimenides and Pherecydes. We have no reliable chronology of Acusilaus, and nor could Eudemus have. In Plato's *Symposium* (178b = 9 B 2) it is said that Acusilaus followed Hesiod (which is confirmed by the

Eudemus' choice of main personages for his *History of Theology* fits quite well with what the *Metaphysics* says about the theologians. Here they comprise a rather definite group of ancient poets (Orpheus, Homer, Hesiod) and the authors of mythical cosmogonies (Pherecydes),[53] who were to be treated separately from the philosophers (φυσικοί).[54] In the first place, theologians belong to a more ancient period;[55] second, they reasoned not rationally but 'mythically'.[56] Accordingly, the philosophers whom Aristotle placed among the physicists and whom we usually call the Presocratics do not figure in the *History of Theology*. And vice versa: none of Eudemus' theologians are mentioned in Theophrastus' *Opinions of the Physicists*. The boundary between these groups coincides to a considerable degree with the contemporary boundary between rational philosophy and mythical theogonies and cosmogonies. We are much indebted to Aristotle that the history of Greek philosophy still starts with Thales and not with Homer or Orpheus.[57] It is worth recalling that, before Aristotle, Hippias of Elis (86 B 6) emphasized affinities rather than differences between philosophers, poets, and 'barbarian' sages and that, since the Hellenistic period, allegorical interpretations transform Homeric poems into a source of all wisdom, including philosophy.[58]

In their choice of the theologians and physicists, Eudemus and Theophrastus not only followed Aristotle's criteria, they obviously aimed at coordinating their own plans. We see the same approach in the distribution of the material between Eudemus' *History of Astronomy* and the astronomical division of

affinity of their principles), and in the *Laws* Epimenides is dated to ca. 500 BC (642d 4 = 3 A 5). Eudemus (or Damascius) could have taken Plato's remarks into account.

53 οἱ περὶ Ἡσίοδον (1000a 9); οἱ ἀρχαῖοι ποιηταί (1091b 3ff.), among them Orpheus (ἀρχή Night), Homer (Ocean, cf. 983b 27), Hesiod (Chaos, cf. 984b 26–28), and finally, Pherecydes and the Persian Magi (1091b 10–11). Aristotle mentions the principles of the Magi, Ormuzd and Ahriman, in his dialogue *On Philosophy* (fr. 6 Rose = fr. 23 Gigon). The same principles are to be found in Eudemus.

54 These two groups are always set apart (*Met.* 1071b 27, 1075b 26, 1091a 34).

55 οἱ παμπάλαιοι καὶ πρῶτοι θεολογήσαντες (*Met.* 983b 29), οἱ νῦν – οἱ πρότερον (1000a 5). To be sure, this should not be taken too literally: Pherecydes was younger than Thales and Anaximander.

56 μυθικῶς σοφιζόμενοι (*Met.* 1000a 18). Mansfeld (Aristotle, 41f.) considers the clarity of statements to have been an important criterion for Aristotle in drawing the line between the two groups. Aristotle, however, accused many of the physicists of vagueness, as well. See Palmer, *op. cit.*, 182ff.

57 Mansfeld, J. Aristotle and the others on Thales, or the beginning of natural philosophy, *Studies*, 126–146.

58 It is noteworthy that Orpheus, Homer, Hesiod, and Pherecydes found their places in the later versions of doxography (see index in Diels' *Doxographi Graeci* and below, 295f.). On the penetration of the Homeric material into doxography, see Diels. *Dox.*, 88f.; Mansfeld. Aristotle on Thales, 122f. On the Oriental origins of philosophy, see the prologue in Diogenes Laertius.

Theophrastus' compendium. While explaining differences between (mathematical) astronomy and physics, Aristotle notes: it seems absurd that a physicist should be supposed to know the nature of celestial bodies, but not to know any of their essential attributes, e.g. that they are spherical, particularly since these problems are indeed discussed in the works of natural philosophers. But an astronomer (μαθηματικός) abstracts from the physical nature of the heavenly bodies and aims to explain their visible motion in the heavens by applying mathematical methods.[59] As a result of this asymmetry in the relationship between physics and mathematical astronomy, the physicists had more freedom than the astronomers to discuss things that were within the competence of other sciences,[60] whereas physical problems (the eternity of the cosmos, the nature of celestial bodies, etc.) were outside the competence of the astronomers.[61] Relying on these criteria, Eudemus limited his *History of Astronomy* to the purely mathematical aspects of this science, while Theophrastus treated opinions belonging both to physical and mathematical astronomy,[62] but *only* those that come from the physicists. Such typical *mathēmatikoi* as Hippocrates of Chios, Archytas, and Eudoxus also expressed their views on problems of physics and physical astronomy,[63] but they are not included in Theophrastus' work.[64] Unlike Eudemus (fr. 149), Theophrastus does not mention such professionals in astronomy as Meton and Euctemon.

Along with the distinctions to be made among the various sciences, we should also take into account the particular aims of each treatise. Theophrastus'

---

59 *Phys.* 193b 22–194a 11. On the difference between physics and mathematical astronomy, see also *APo* 78b 36f.; *PA* 639b 7; *Met.* 1073b 3f. The Stoics held similar views (D. L. VII, 132–133). Posidonius modifies Aristotle's theory (fr. 18 E.-K.; see below, 289f.). See also Philop. *In Phys.*, 218–222.

60 Of course, with due reference to *mathēmatikoi*, as does Aristotle himself (*Cael.* 291a 29–b9, 297a 3, 298a 15; *PA* 639b 7; *Met.* 1073b 3). On the same asymmetry in reference to physics and medicine, see above, 128.

61 According to Theophrastus (*Met.* 9b 25f.), astronomy deals with the motion of celestial bodies, their size and form, and the distance between them.

62 In the first one we can include, e.g., the sections on whether the cosmos is animated and eternal; what parts it consists of, whether there is a void outside the cosmos, what the nature of the moon is, of the sun, etc.; in the second, the questions concerning the form of celestial bodies, their order, size, etc. See the corresponding sections in Aëtius' book II (*Dox.*, 268).

63 See 1) Hippocrates and his student Aeschylus (Arist. *Mete.* 342b 29f. = 42 A 5); this opinion about the comets (though without Hippocrates' name) entered Aëtius' doxography (III,2.1) through Posidonius (*Dox.*, 230f.); 2) Archytas (Arist. *Met.* 1043a 19 = 47 A 22; Eud. *Phys.* fr. 60, 65 = 47 A 23–24; 47 A 25); 3) Eudoxus (fr. 287–288 Lasserre).

64 Eudoxus' opinion on the Nile's floods with reference to Egyptian priests (*Dox.*, 386.1f.) entered the doxography after Theophrastus (*Dox.*, 228f.; cf. Eudox. fr. 287–288 Lasserre); the same is true of the reference to Eudoxus and Aratus (*Dox.*, 347.21f.). See below, 295f.

compendium was devoted specifically to the *doxai* of the physicists as a *distinctive group* that – in Aristotle's view and that of his students – differed from the other groups, such as theologians, mathematicians, and physicians.[65] For this reason, we do not find theologians even in the section περὶ θεοῦ (Aët. I, 7), nor mathematicians in the astronomical part, nor doctors in the embryological part. In contrast, Eudemus' histories of geometry, astronomy, and arithmetic do not deal with the opinions of mathematicians as a specific group, but with *mathematical discoveries*, which could also have been made by those whom the Peripatetics considered physicists – Thales, Anaximander, Anaxagoras, and others. In the same way, as the title implies, Meno's *Medical Collection* contained not physicians' opinions, but *medical theories* about the origins of diseases. For this reason, Meno could legitimately include in his work ideas on this subject coming from Hippon, Philolaus, and Plato. Finally, Eudemus' *History of Theology*, though similar in title and in the chronological arrangement of material to his histories of sciences, was in its criteria for the selection of material closer to the *Physikōn doxai*, because it dealt with the specific group that Aristotle called the theologians.

3. History in the Lyceum

Close interconnection between Aristotle's views on specific sciences and Peripatetic writings on the history of these sciences; clear evidence of cooperation between Eudemus, Theophrastus, and Meno; the absence of apparent signs of duplication – all this supports my hypothesis that here we are dealing with a project that was rationally planned and implemented. Meanwhile, in recent decades the historical and even the historiographical[66] character of these writ-

---

65 The traditional understanding of the title of Theophrastus' work as Φυσικῶν δόξαι (*Opinions of the Physicists*), rather than Φυσικαὶ δόξαι (*Physical Opinions*), as Mansfeld proposes (Doxography and dialectic, 3057 n. 1; idem. *Physikai doxai* and *Problēmata physica* from Aristotle to Aëtius (and beyond), *Theophrastus: His psychological, doxographical, and scientific writings*, ed. by W. Fortenbaugh, D. Gutas, New Brunswick 1992, 63–111) is supported by the fact that the expression δόξα(ι) τῶν φυσικῶν (τῶν φυσιολόγων, τῶν περὶ φύσεως) is found both in Aristotle (*Phys.* 187a 27; *Met.* 1062b 21, 25; cf. δόξαι τῶν ἁρμονικῶν, Aristox. *Elem. harm.*, 7.3) and in his commentators (e.g. Alex. *In Met.*, 72.2, 652.30, 719.8; Themist. *In Phys.*, 211.29; Simpl. *In Phys.*, 148.28, 355.20, 358.12; *In Cael.*, 561.1; Philop. *In Phys.*, 26.23, 89.7, 108.15; Olymp. *In Mete.*, 150.28; see also Strab. II, 5.2.22–24 = Posid. fr. 3c Theiler; Euseb. *Prep. evan.* XV, 340.23). Meanwhile, φυσικαὶ δόξαι is attested neither in the tradition of the Lyceum, nor in Aristotle's commentators (cf. φυσικὴ δόξα in Olymp. *In Mete.*, 138.29). Mansfeld, J. Deconstructing doxography, *Philologus* 146 (2002) 279f., does not adduce any indisputable example of φυσικαὶ δόξαι (in plural), whereas φυσικὴ δόξα in some of his examples means 'natural', and not 'physical opinion'.

66 By historiography I mean a general field including not only history in the proper

ings has been questioned or disputed. For example, in their works on doxography, J. Mansfeld, D. Runia, and H. Baltussen clearly tend to reduce the historical orientation of Theophrastus' work to a minimum or even to deny it in favor of a systematic one.[67] According to them, the doxography is a systematically organized collection of 'physical opinions', born out of Aristotelian dialectic and designed for dialectical discussions held in the Lyceum;[68] thus, it would be an obvious anachronism to call it 'the history of philosophy'. From this viewpoint, Peripatetic doxography can be compared to a contemporary database, to be mined as needed in the course of theoretical discussion or research. C. Eggers Lan insisted that Eudemus did not write a history of the exact sciences, but "classified authors according to geometric topics".[69] Earlier, Wehrli had related Eudemus' fragment on the principles of the theologians not to the *History of Theology*, but to a systematic treatise.[70] To find a similar goal for Meno's *Medical Collection* is even easier.[71] As a result, the historiographical project becomes systematic or simply disintegrates into separate works hardly connected with each other.

The attempts to separate Theophrastus' doxography from Eudemus' history of exact sciences, to deprive it of its historical sense, and to consider it only as an application of Aristotle's dialectic do not seem to me convincing,[72] nor do recent works questioning the historical character of Eudemus' writings.[73] In spite of the significant differences in the methods the individual Peripatetics employed, it is precisely a historical – and not dialectical or systematic approach – that unites various parts of the project into one meaningful whole.[74]

---

sense, but also genres that cannot always be *directly* related to it, such as ancient Greek biography and doxography.

67 Mansfeld, Runia. *Aëtiana*; Mansfeld, J. Sources, *The Cambridge companion to early Greek philosophy*, ed. by A. A. Long, Cambridge 1999, 22–44; Runia, D. What is doxography?, *AHM*, 33–55; Baltussen. *Theophrastus.*

68 Mansfeld. Doxography and dialectic, 3063; idem. Doxographical studies, Quellenforschung, tabular presentation and other varieties of comparativism, *Fragmentsammlungen*, 22.

69 Eggers Lan, C. Eudemo y el 'catálogo de geómetras' de Proclo, *Emerita* 53 (1985) 130.

70 See Wehrli's commentary to Eud. fr. 150. Cf. above, 130 n. 51.

71 E.g. it could be related to Aristotle's work περὶ νόσου καὶ ὑγιείας (Manetti, *op. cit.*, 129).

72 Zhmud, L. Revising doxography: Hermann Diels and his critics, *Philologus* 145 (2001) 219–243.

73 Mejer, J. Eudemus and the history of science, *Eudemus of Rhodes*, 243–261; Bowen, A. C. Eudemus' history of early Greek astronomy, ibid., 307–322.

74 "Aristotle's own great achievement in the field of history during his later years and the parallel works of his disciples organized by him show that the investigation of the detail occupied his mind on a large scale ... This sort of historical interest cannot be explained any longer as an outgrowth of his dialectical method ... We must not separate Aristotle's interest in the history of philosophy from his historical research in

This conclusion will appear even more obvious if we consider this project against the background of other historiographical genres practiced in the Lyceum. Therefore, we will venture into a more detailed analysis of Aristotle's attitude toward history in general and toward the history of knowledge in particular, as well as the goals he tried to achieve by directing his students to the history of the theoretical sciences.

In contrast to his approach to mathematics, which attracted Aristotle as a philosopher but never became an object of his independent research,[75] when dealing with history and natural sciences he revealed a keen interest in the empirical investigation of particular facts, without, however, forgetting his theoretical tasks: to explain the 'causes' of things. Being fully aware of the historical character of such human accomplishments as the state, art, philosophy, and science, Aristotle endeavored to reveal the inner logic of their development. Even if he does not entirely meet the modern requirements of history, Aristotle did much more for its development than any other ancient philosopher,[76] both through his own studies and through the historiographical works of his students.

Certainly, Aristotle was hardly a "historian in the modern sense of this word".[77] The question, however, is whether "a historian in the modern sense of the word" is the only type of historian possible. What will remain of Jacoby's *Fragmente der griechischen Historiker*, if "the modern notions of an objective historical research" are applied? From the viewpoint of the present-day criteria, Eudemus' works belong to the history of science only with some reservation. Even more reservations can be held concerning Theophrastus' doxography, especially concerning Meno's doxography. In applying rigid criteria to the first specimen of newly-born genres, however, we should be aware of the limitation of this procedure, which, though quite legitimate, is hardly the only correct one. On the contrary, the historical approach to the project of the Lyceum shows that, in spite of all the differences between the Peripatetic historiography of science and philosophy and its contemporary counterpart, the latter is deeply rooted in the ancient tradition, which in turn begins with Aristotle and his

---

all these other fields of civilization." (Jaeger, W. Rec.: Cherniss, H. *Aristotle's Criticism of Presocratic Philosophy*, *AJP* 58 [1937] 354).

75 See above, 111 f. Hussey, E. Aristotle and mathematics, *Science and mathematics in ancient Greek culture*, ed. by C. J. Tuplin, T. E. Rihll, Oxford 2002, 217–229.

76 On Aristotle's contribution to the development of historical research, see von Fritz. Die Bedeutung des Aristoteles für die Geschichtsschreibung, 91 ff.; Weil, R. *Aristote et l'histoire*, Paris 1960; idem. Aristotle's view of history, *Articles on Aristotle 2. Ethics and politics*, ed. by J. Barnes et al., London 1977, 202–217; Huxley, G. On Aristotle's historical methods, *GRBS* 13 (1972) 157–169; De Ste. Croix, *op. cit.*; Blum, *op. cit.*, 20 ff.

77 Baltussen, H. A 'dialectical' argument in *De anima* A 2–4, *Polyhistor. Studies in the history and historiography of ancient philosophy presented to J. Mansfeld*, ed. by K. Algra et al., Leiden 1996, 335 f.

school. Thus, without identifying Peripatetic historiography with modern historiography, we have every reason to compare them, particularly since we are dealing with the development of one and the same phenomenon.

To realize better the importance that the Peripatetics attached to ἱστορία, we should bear in mind the following. First, the Aristotelian theory of science stresses the empirical origin of any knowledge. The question ὅτι, 'that (something is the case)', i.e., the collection and description of facts, not only precedes the question διότι, 'why (something is the case)', i.e., the explanation of general or particular causes, but actually makes it possible in the first place.[78] Thus, any scientific explanation is based on the facts established by observation (φαινόμενα) and correspondingly arranged beforehand.[79] In this sense, even a purely descriptive work is a necessary part of scientific procedure inasmuch as it is the prerequisite for subsequent theoretical analysis; and the Peripatetics, as we know, wrote hundreds of such works. In natural sciences, the questions about facts and about their causes can be asked within the framework of two different types of research, empirical and theoretical, which nevertheless belong to the same science, e.g. zoology (physics).[80] What the Peripatetics related to the field of natural history (φυσικὴ ἱστορία),[81] however, was not just a loosely arranged collection of facts lacking any analysis. For example, two of Theophrastus' works, *Historia plantarum* and *De causis plantarum*, are devoted to research on ὅτι and on διότι respectively. In *Historia plantarum*, however, we find not merely assembled data but botanical classification, morphology, and taxonomy.

Second, for Aristotle and the Lyceum as a whole, natural history was not yet rigidly separated from the history of human deeds and events, ἱστορίαι περὶ τῶν πράξεων.[82] Comparing the contents of Theophrastus' (fr. 196a FHSG), Aristoxenus' (fr. 131), and Hieronymus of Rhodes' (fr. 35–36) identically en-

---

78 ὅτι μὲν γὰρ οὕτω ταῦτα συμβαίνει, δῆλον ἐκ τῆς ἱστορίας τῆς φυσικῆς, διότι δέ, νῦν σκεπτέον (*De inc. an.* 704b 9). See also *HA* 491a 7–14, *PA* 646a 8–12.

79 E.g. *APo* 89b 29, 93a 16f.; *PA* 639b 7f., 640a 14f. and esp. *HA* 491a 7–14. In astronomy, ἀστρολογικὴ ἐμπειρία provides the researcher with the evidence and principles, while ἀστρολογικὴ ἐπιστήμη bases its proofs on these grounds (*APr* 45a 17f.). See Kullmann, W. *Wissenschaft und Methode*, Berlin 1974, 204ff.

80 See above, n. 78.

81 Arist. *HA* 650a 32, *De inc. an.* 704b 10; Aristotle often called his empirical studies of animals ἱστορίαι περὶ ζῴων, e.g. *GA* 716b 31, 717a 33, 728b 14, 740a 23; cf. Theophr. *CP* I,1,1–2, I,5,3, I,9,1. As an empirical type of research, φυσικὴ ἱστορία has to be distinguished from περὶ φύσεως ἱστορία that traditionally referred to natural philosophy (φυσικὴ ἐπιστήμη in the Aristotelian terms) and included the study of general regularities. Περὶ φύσεως ἱστορία (*Cael.* 268a 1) means the same as περὶ φύσεως ἐπιστήμη (298b 2), and ἡ τῆς ψυχῆς ἱστορία (*De an.* 402a 4) is characterized as purely theoretical research; see also *PA* 639a 12, 641a 29 and Theoph. fr. 224–225, 230 FHSG.

82 Arist. *Rhet.* 1360a 36. Louis, P. Le mot ἱστορία chez Aristote, *RPh* 29 (1955) 39–44; Weil. *Aristote*, 90f.

titled Ἱστορικὰ ὑπομνήματα, we can easily see that Theophrastus deals with natural history, whereas his colleagues in the Lyceum study historico-biographical material. Like natural history, political history describes particular things, τὰ καθ' ἕκαστον, such as what Alcibiades did or suffered.[83] But in contrast to natural history, political history does not possess its own theoretical counterpart, though the facts and evidence collected by history appear useful for practical sciences, such as politics.[84] Being restricted *mostly* to the particular, history does not explore general causes and, therefore, from the viewpoint of philosophy, is not a true science. This is the sense in which the well-known words that poetry is more 'philosophical' and 'serious' than history (*Poet.* 1451 b 2f.) are often interpreted.[85] Whatever Aristotle meant by this remark, it is true that none of the passages in which he discusses sciences includes history, in contrast to grammar, the science that studies all (articulated) sounds (*Met.* 1003b 19f.). One of the reasons for this marginal position of history is that a few Aristotle's remarks on this subject refer to the traditional type of political history[86] rather than to history as it was practiced in the Lyceum. Otherwise, it remains incomprehensible why Aristotle and his students paid so much attention to studies that qualify as historical both from the ancient and from the modern point of view.

Thus, when trying to understand why and how Aristotle and his students studied the history of knowledge, we should rely not only on their theoretical views on ἱστορία, but on their actual practice as well. Let me start with some parallels. Along with theoretical sciences, the other two types of sciences, practical and productive – rhetoric, poetics, and music – also became subjects of historical and antiquarian studies. In his Τεχνῶν συναγωγή, Aristotle considered the history of rhetoric with special attention to the *prōtoi heuretai* and their discoveries. He started with the founders of rhetoric, Corax and Tisias, and ended, as it seems, with Isocrates.[87] Such an approach, which we know from Glaucus' book *On the Ancient Poets and Musicians*, was also characteristic of Heraclides Ponticus' Συναγωγὴ τῶν ἐν μουσικῇ (fr. 157–163) and the first book of Aristoxenus' *On Music* (fr. 78–81, 83). Aristotle's dialogue *On the Poets* is a work exploring the history of poetry. Here much attention is given to

---

83 Arist. *Poet.* 1451b 3f.; cf. ἱστορία περὶ ἕκαστον, contrasted to the study of the causes (*HA* 491a 7–14).

84 A politician and a lawmaker should know both the past and the laws of other peoples; that is why the works in geography and history are useful (*Rhet.* 1360a 31f.).

85 See e.g. Zoepffel, R. Historia und Geschichte bei Aristoteles, *AHAW* no. 2 (1975) 37. Cf. Fritz, K. von, *Gnomon* 52 (1977) 345f.; Huxley, G., *CR* 28 (1978) 89f.; Kinzl, K., *Gymnasium* 85 (1978) 99f.

86 *Rhet.* 1360a 37; 1409a 28; *Poet.* 1451b 3, 6, 1459a 17f.

87 Fr. 136–141 Rose = fr. 123–134 Gigon. See Leo, *op. cit.*, 49, 99f.; Blum, *op. cit.*, 46. Cf. *SE* 183b 15f., where this history is presented in compressed form: at first the anonymous *prōtoi heuretai* of rhetoric, followed by Tisias, Thrasimachus, Theodorus, and others.

the founders of various genres and their relative chronology.[88] Note that these outlines are devoted to the disciplines that Aristotle studied in his systematic treatises, *Rhetoric* and *Poetics* and Aristoxenus in his *Elements of Harmonics*; historical evidence collected in the Τεχνῶν συναγωγή was later used in the *Rhetoric*.[89]

At the same time, not every historical or antiquarian work of Aristotle and his students has to be directly related to their systematic treatises or conceived as a preliminary for subsequent theoretical analysis. Some of these works do not presuppose any specific theoretical goal, others are only indirectly related to the systematic writings, still others were written when the corresponding systematic treatise (or treatises) had been already finished. Following the first attempts made by Hippias and Hellanicus, Aristotle compiled lists of Olympic victors, the victors in sport and musical agones in Delphi, and the victors in dramatic agones at the Dionysia and Lenaea.[90] He also compiled (probably with the help of his students) the so-called Διδασκαλίαι, an extensive list of all the tragedies, comedies, and satyr plays performed at these artistic festivals, dating them by the names of the Athenian archons.[91] Whereas the lists of Olympic and Pythian victors were compiled irrespective of any further theoretical task, the Διδασκαλίαι was used in the study of the Attic tragedy in Aristotle's *Poetics*, though significance of this collection cannot be reduced to that of preparatory work, subordinate to predetermined theoretical task.

Another, more extensive and time-consuming Peripatetic project, preparing the 158 historical outlines of the constitutions of the various Greek cities, demonstrates a rather complicated relationship between historical and systematical studies. Books IV–VI of Aristotle's *Politics* rely on this collected and arranged material, yet the other books are mostly theoretical in character and were written before the project started. The surviving Aristotelian *Athenian Constitution* consists of two parts: the first gives an outline of the development of the Athenian constitutional system (chapters 1–41), the second describes the main principles of its functioning (chapters 42–69). It should be stressed that *Athenian Constitution* is a historical work of independent value addressed to a broad audience, and not just a dossier.[92] Certainly, historical subject matter and narrative form do not exclude subsequent systematization, but rather, in the case of the constitutions, presuppose it (Arist. *EN* 1181b 12ff.). It is revealing, however, that *Athenian Constitution* has been written after the *Politics*, ca. 329–322, and that quite often it gives more accurate and precise treatment of the historical events mentioned in the *Politics*. The preserved fragments of other constitutions show an obvious predominance of separate *Geschichten*, i.e., interesting

---

88 Fr. 70–77 Rose = fr. 14–22 Gigon; see *prōtos heuretēs* in fr. 14, 15, 17 Gigon.

89 Blum, *op. cit.*, 46; Gigon. *Aristotelis fragmenta*, 390.

90 D.L. V, 26 No. 130–131, fr. 615–617 Rose = fr. 408–414 Gigon; D.L. V, 26 No. 134–135. Weil. *Aristote*, 131f.; Blum, *op. cit.*, 23ff.

91 D.L. V, 26 No. 137 = fr. 415–462 Gigon. Weil. *Aristote*, 137f.; Blum, *op. cit.*, 24ff.

92 Dovatour, A. I. *Aristotle's* Politics *and* Polities, Moscow 1965, 296 (*in Russian*).

stories.[93] Even if this correlation is accidental, there is no doubt that Aristotle recognized an independent value and interest in these stories and hardly regarded them only as material for further generalizations. On the other hand, as the title of the whole collection suggests, the constitutions were classified in accordance with Aristotle's view of various forms of government,[94] which points out a conceptual framework of the project.

Aristotle's students demonstrate an even keener interest in historical studies. Callisthenes, Aristotle's grand-nephew and student who helped him to compile the list of Pythian victors, was a historian, as were two other Peripatetics, Clytus of Miletus and Leon of Byzantium.[95] Theophrastus' fellow student and friend Phanias of Eresus wrote historical works *Pritanes of Eresus* (fr. 17–19) and *On the Sicilian Tyrants* (fr. 11–13). The student of Theophrastus, Praxiphanes, dealt with Thucydides in his Περὶ ἱστορίας (fr. 18). The content of Theophrastus' own Περὶ ἱστορίας (D. L. V, 47) is unknown, but it is more likely that this book was on 'history' than on 'research'.[96] Aristoxenus' biographies are to some extend connected to the theoretical views of the Lyceum,[97] but more often than not they reveal his personal rather than philosophical preferences.

History as a purely narrative or descriptive genre, restricted to the particular and detached from considering general regularities, was not the only option for the Lyceum. Along with Eudemus, Dicaearchus is an especially interesting case, inasmuch as in his important works he developed Aristotle's views on the historical progress.[98] His *Life of Hellas* (fr. 47–66) was the first general cultural history, modeled after contemporary universal history and dealt, apart from particular events and individuals, with the general stages of the development of civilization. Dicaearchus for the first time introduces and causally explains the transition from gathering to cattle breeding and further to agriculture (fr. 47–51).[99] His view on the moral decline as inseparable from the economical progress can be traced to Plato.[100] Dicaearchus' other work, *On the Destruction of the People* (fr. 24), seems to be related to Aristotle's general conception of

---

93 See e.g. fr. 504, 512, 532, 549, 558, 583 Rose. Dovatour, *op. cit.*, 149.

94 "Constitutions of 158 cities arranged by the type (κατ' εἴδη), democratic, oligarchic, aristocratic, tyrannical" (D. L. V, 26). κατ' εἴδη was suggested by P. Moraux, κατ' ἰδίαν by I. Düring.

95 See *Die Philosophie der Antike*, Vol. 3, 566.

96 His praise of Herodotus' and Thucydides' style (fr. 697 FHSG) may come from this work (Regenbogen, O. Theophrastos, *RE Suppl.* 7 [1940] 1526). See also Wehrli, F. Praxiphanes, *Die Philosophie der Antike*, Vol. 3, 603.

97 Leo, *op. cit.*, 99 f., 102 f.; Dihle, A. *Die Entstehung der historischen Biographie*, Heidelberg 1987.

98 Zhmud, L. Dikaiarchos aus Messene, *Die Philosophie der Antike*, Vol. 3, 568 ff. On Dicaearchus' historical approach to philosophy, see White, S. A. *Principes sapientiae*: Dicaearchus' biography of philosophy, *Dicaearchus of Messana*, 195–236.

99 Cf. *VM* 3; Thuc. I, 2 ff.

100 Schütrumpf, E. Dikaiarchs Βίος Ἑλλάδος und die Philosophie des vierten Jahrhunderts, *Dicaearchus of Messana*, 269 ff.

history, according to which humanity is eternal, while different natural disasters separate one civilization from another.[101] Having discussed various natural catastrophes, Dicaearchus claims that more people have been destroyed by the "attack of people" (wars, revolts, etc.) than by any other evil. His conclusion that the people mostly help as well as damage the other people is close to the view, expressed in Aristotle's *Politics* (1253a 31ff.).

Thus, relationship between history and theory cannot be conceived as strictly unilateral, insofar as many historical works of the Peripatetics were not simply chronologically arranged collections of facts intended for further theoretical analysis, but proceeded from certain philosophical premises and relied on Aristotle's conception of the progress of civilization and its separate branches (τέχναι, philosophy, sciences, etc.).

4. The aims of the historiographical project

The immediate tasks of the historiographical project were collection, systematization, and preliminary analysis of the evidence, related to the historical development of the theoretical sciences. But whereas these immediate tasks are quite securely reconstructed on the basis of sources available to us, the further use of the collected and systematized material remains a matter of speculation. The problem is that this project falls in the last decade of Aristotle's life, and what was to follow afterward remained apparently unrealized owing to his escape from Athens and sudden death. Neither his preserved works, nor fragments and titles of the perished writings indicate that he used the evidence collected and arranged by his students.[102] Hence, we can only guess what kind of new knowledge did he expect to discover by directing his students to the history of philosophy and science. Did Aristotle, relying on this material, intend to revise some particular theories in physics or theology? Did he regard the project as relevant to his theory of science? Or did the establishment and systematization of the facts related to the history of knowledge possess in his eyes a value of their own, irrespective of any further use?

The problem does not get any easier if we abandon the idea of the common project initiated by Aristotle and treat the respective works of the Peripatetics separately. What were then Theophrastus' and Eudemus' objectives in collecting and systematizing the opinions of the physicists and theologians and the discoveries of the mathematicians? Did they intend to build some further theories on this material or to use it in their systematical works, in the way Aristotle's *Politics* used the evidence of the constitutions? If so, clear traces of such

---

101 *Cael.* 270b 16–24; *Mete.* 339b 25–30; *Met.* 1074b 10–13; *Pol.* 1269a 5ff., 1329b 25–33.

102 Cf. Gigon, O. Die ἀρχαί der Vorsokratiker bei Theophrast und Aristoteles, *Naturphilosophie bei Aristoteles und Theophrast*, ed. by I. Düring, Heidelberg 1969, 114ff.; Mansfeld. Aristotle, 73 n. 29.

theories or works are not to be found in their legacy. It might be the case, therefore, that no such theories or works were really intended. Indeed, granted that the objectives of the Peripatetics were mostly historical, we do not need to postulate any specific 'final goal', external to the historiographical project. Both the project as a whole and its separate parts possess a genuine value of their own, imparting a historical meaning to contemporary philosophy and science as the last stage in the long quest for truth.

As for the further use of collected and systematized material, it could have been intended for various historiographical and theoretical purposes. Although we cannot pinpoint a specific theoretical goal for the whole project or for its separate parts, we can at least exclude the least plausible hypotheses suggested on this point. To them belongs, first of all, the interpretation of Aristotelian dialectic as simultaneously both a source and a goal of the physical doxography.[103] Nor is the interconnection between the systematic *method* of doxography and its systematic *purposes* obvious.[104] In his theoretical works, Aristotle often used historical material preliminarily arranged by him chronologically. The chronological method of organizing the material – the method characteristic, although in varying degrees, of all parts of the Peripatetic project – can be regarded as an important indicator of its historical orientation. Meanwhile, from the point of view of Aristotelian ἱστορία, the historical and the systematic approaches hardly contradict each other; rather, they are different methods of bringing facts into a system. The specific feature of history was (and still is) that it allows and even encourages the chronological principle of organizing facts, whereas natural history employed other methods. In the contemporary humanities, chronology and systematics often complement each other: the history of literature unites literary works by genres, the history of philosophy groups thinkers according to schools, and the order of consideration may not agree with the chronology of individual authors and works. It is easy to imagine a history of Greek culture consisting of chapters on religion, mythology, literature, etc., or an economic history of Rome with chapters on trade, agriculture, slavery, etc. The extent to which each chapter can be organized chronologically depends on the material and intentions of the author. Therefore, the fact that Eudemus employs only a chronological approach, while Theophrastus and Meno combine this with a systematic approach, does not undermine the historical orientation of doxography.

Both before and after Aristotle, scientists, historians, and philosophers described the opinions of their predecessors and disputed them[105] without attach-

---

103 See above, 133f., 134 n. 72. On the origin of the Peripatetic doxography, see Zhmud, L. Doxographie in ihrer Beziehung zu den anderen Genres der antiken Philosophiegeschichte, *Die Philosophie der Antike*, Vol. 1: *Frühgriechische Philosophie*, ed. by H. Flashar et al., Basel 2007 (*forthcoming*).

104 Cf. Runia. What is doxography?, 51.

105 Hdt. II, 20–23 presents one of the earliest doxographical overviews on the Nile's floods.

ing to this procedure any independent value that might be described in terms of historical interest. The doxographical overview at the beginning of the *Metaphysics* undoubtedly reveals such an interest.[106] Certainly, the material of Aristotle's doxographical overviews, the opinions of physicists, was directly related to his scientific and philosophical views, so that it is quite natural to expect here much aberration and subjectivism. But even these features reflect, to a large extent, his *historical* views, rather than prove that ideas of his precursors were important to him *only* insofar as they could provide material for the construction of his theories.[107] As W. Jaeger acutely observed, Aristotle "was the first thinker to set up ... a conception of his own position in history"; he invented "the notion of intellectual development in time and regards even his own achievements as the result of an evolution dependent solely on its own laws."[108] From this perspective, the opinions of the Presocratics were, naturally, regarded as a preliminary stage for Aristotle's own theories. No wonder Theophrastus interpreted them in Peripatetic terms, quite often making the earlier thinkers answer questions that they had never formulated themselves. This prompts us to treat Theophrastus' interpretations with much caution, but hardly casts doubt on the historical orientation of his *Physikōn doxai.*

Theophrastus' *Physikōn doxai* and especially Eudemus' history of the theoretical sciences were much less related to the problems of their own theoretical works than Aristotle's doxographical overviews. Eudemus was neither a mathematician, nor an astronomer, nor a theologian, and could not regard the heroes of his histories as his predecessors. Theophrastus was a φυσικός, but his own *Physics*, if we judge from the preserved fragments, followed Aristotle's *Physics*, i.e., developed contemporary φυσικὴ ἐπιστήμη, leaving aside the φυσικῶν δόξαι that he had collected.[109] It is revealing that Democritus is nearly the only Presocratic mentioned in the fragments of Theophrastus' *Physics* (fr. 238 FHSG), except for the controversial reference to Xenophanes.[110] Therefore, the thesis that Theophrastus systematized the opinions of

---

106 See below, 154f.

107 As H. Cherniss thought (*Aristotle's criticism of Presocratic philosophy*, Baltimore 1935, 347ff.).

108 Jaeger. *Aristotle*, 3.

109 Though it is possible that Theophrastus' *Physics* was written before *Physikōn doxai*, doxographical passages in his systematic works are generally very rare: Gottschalk, H. Rec., *Gnomon* 39 (1967) 20. Cf. below, 144 n. 115, 158 n. 166. Eudemus' *Physics*, although it contains many such passages (see below, 152 n. 141), followed Aristotle's *Physics* even more closely than Theophrastus', sometimes nearly paraphrasing it. It is unlikely that Eudemus' used Theophrastus' doxography.

110 Θεόφραστος ἐν τοῖς Φυσικοῖς γέγραφεν (fr. 232 FHSG). This reference can be either from Theophrastus' *Physics* or from his *Physikōn doxai.* See Steinmetz, P. *Die Physik des Theophrastos von Eresos*, Bad Homburg 1964, 334ff.; Sharples, R.W. Theophrastus on the heavens, *Aristoteles Werk und Wirkung*, ed. by J. Wiesner, Vol. 1, Berlin 1985, 577–593; Mansfeld. *Studies*, 147ff.; Runia, D. Xenophanes or

the Presocratics in order to return to them once again in his theoretical discussion of the respective physical problems finds no corroboration.

The following also testifies against this thesis: from the Peripatetic point of view, the problems presented in Theophrastus' doxography are mostly among those already *solved*. It is hard to imagine that Aristotle commissioned his student to collect the opinions of Anaximenes, Xenophanes, Heraclitus, Anaxagoras, and others on the moon's eclipses in order to rethink this question again. Most of his physical treatises were written before Theophrastus started to collect material for the *Physikōn doxai*, and it is unlikely that Aristotle intended to rewrite them after his student had finished his work.

Another confirmation that Theophrastus' doxography deals mainly with already solved physical problems comes from Aristotle's little-known treatise *De inundatione Nili*, preserved in an abridged Latin translation of the thirteenth century.[111] The abridged version is mainly doxographical; to what extent this corresponds to the original version, which included three books, is hard to say. At the beginning of this work, Aristotle poses the problem: "Why do the Nile's floods, unlike those of all other rivers, come in the summer?" Then he cites and criticizes his predecessors' opinions: in particular, those of Thales, Diogenes, Anaxagoras, Nicagoras of Cyprus, and Herodotus.[112] In the doxographical part of the treatise our attention is especially attracted by one detail, typical of the first chapter of the *Physikōn doxai*: Aristotle gives not only the names of the authors, but also their patronymics and birthplaces.[113] Yet in contrast to doxography, the ending of this work contains a solution to the problem: "The problem does not exist anymore", notes Aristotle, and, referring to observations, claims that the Nile's floods are caused by the Ethiopian seasonal rains

---

Theophrastus? An Aëtian *doxographicum* on the sun, *Theophrastus of Eresus*, 112–140.

[111] On the authorship and history of this treatise, see Diels, *Dox.*, 226f.; Partsch, J. Des Aristoteles Buch "Über das Steigen des Nil", *ASGW* 27 (1909) 553–600; Balty-Fontaine, J. Pour une édition nouvelle du "Liber Aristotelis de inundatione Nili", *Chronique d'Égypte* 34 (1959) 95–102; Bonneau, D. Liber Aristotelis De inundatione Nili, *Etudes de Papyrologie* 9 (1971) 1–33; Bollack, M. *La raison de Lucrèce*, Paris 1979, 539f. Bollack counters Steinmetz's attempt (*op. cit.*, 278ff.) to ascribe this treatise to Theophrastus. Editions of the text: Arist. fr. 248 Rose; *FGrHist* 646 F 1; Bonneau, *op. cit.*, 3–7; Jacoby (*FGrHist* 646 T 1–2) collects references by ancient authors to the Greek text of Aristotle.

[112] The theories of Euthymenes of Massalia, Oenopides, and possibly Ephorus and Plato are presented anonymously. In the chapter on Nile's floods (IV, 1), Aëtius cites the opinions of Thales, Euthymenes, Anaxagoras, Democritus, Herodotus, Ephorus, and Eudoxus. Cf. above 141 n. 105.

[113] Thales, son of Examyes, from Miletus (3), Diogenes, son of Apollothemis, from Apollonia (4), Anaxagoras, son of Hegesibulus, from Clazomenae (5). Nicagoras' patronymic is not mentioned (9), and Herodotus is presented only by name (10). Cf. below, 161.

that filled the river to overflowing (sec. 12).[114] It is obvious that he did not turn to the history of this problem in order to solve it on the basis of earlier opinions. He addressed this issue *after* gaining access to the new empirical evidence that finally settled it. The *Physikōn doxai* does not contain 'correct answers', i.e., the solutions to the problems posed by the Presocratics, because these answers belong to φυσικὴ ἐπιστήμη and are given in the respective theoretical works of Aristotle and Theophrastus; moreover, including them would have made the already voluminous compendium still larger. But the principal approach to the *doxai* remains the same as in *De inundatione Nili*: their collection does not open a discussion on the specific physical problem, but gives *a historical retrospective* of this discussion after the problem itself appears to have been solved.[115]

Unlike the case with physics, in mathematics and mathematical astronomy Aristotle and his disciples could not claim the independent solution of problems; at best, they could register those that contemporary specialists considered to have already been solved. Nor did the principles of the theologians hold any but purely historical interest for them. As for the causes of diseases, the many remarks on this subject scattered in Aristotle's works seem to suggest that he thought the main cause was the imbalance between the natural qualities (warm, cold, wet, and dry) as a result of bad nutrition, overstraining, or some external factors.[116] In its general form, this doctrine had been proposed by Alcmaeon (24 B 4) and was later developed in the works of the Hippocratics and the physicians of other schools, such as Philistion of Locri (An. Lond. XX, 25–40). In Meno's medical doxography, a similar point of view was represented by a group of physicians who explained the diseases proceeding from the four elements (ibid. XIV, 9), connected, in turn, with the four qualities. Another group, twice as large as the first, saw the cause of the diseases in the so-called residues, περιττώματα (ibid. XIV, 7). It is quite probable that Aristotle, who did not consider himself a specialist in medicine, tended in this case to rely on

---

114 The Greek quotation from Aristotle, οὐκέτι πρόβλημά ἐστιν, preserved in an anonymous biography of Pythagoras cited by Photius (*Bibl.* 242, 441 a 34 = *FGrHist* 646 T 2 a), agrees with the Latin translation: jam non problema videtur esse (Partsch, *op. cit.*, 574). Aristotle apparently relied on the results of the expedition organized by Alexander to solve the problem of the causes of the Nile's floods. This allows us to date the treatise to 330–327 BC (Bonneau, *op. cit.*, 21 f.).

115 The order of the subjects in the chapter on meteorological *doxai* (Aët. III, 3–7) agrees with the sequence known from Theophrastus' *Meteorology* (Daiber, H. The *Meteorology* of Theophrastus in Syriac and Arabic translation, *Theophrastus of Eresus*, 166–293). Typically, no names of the Presocratics appear in the *Meteorology* itself; their theories are integrated in Theophrastus' own doctrine. Cf. above, 142 n. 109.

116 Tracy, T. *Physiological theory and the doctrine of the mean in Plato and Aristotle*, The Hague 1969. On excess and defect as causes of diseases, see also Manetti, *op. cit.*, 126f.

the opinions of the professionals more than he normally did in other fields. Still, the collection of their opinions here, as everywhere, answered the same purpose as Theophrastus' doxography, showing the difficult way to the 'truth' that was already at Aristotle's disposal. [117]

Much of the data collected by Theophrastus and Eudemus must have been known to Aristotle long before. The very fact that these data were mentioned and analyzed by the head of the Lyceum was for the Peripatetics one of the reasons to pay attention to them.[118] Theophrastus, for his part, proceeded from his own numerous works dealing with individual Presocratics.[119] Important, therefore, was not only the collection of separate facts from which to draw conclusions, but the totality of them organized to reveal the intrinsic logic of scientific and philosophical development. The historiographical project presupposed a certain completeness of evidence, which indeed was achieved in most cases. Apart from the physicists mentioned in Theophrastus, we know almost no one else from the relevant period.[120] The same is true for the mathematicians: Eudemus gives the names of practically all the geometers up to his own time. His list of Greek 'theologians' is indeed complete, and his account of the Oriental theogonies is highly valuable.[121] Almost half of the names given by Meno are not found in other sources. Theophrastus and Meno missed neither those whom Aristotle obviously held in low regard (like Hippon), nor those whose names he preferred not to mention at all (like Philolaus).[122] As a result, here, as in the corpus of the constitutions, many more data were collected than the most detailed analysis of every particular problem actually needed, which testifies to the significance of the facts themselves. In Aëtius we find, e.g., 15 different answers to

---

117 See the similar conclusion in Manetti, *op. cit.*, 129.

118 For parallels between Aristotle's mathematical passages and Eudemus' *History of Geometry*, see below, 197 f., 202 f. The material of the *History of Theology* overlaps that of the *Metaphysics* and *On Philosophy* (see above, 131 n. 53). On Theophrastus' dependence on Aristotle, see McDiarmid, J. B. Theophrastus on Presocratic causes, *HSCPh* 61 (1953) 85–156; Mansfeld. Aristotle.

119 He wrote about Anaximander, Anaximenes, Anaxagoras, Empedocles, Archelaus, Democritus, Diogenes, and Metrodorus (D. L. V, 42–44, 49). Besides, he could have used the material of Aristotle's monographs on Xenophanes, Alcmaeon, and Melissus (D. L. V, 25).

120 Interestingly, Theophrastus wrote a special work on Hippocrates of Chios' student Aeschylus (D. L. V, 50 = 137 No. 42 FHSG) but seemed not to mention him in the *Physikōn doxai* (cf. above, 132 n. 63); on Menestor, see below, 158 n. 166.

121 Casadio, G. Eudemo di Rodi: Un pioniere della storia delle religioni tra Oriente e Occidente, *WS* 112 (1999) 39–54.

122 On Hippon, cf. 38 A 6 (= Arist. *Met.* 984 a 3) and A 3–4, 10, 13–14, etc. (from Theophrastus), A 11 (from Meno). Aristotle only once refers to Philolaus' oral dictum (*EE* 1225 a 30) and ascribes his astronomical system to anonymous 'Pythagoreans'. Theophrastus and Meno attribute specific theories to Philolaus (44 A 16–23, 27).

the question of the nature of the sun (II, 20);[123] none of them concurs with the solution offered by Aristotle and Theophrastus (the sun consists, like the other celestial bodies, of the fifth element, ether).[124] What was the purpose of collecting them? It is very unlikely that they intended to turn to this problem again after enriching themselves with new – or rather with old – knowledge. This collection was mainly of historical interest, showing the difficult path to the truth that was finally revealed in Aristotle's and Theophrastus' physical teaching, i.e., outside of the *Physikōn doxai*.[125]

Aristotle attributed to facts an independent value. These acquire even greater significance when, properly selected and arranged, they help in the search for the 'causes' (διότι). Although explaining the 'causes' was not among the immediate tasks of the historiographical project, this does not mean that the Peripatetics limited their studies to particular facts. The discoveries of the scientists and the opinions of the philosophers, doctors, and theologians not only paved the way for conclusions of a general character, they themselves had already been selected and arranged in accordance with Aristotle's theoretical views on science and its development. This means that the conclusions, as so often, were known in advance. Thus, e.g., at the beginning of his *History of Geometry*, Eudemus states a general rule of cognitive evolution from αἴσθησις to λογισμός and further to νοῦς (fr. 133), and this is not the only case in which philosophical ideas were applied to historical material (5.4).

Speaking of the original objectives of the Peripatetic project, one has to bear in mind one important circumstance. In terms of their literary form, neither Eudemus' histories, nor Theophrastus' doxography can be regarded as 'esoteric' writings, *pragmateiai*, intended only for use in the Lyceum. In spite of the certainly not easily digestible subject matter of these works, both their form and their subsequent fate strongly imply that, just like the *Constitutions*, they were addressed to a wider audience than the Peripatetic community.[126]

---

123 Thales, Anaximander, Anaximenes, Xenophanes, Hecataeus of Miletus, Parmenides, Heraclitus, Anaxagoras, Empedocles, Philolaus, Democritus, Antiphon, Diogenes, Metrodorus, Plato.

124 Contrary to Steinmetz (*op. cit.*, 116ff., 161ff.), it is hardly the case that Theophrastus abandoned the idea of the heavenly ether. See Sharples, R.W. *Theophrastus of Erasus. Commentary*, Vol. 3.1: *Sources on physics*, Leiden 1998, 85ff. and fr. 158, 161a FHSG.

125 Gigon, O. Die Geschichtlichkeit der Philosophie bei Aristoteles, *Archivio di filosofia* 23 (1954) 117, aptly called Theophrastus' *Physikōn doxai* "geschichtliche Ergänzung zur eigenen Physik".

126 On Eudemus' *History of Geometry* as a literary work, see Becker, O. Zur Textgestaltung des Eudemischen Berichts über die Quadratur der Möndchen durch Hippokrates von Chios, *Q & St* B 3 (1936) 416f. Eudemus' histories were his only works known in the Hellenistic period (5.1). According to a plausible reconstruction, *Physikōn doxai* was available to Epicurus already ca. 306 (Sedley, D. *Lucretius and the transformation of Greek wisdom*, Cambridge 1998, ch. 6).

5. Eudemus' history of science

Let us turn now from the tasks of the Peripatetic project to the various forms in which its specific parts came to be realized. What are the differences between Eudemus', Theophrastus', and Meno's approaches to their branches of knowledge and what are the reasons for these differences? Being of particular interest to us and serving as a starting point, Eudemus' history of science was the most historical part of the project. The history of theology can be placed somewhere between the history of science and a much more systematically organized doxography. Is it explained by the specifics of the material itself, the differences in approach to mathematics, physics, and theology, or some other reasons? What made chronology the main principle of the organization of material in Eudemus' historiography of science? Was Γεωμετρικὴ ἱστορία the history of advancing knowledge, or rather, as Eggers Lan believed, "a classification of authors by geometrical subjects"?

Let us start with the titles of Eudemus' historical treatises cited in several authors. The most exact variant is given by Simplicius: Γεωμετρικὴ ἱστορία (fr. 140) and Ἀστρολογικὴ ἱστορία (fr. 148); Porphyry quotes Ἀριθμητικὴ ἱστορία (fr. 142). The list of Theophrastus' works (251 No. 2 FHSG) includes (Eudemus') Περὶ τὸ θεῖον ἱστορία. But what does Γεωμετρικὴ ἱστορία actually mean: 'geometrical research', 'inquiry into geometry', or, still, the 'history of geometry' proper? It is obviously not a mathematical treatise: Eudemus did not consider mathematical problems as such, but the way they were solved, in historical succession, by different mathematicians. (This holds good for the history of theology as well.) The subject of his study accordingly coincides with the subject of the history of science as we understand it now. The titles of Eudemus' works are, therefore, not as close to ἱστορία as 'research' – the meaning it takes in the titles of Aristotle's and Theophrastus' treatises (Ἱστορία ζῴων and Περὶ φυτῶν ἱστορία) – as they are to ἱστορία in a more narrow sense, attested already in Herodotus (VII, 96) and understood usually as a "written account of one's inquiries, narrative, history".[127] There is no doubt that in Eudemus' times ἱστορία *could* mean what we now call history[128] and further that such rendering corresponds best to the historical character of his writings. But if Eudemus' book had been entitled Περὶ γεωμετρίας or had remained untitled, it would not make any important difference for defining its genre. The fact that Thucydides' work had no title and the author himself never used the word ἱστορία does not prevent us, any more than it did the Greeks themselves, from relating it to the historical genre. In the case of Eudemus, we also have a

---

[127] *LSJ* s.v. II; Hornblower, S. *Thucydides*, Baltimore 1987, 9.

[128] On the usage of ἱστορία and ἱστορικός in the sense of 'history' and 'historian', see Arist. *Rhet.* 1359b 30f., 1360a 30–36; *Poet.* 1451b 1–7, 1459a 21–24; Anaximenes (*FGrHist* 72 F 3, 9); Louis, *op. cit.*, 40f. Eudemus' older contemporary Ephorus entitled his universal history Ἱστορίαι; he was followed by Duris of Samos (born ca. 340). On Theophrastus' and Praxiphanes' Περὶ ἱστορίας see above, 139.

chance to verify how far our own notions correspond to the ancient ones. The *Catalogue of geometers*, which goes back to Eudemus, ends with a revealing sentence: "those who have written histories (of geometry) bring to this point their account of the development of this science."[129] These words clearly show that ancient readers, too, regarded Γεωμετρικὴ ἱστορία as a historical account of the progress of geometry.

The subject of Eudemus' works on the history of science was the development of three *mathēmata* – geometry, arithmetic, and astronomy – from their emergence at the beginning of the sixth century to the middle third of the fourth century; nor did the author fail to mention their oriental origins. The last mathematicians named in the *History of Geometry* belong to the generation of Eudoxus' students. This agrees with the chronological framework of the *History of Astronomy*, which begins with Thales (fr. 143) and ends with Eudoxus' student Callippus (fr. 149), who was Eudemus' contemporary. In the *Catalogue*, all the names of the mathematicians are given in chronological order, often with indications of who was older and younger, who was whose student, etc. Besides, we know that Eudemus considered the Pythagoreans in the first book of the *History of Arithmetic* (fr. 142), Hippocrates' quadrature of the lunes in the second book of the *History of Geometry* (fr. 140), and Eudoxus' and Callippus' theories in the second book of the *History of Astronomy* (fr. 148–149).

Wehrli, in contrast, believed that Eudemus' material was not arranged in chronological order, following the succession of the mathematicians, but rather in terms of the history of problems.[130] This principle is indeed convenient for the analysis of the approaches of several generations of mathematicians to the solution of the same problem; Pappus often used it when dealing with famous problems of the past, such as duplicating the cube, etc. The material of the early Greek mathematics accessible to Eudemus, however, was too various and abundant to be reduced to a thin thread of problems running throughout its history, which made the historico-problematic approach to it thoroughly inadequate. Nor does this approach seem very convenient for the historical treatment of whom such and such mathematical discovery belongs to, which was, as I will show, one of Eudemus' main goals. It is obvious that such figures as Thales, Mamercus, Pythagoras, and Oenopides interested him owing to their discoveries, and not by virtue of the fact that they had been working on the same problems. Oenopides' discoveries in geometry, by the way, can hardly qualify as a "maßgebender Gedanke" (5.4); Eudemus probably mentioned them for the simple reason that he knew that Oenopides made them. As for Mamercus, Eudemus could hardly know anything about him apart from the fact that he was

---

129 οἱ τὰς ἱστορίας ἀναγράψαντες μέχρι τούτου προάγουσι τὴν τῆς ἐπιστήμης ταύτης τελείωσιν (Procl. *In Eucl.*, 68.4f. = Eud. fr. 133).

130 "Problemgeschichtliche Anordnung" (Wehrli. *Eudemos*, 119). "Der Stoff war nach Auftreten und Entwicklung der maßgebenden Gedanken, nicht nach Autoren geordnet" (ibid., 113). The last definition repeats Leo's words (*op. cit.*, 100), which referred to doxography, rather than to the history of science.

renowned as a geometer; it is only in a historical context that such a figure could have been mentioned.

It would not be surprising if one and the same mathematician, e.g. Hippocrates, appeared to be mentioned by Eudemus in connection with the problem of doubling the cube along with Archytas and Eudoxus, while figuring at the same time along with Antiphon in connection with squaring the circle. Such repeated references are typical for any history of science. Apart from that, Eudemus could well have deviated from the chronological principle of organizing the material while discussing problems that had preoccupied several generations of geometers. On the whole, however, as we see from the *Catalogue*, he proceeded from generation to generation, from teachers to their disciples, rather than from one problem to another.[131] It is difficult, e.g., to suppose that Eudemus should have considered together the three authors of *Elements* he mentions, rather than in chronological order. In exactly the same way, the *History of Astronomy* was not organized in accordance with the problems (the moon's eclipses, the order of the planets, the position of the earth, and so on), but in accordance with the *prōtos heuretēs* principle. Eudemus preferred the chronological arrangement even in the *History of Theology*, where much more limited material referred to the history of one single problem, the principles of the theologians, and easily allowed systematization by their number or type.

Since the formula πρῶτος εὑρετής is found, full or abridged, in practically every fragment of Eudemus' works on the history of science, they can be regarded, in their entirety, as a detailed answer to the question: 'who discovered what?'. In seven of the nine fragments of the *History of Geometry* that reached us under the name of Eudemus, we find either πρῶτος, or εὕρημα (εὕρεσις, εὑρίσκω) or both of them in combination.[132] Clear traces of this terminology have survived in the *Catalogue* as well: πρῶτον εὑρῆσθαι in connection with the invention of geometry in Egypt (*In Eucl.*, 64.18), πρῶτος and εὗρεν on Thales (65.7f.), ἀνεῦρε on Pythagoras (65.21), εὑρών and πρῶτος on Hippocrates (66.4f.), εὑρεῖν on Leon (66.22), πρῶτος on Eudoxus (67.2f.), and ἀνεῦρε on Hermotimus (67. 20f.). No less revealing are those testimonies on geometers that can be safely related to the *History of Geometry*.[133] Further,

---

131 Heath. *Elements* I, 38; Edelstein, *op. cit.*, 95. Pappus (*Coll.* IV, 272.15f.) and Eutocius (*In Archim. De sphaer.*, 57.13f.), on the contrary, could consider different solutions of the same problem without regard to their authors' chronology or even their names. See Knorr. *TS*, 77ff., 213ff.

132 τοῦτο τὸ θεώρημα εὑρημένον ὑπὸ Θαλοῦ πρώτου (fr. 135), εὕρεσις (fr. 136), εὑρήματα τῶν Πυθαγορείων (fr. 137), Οἰνοπίδου εὕρημα (fr. 138), Ἱπποκράτης καὶ Ἀντιφῶν ζητήσαντες ... εὑρήκασιν (fr. 139), ὑφ' Ἱπποκράτους ἐγράφησαν πρώτου (fr. 140), Ἀρχύτου εὕρησις (fr. 141).

133 See below, 170ff. In the material on Thales, πρῶτος and εὕρεσις are mentioned (*In Eucl.*, 250.20f.); in that on the Pythagoreans, εὕρημα (*Schol. in Eucl.*, 273.3–13, twice); on Oenopides, πρῶτος (*In Eucl.*, 283.7f.); on Hippocrates, πρῶτος and εὗρεν in Proclus (213.7f.) and πρῶτος in Eratosthenes (Eutoc. *In Archim. De*

πρῶτος εὑρετής is mentioned in five of the seven fragments of the *History of Astronomy*, whereas an excerpt of this work, analogous to but much shorter than the *Catalogue*, is entitled: τίς τί εὗρεν ἐν μαθηματικοῖς;[134]

We have noted already that the search for discoverers, which had mightily stimulated the formation of the history of culture, remained topical in the Lyceum as well. Some of the Peripatetics paid a direct tribute to the genre of heurematography in works bearing the standard title of *On Discoveries*.[135] Nor did Aristotle ignore this subject matter himself.[136] Considering that in philosophical biography discoveries are a recurrent topic, it could well go back to Aristoxenus, the founder of the genre.[137] His biography of Pythagoras ascribes to the philosopher the introduction in Greece of measures and weights and the identification of the Evening and the Morning Stars with Venus (fr. 24). A fragment of his work *On Arithmetic* says that Pythagoras advanced the science of numbers discovered by the Egyptian god Thoth (fr. 23). Extracts from other biographies by Aristoxenus (who wrote on Socrates, Plato, and Archytas) unfortunately do not contain any similar information, but his work *On Music* is brimming with references to various discoveries (fr. 78–81, 83), as is the history of music by Heraclides of Pontus (fr. 157–159, 163). Dicaearchus developed this subject in his works on musical agones (fr. 75–76, 85), and in his *Life of Hellas* did not fail to mention the discovery of horse-breeding by the Egyptian king Sesostris (fr. 57). Still more important is the subject of the first discoverers in the histories of various τέχναι written by Aristotle; it is presented in doxography as well.[138] Persistent interest in cultural novelties and their authors is thus typical of the majority of the historically oriented genres practiced in the Lyceum.

The influence the early heurematographic tradition exercised on these genres had various forms and gradations. While Peripatetic heurematography provides an example of the direct continuity of genre, individual references to discoveries in the context of a biography or a systematic treatise testify rather to the thematic continuity. The most interesting case seems to be where the prin-

---

sphaer., 88.18–23); on Democritus, πρῶτος (Archim. II, 430.5f.); on Archytas, πρῶτος (D. L. VIII, 83); and on Eudoxus, ἐξηύρηκεν πρῶτος (Archim. II, 430.1f.).

134 Fr. 144, 145, 146, 147, 148; Ps.-Heron. *Def.*, 166.23–168.12. = Eud. fr. 145.

135 Heraclides Ponticus (fr. 152), Theophrastus (fr. 728–734 FHSG), Strato (fr. 144–147).

136 Fr. 382, 479, 501, 600, 602 Rose = fr. 924 Gigon; see Eichholtz, *op. cit.*, 24f.; Wendling, *op. cit.*

137 Leo, *op. cit.*, 46f., 99f.; see above, 35 n. 60.

138 See above, 137f. Aristotle regarded Thales as the founder of physics (*Met.* 983b 20), Empedocles of rhetoric, Zeno of dialectic (fr. 65 Rose), and Socrates of ethics (*Met.* 1078b 17). Particularly numerous are the mentions of πρῶτος in the historical overview of the ἀρχαί in *Metaphysics* A: Hesiod or Parmenides (984b 23, 31), Empedocles (985a 8, 29), Pythagoreans (985b 23), and Xenophanes (986b 21). Theophrastus follows and develops these ideas (fr. 225, 226a, 227d–e, 228a FHSG).

ciple of *prōtos heuretēs* becomes the constitutive feature of a historical treatise, for example in Aristotle's Τεχνῶν συναγωγή (and, earlier, in Glaucus of Rhegium) or, after him, in Eudemus' history of science.[139] It is important to point out, though, that this principle *as such* is by no means identical to the historical approach. No wonder then that the Peripatetic heurematography features traditional discoverers, rather than real innovators of the historical epoch, let alone scientists.[140] The attention is invariably focused on cultural innovations as such, constituting a list that is not ordered systematically or chronologically. Unlike heurematography, the history of various arts and sciences (rhetoric, poetry, or mathematics) attempts to show the dynamic of their development and does not base the account on the list of discoveries, but on the chronology of their authors, thus giving it a historical perspective.

In the history of science, Eudemus could employ a chronological approach much more consistently than his colleagues, owing mostly to the character of his material. In fact, in this period, the cumulative development of the exact sciences was much more obvious than that of physics or medicine. The discoveries of mathematicians necessarily depended on what had been achieved before them: Hippocrates and Theaetetus relied on Pythagorean mathematics, Archytas and Eudoxus developed the theories of Hippocrates, etc. A mathematician could base his research on a solid foundation created by his predecessors and move further in his quest for the truth faster than the others. To be sure, Eudemus records in details some unsuccessful attempts to solve mathematical problems, e.g. Antiphon's squaring of the circle (fr. 139–140). It is, however, in the nature of mathematics that its history includes many more victories than failures, especially in comparison with other sciences. No wonder that, in the history of early Greek geometry, Eudemus encountered fewer cases in which many scientists *failed* to solve one and the same problem than in the history of physics. Each of the geometers mentioned by Eudemus could claim credit for real discoveries that allowed them to be listed among the *prōtoi heuretai*.

In the *History of Astronomy*, Eudemus, judging from the preserved fragments, also focused on the most important discoveries, starting from Thales' prediction of the solar eclipse (fr. 143–144) and ending with Callippus' modification of Eudoxus' system (fr. 149). Unlike the astronomical division of Theo-

---

[139] Leo, *op. cit.*, 47f., 49, even believed that the principle of *prōtos heuretēs*, developed by Aristotle in Τεχνῶν συναγωγή, came to be accepted as a method of research and presentation of material by all of his students, including Theophrastus.

[140] In Theophrastus (fr. 728–732, 735 FHSG) we find Prometheus, Demeter, Cadmus, Palamedes, the Oriental primogenitors of crafts, and Delas, one of the Phrygian Dactyls, with whom heurematography actually starts (see above, 24f.). But even in cases where historical figures are mentioned, the context remains traditional (fr. 733–736). Heraclides' only fragment ascribes the invention of coins to Pheidon of Argos (fr. 152). Strato argues against Ephorus, who was too enthusiastic about Scythian inventors, and discusses the authorship of the saying 'nothing beyond measure' (fr. 144–147).

phrastus' doxography, the *History of Astronomy* was selective and did not pretend to completeness; this allowed Eudemus to present the development of astronomy by charting the figures of the first discoverers in succession. Accordingly, he was not concerned with opinions, but with discoveries, whose importance was evaluated by the criteria of contemporary astronomy. From the earlier periods of astronomy, he selected those ideas that his own time considered to be true, or at least those that appeared significant in the progressive development of this science. It is revealing that the histories of geometry and astronomy trace the development of these sciences up to Eudoxus' students, i.e., two generations further than the *Physikōn doxai*, which concludes with Plato. This means that Theophrastus' point of departure was Aristotle's physical teaching. It is from this position that he treated and criticized all the previous opinions, whereas Eudemus relied on the criteria shared by the community of contemporary mathematicians and astronomers. Although both considered natural philosophy (φυσικὴ ἐπιστήμη) to be a science in the same way as mathematics, their historiographical writings reveal fundamental distinctions between philosophy and science, in spite of their theoretical views.

At the same time, I would not account for Eudemus' historical approach solely by recourse to the differences between the exact sciences and physics and to the peculiarities of their development. Eudemus' interest in the history of ideas is manifest outside mathematics as well. The fragments of his *Physics* contain an uncommonly large number of doxographical digressions, where ideas of Parmenides, Zeno, Melissus, Anaxagoras, Empedocles, the Pythagoreans, Archytas, and Plato are considered and/or criticized.[141] All these names (except for Archytas) can be found in Aristotle's *Physics* as well, but Eudemus, it seems, paid much more attention to historical details. Thus, he says at the beginning of his *Physics* (fr. 31) that Plato was the first to call the principles (ἀρχαί) στοιχεῖα and (if the rest of the fragment comes from Eudemus) that it was Aristotle who further found the concept of ὕλη.[142] Eudemus' *History of Theology* is obviously indebted to Aristotle's dialogue *On Philosophy*, whose first book shows the development of philosophy from the Persian Magi (whom Aristotle believed to be more ancient than the Egyptians) to Plato.[143] Although in discussions on the principles of theologians the topic of discoveries was rather irrelevant, Eudemus arranged their names in chronological order, which he might have considered most natural.

---

141 Fr. 31, 35–47, 53–54, 60, 65, 67, 75, 78, 82, 89, 110–111, 118. Gottschalk, H. Eudemus and the Peripatos, *Eudemus of Rhodes*, 25 ff.

142 See below, 232 n. 18.

143 The Magi and Egyptians (fr. 6 Rose = fr. 23 Gigon, with a parallel reference to Eudemus), Orpheus (fr. 7 a Rose = fr. 26 Gigon), the Seven Sages (fr. 3 a–b Rose = fr. 28–29 Gigon). Cf. above, 113 n. 154.

6. Doxography: between systematics and history

A different lot fell to the principle of *prōtos heuretēs* in doxography. It was predetermined, in many respects, by the purposes of Theophrastus' work and the character of his material. Doxography aimed to describe *all* the (relevant) *doxai*, and not only the true ones. In a word, its subject was the historically recorded opinions of physicists. This task dictated another method of arranging the material, differing from that employed in the history of science. Apart from the opinions true from the viewpoint of Peripatetic physics, which could be qualified as discoveries and presented in chronological order, Theophrastus had to register a great deal of *wrong* opinions. In physics, as in medicine, studied by Meno, firm proofs were often lacking and it was difficult to distinguish true opinions from false ones. The ideas of an earlier thinker could seem sounder than those suggested later, so that a 'progressive' scheme of things did not always work. Furthermore, Theophrastus had to deal with many *identical* opinions on the same problem, e.g. whether the cosmos is eternal or not.[144] In the history of geometry, such a situation was hardly possible: after a theorem has been proven, one can try for a more elegant or simpler proof, but no one would simply state *the same* as it regularly happened in physics. Thus, Eudemus gives several successive solutions of the problem of doubling the cube, which from the mathematical point of view are really different. In the history of astronomy, for any important discovery (e.g. that the moon reflects the sun's light or that the angle of the obliquity of ecliptic is equal to 24°), Eudemus apparently registered its immediate author alone, without mentioning all those who shared this view. Finally, Peripatetic physics covered a much wider range of problems than any mathematical science: it included matters that concern physics, astronomy, and meteorology, as well as psychology, physiology, embryology, and even geography (on the Nile's floods, Aët. IV, 1)

As a result, the number of various *doxai* – opposite, similar, or identical – that were included in the doxography, as well as the number of their authors, greatly exceeded the relatively limited material that Eudemus had worked on. Together with the distinctive features of the physical opinions that had to be fitted to the Procrustean bed of the Peripatetic categories, this factor largely predetermined the complicated structure of the *Physikōn doxai*, which combined several principles of arranging material. Some of them were used in the earliest doxographical accounts, others were developed by Aristotle. Herodotus, as far as we can judge, used the chronological principle,[145] Hippias ar-

144 Aët. II, 4 (εἰ ἄφθαρτος ὁ κόσμος): Anaximander, Anaximenes, Anaxagoras, Archelaus, Diogenes, Leucippus: the cosmos is perishable; Xenophanes, Parmenides, Melissus: the cosmos is eternal.

145 His doxographical overview of the theories explaining the causes of the Nile's floods (II, 20–23) contains the opinions (without mentioning any names) of Thales, Euthymenes of Massalia (rendered by Hecataeus), and Anaxagoras. Cf. *FGrHist* 1 F 302; 647 F 1, 5 and 2, 1–3 (= Aët. IV,1.1–3); Jacoby, F. Euthymenes von Massalia, *RE* 6

ranged the ideas in accordance with their supposed relationship and similarity, while Gorgias and Isocrates classified the material by the character and/or the number of the principles admitted by the given philosopher.[146] In *Physics* I, 2 Aristotle characterized the Presocratics' principles according to the scheme that goes back to Plato's method of division (διαίρεσις): there must be either one principle or many; if one, it must be either motionless or in motion; if many, then either limited (two, three, four, etc.) or unlimited plurality; if unlimited, then either one in kind or different in kind. Since Aristotle cites but a few names here (Parmenides, Melissus, Democritus), it is obvious that the systematic aspect of the doctrine of principles was more interesting to him than the historical one. In the doxographical section of *De anima* (I, 2), he pointed out that for the study of the soul "it is necessary to consult the views of those of our predecessors who have declared any opinion on this subject" (403b 20f.).[147] In this context, many more names and opinions naturally appear,[148] whose individual features make the consistent application of the diaeretical scheme practically impossible.[149] Although Aristotle uses different systematic methods at once (or, rather, owing to this), the results of his systematization do not look very convincing here, while there is no chronology or indications of the historical links between separate theories.

The doxographical survey in *Metaphysics* A 3–7, which, along with the treatise *On the Nile's Floods*, can be regarded as one of the most important models for the *Opinions of the Physicists*,[150] appears completely different. In

(1907) 1507–1511; Gigon, O. *Der Ursprung der griechischen Philosophie*, Basel 1945, 48ff.; Lloyd, A.B. *Herodotus Book II. Commentary 1-98*, Leiden 1976, 91ff., 98ff.; Bollack, *op. cit.*, 539f.; Brodersen, K. Euthymenes aus Massalia, *DNP* 4 (1998) 318–319.

146 Mansfeld. Aristotle, 55ff.; Zhmud. Doxographie.

147 Cf. *Met.* 983b 1f. and the characteristic note made in another doxographical passage: "It is what we are all inclined to do, to direct our inquiry not by the matter itself, but by the views of our opponents." (*Cael.* 294b 7–9).

148 Democritus, Leucippus, the Pythagoreans, Anaxagoras, Homer, Empedocles, Plato, Thales, Diogenes, Heraclitus, Alcmaeon, Hippon, Critias.

149 Aristotle starts from two main principles: the soul is a source of motion and mind (403b 24f.), which can be combined (Diogenes, e.g., admitted both, 405b 21f.); then he adds to them the third, binary principle, that of corporeality/uncorporeality (404b 30f.), so that at the end of his overview he mentions three of them (405b 11f.). These principles can be reduced, in turn, to the ἀρχαί of every thinker (water, air, fire, etc.), but there are a few exceptions. One is admitted by Aristotle himself: Critias derived soul from blood (405b 5f., 13), the others ignored (Thales did not consider water to be the source of a soul; nothing is said at all on Alcmaeon's ἀρχαί, 405a 19f., 29f.). The difference between the monists and pluralists is noticed (404b 9f., 405b 17), but does not play any particular role in the account. Mansfeld. Aristotle, 37ff., believes that Aristotle combined two principles here: 1) by related ideas, which goes back to Hippias; 2) by the number and nature of the ἀρχαί.

150 See already Zeller, E. Über die Benützung der aristotelischen Metaphysik in den

this survey, which traces the development of the notions connected with the four causes, the main principle for presenting the opinions is by the type of causes (first comes the material cause, then the efficient, etc.). But from the very beginning, this principle is combined with the historical one,[151] since all the early physicists, including Thales, the ἀρχηγός of natural philosophy (983b 20), and some of the later ones as well, admitted only the material cause. In the section on material causes, the monists' opinions are grouped according to the similarity of their elements: Thales and Hippon suggested water; Anaximenes and Diogenes, air; Hippasus and Heraclitus, fire.[152] The monists were followed by Empedocles, who added the fourth element, earth, to the three already known ones, and, later, Anaxagoras, who considered the number of elements to be infinite. Here Aristotle adds an important chronological reason: though Anaxagoras was older than Empedocles, his philosophy was later;[153] as a result the succession 'one element – many elements – an infinite number of elements" acquires a historical meaning.

Under the pressure of facts and the truth itself, Aristotle continues, philosophers, namely Anaxagoras (984b 18) and Empedocles (985a 5),[154] turned from material causes to causes of motion.[155] Immediately after these, however, he names Leucippus and Democritus,[156] who admitted material causes only. This lack of consistency is explained, first of all, by the fact that the Atomists lived later than the majority of the philosophers mentioned previously.[157] Another chronological remark connects Leucippus and Democritus with the Py-

---

Schriften der älteren Peripatetiker (1877), *Kleine Schriften*, Vol. 1, Berlin 1910, 197ff. See also McDiarmid, *op. cit.*, 91ff. Cf. above, 143f.

151 Kienle, W. von. *Die Berichte über die Sukzessionen der Philosophen in der hellenistischen und spätantiken Literatur* (Diss.), Berlin 1961, 51f.; Gigon. Die ἀρχαί der Vorsokratiker, 121f.

152 The presence in this section of material from Hippias' work, on Thales and Homer in particular (Patzer, *op. cit.*, 33ff., 40), might have been an additional reason for putting 'related' elements together. It should be noted that the earlier thinker is mentioned first in each of the three pairs.

153 984a 12: τοῖς δ' ἔργοις ὕστερος. Alexander understood these words to mean that Anaxagoras was worse than Empedocles, but they are more likely to have a temporal sense (Zeller, E. *Die Philosophie der Griechen*, 6th ed., Leipzig 1919, 1261 n.2; Ross, *op. cit.*, 132; Mansfeld. *Studies*, 300ff.).

154 The latter is called πρῶτος twice (985a 8, 29), in connection with various aspects of his theory.

155 Among other candidates for introducing this cause, whose authorship Aristotle doubted, he names Hermotimus of Clazomenae (who lived before Anaxagoras), Hesiod, and Parmenides. Aristotle postpones considering the question τίς πρῶτος (984b 31), never to return to it.

156 Democritus is called the ἑταῖρος of Leucippus (985b 4–5), and this is the first indication of the teacher-student relationship, which proved of great importance for the history of philosophy.

157 Ross, *op. cit.*, 28; von Kienle, *op. cit.*, 52.

thagoreans, who come next in his account: the latter lived "at the same time and even earlier" than the former. The principles of these Pythagoreans, who "were the first to advance mathematics" (985b 23–24), were numbers, while other (probably, later) Pythagoreans named ten pairs of opposite principles.[158] Aristotle concludes that Alcmaeon speculated along the same lines, and that either he derived this view from them, or they derived it from him (986a 27f.). It is revealing that Aristotle does not limit himself to indicating the similarity between the doctrines, but tries to establish who influenced whom. The words that come next (986a 29–30), "for Alcmaeon was contemporary with the old age of Pythagoras" (καὶ γὰρ ἐγένετο τὴν ἡλικίαν Ἀλκμαίων ἐπὶ γέροντι Πυθαγόρᾳ), if they are in fact Aristotle's own,[159] show that he was inclined, though not without hesitation, to date Alcmaeon before these (later) Pythagoreans, so that Alcmaeon is said to influence them, rather than vice versa.[160]

Having reached the penultimate stage of his overview in his account of the pluralists, Aristotle goes back to the 'metaphysical' monists, namely, the Eleatics. In the first place he names Xenophanes (πρῶτος τούτων ἑνίσας, 986b 21), with Parmenides "who is said to have been his student" and Melissus coming next. Finally, "after these systems came the philosophy of Plato" (987a 30f.), whose similarity to the teaching of the Pythagoreans Aristotle is never tired of emphasizing, without, however, forgetting to mention Cratylus, Heraclitus, and Socrates, who influenced Plato in his youth (cf. 988a 15–17). Plato, according to Aristotle, added to the two causes the third, formal one, himself admitting only two. As for the fourth cause, the final one, no one used to speak of it clearly.

Let us sum up the distinctive historical features of Aristotle's overview. 1) Chronological sequence. Without being the only or even the main method of arranging the material, it is, nevertheless, constantly in the foreground of Aris-

---

158 986b 22f.: limit – unlimited, even – odd, etc. In fact, the table of opposites goes back to Speusippus (Cherniss. *Aristotle's criticism*, 391; Burkert. *L&S*, 51f.; Tarán. *Speusippus*, 33ff.; Zhmud. Philolaus, 261ff.). Cf. Arist. *Met.* 1092a 35, 1087b 4, b 25, 1085b 5.

159 These words, absent from one of the manuscript traditions (A^b) and from Alexander's commentary, survived in a more reliable tradition (EJ), as well as in the commentary by Asclepius (*In Met.*, 39.21). Ross, who generally preferred EJ (*op. cit.*, clxv), nevertheless considered these words to be a later interpolation. Wachtler, J. *De Alcmaeone Crotoniata* (Diss.), Leipzig 1896, 3f. analyzed this passage in detail, arguing very convincingly for its authenticity. See also Guthrie, *op. cit.*, 342; Zhmud. *Wissenschaft*, 75.

160 Irrespective of the authenticity of these words, the conclusion that Alcmaeon lived earlier follows from the fact that he spoke vaguely (ἀδιορίστως), while the Pythagoreans "declared both how many and which their contrarieties are" (986a 34–b 2). There are two more places where Aristotle clearly indicates that he considers these Pythagoreans to be a later school of thought: πρῶτοι and οἱ Ἰταλικοί (987a 5, a 10), οἱ πρότερον and οἱ ἄλλοι (987a 28).

totle's attention. 2) Frequent references to the *prōtos heuretēs* of various ideas. 3) Direct references to the teacher-student relationship (Xenophanes and Parmenides, Leucippus and Democritus). 4) Where the principles held by two philosophers are identical, the earlier is mentioned first (Thales and Hippon, Anaximenes and Diogenes, Hippasus and Heraclitus); if the principles are similar, Aristotle solves the problem of priority by resorting to chronological arguments (Alcmaeon and the Pythagoreans). 5) Attention is paid to the influence of earlier thinkers on later ones (Pythagoreans and Plato). 6) There are some references to places of birth.[161] 7) Finally, one can recognize in this overview the beginnings of the future arrangement in schools (Ionians, Pythagoreans, Atomists, Eleatics).

Theophrastus developed and more consistently applied all these features, along with the systematic grouping of the *doxai*, especially in the chapter on the first principles.[162] Let us remind the reader that the *Physikōn doxai* are known mainly from the following sources. The first is a work by Aëtius (a revised version of an earlier compendium, *Vetusta placita,* which, in turn, is a revised version of Theophrastus), reconstructed by Diels from Ps.-Plutarch, Stobaeus, and other later doxographers. The second source is the fragments, quoted by Simplicius, of the chapter Περὶ ἀρχῶν (Aët. I, 3), taken most probably directly from Theophrastus.[163] The third is Theophrastus' *De sensibus*, a long fragment that originally was a division of the *Physikōn doxai* related to the doctrines on the five senses. What precisely the general systematic structure of Theophrastus' compendium was, remains unknown. We can, however, get an idea of it from the composition of *Vetusta placita* as reconstructed by Diels. Its first part deals with the fundamental physical principles and categories, the second with cosmology and astronomy, the third with meteorology, the fourth with the earth, the sea, and the Nile's floods, the fifth with the soul (psychology and physiology), the sixth with the body (physiology and embryology).[164] Except for the first part, this structure corresponded on the whole both thematically and to a certain extent compositionally to many Presocratic writings, starting at least from Alcmaeon. In principle, the subject matter of the *Physikōn doxai* covered the whole of what the Peripatetics under-

---

161 Hippon, Diogenes, Hippasus, Heraclitus, Anaxagoras, Hermotimus, Alcmaeon. *Mete*. 365a 14f. gives the birthplaces of Anaxagoras, Anaximenes, and Democritus, as well as their relative chronology. *Cael*. 294a 22f. mentions Xenophanes of Colophon and Thales of Miletus. Occasional indications of the birthplace of a thinker are found still more often. Cf. above, 143 n. 113.

162 Diels, H. Leucippos und Diogenes von Apollonia, *RhM* 42 (1887) 7.

163 Diels believed that Simplicius borrowed these fragments from Alexander (*Dox*., 108f.; McDiarmid, *op. cit*., 90f.), but see Reinhardt, K. *Parmenides und die Geschichte der griechischen Philosophie*, Bonn 1916, 92 n. 1; Regenbogen. Theophrast, 1536; von Kienle, *op. cit*., 66f.; Steinmetz, *op. cit*., 341.

164 *Dox*., 181. The question of how precisely these parts corresponded to the 16 (or 18) books of Theophrastus remains open.

stood by physics,[165] though many of the problems of zoology and botany studied by Aristotle and Theophrastus are lacking here, since the early physicists either did not touch upon them at all or paid very little attention to them, which made a representative collection of their opinions impossible.[166]

The six parts of *Vetusta placita* were divided into chapters dealing with individual problems and following each other in logical order. In the part on the soul, e.g., opinions on the five senses in general were duly followed by those on the individual senses: sight, hearing, the sense of smell and taste being taken separately (*Dox.*, 182). In *De sensibus*, however, the material is not arranged in accordance with the five senses following in succession, but with the individual thinkers' theories on all the senses. This reflects Theophrastus' attention not so much to the opinions as such, but rather to the doctrines of concrete thinkers, in which he tried to emphasize their common as well as individual features. At first, he divides them into two groups: the first follows the principle 'like by like', the second sticks to the opposite principle (1). Individual physicists' doctrines are then critically exposed according to this division, though not consistently enough. The first group includes the teachings of Parmenides, Plato and Empedocles (3–24), ranged in the order of the growing complexity and elaborateness of their theories (*Dox.*, 105). Yet they are not followed by the proponents of the opposite principle, but by all the others, arranged in chronological order: Alcmaeon, Anaxagoras, Clidemus, Diogenes, and Democritus (25–58).[167]

The multi-level structure of the *Physikōn doxai* can be preliminarily characterized as follows. On the whole, the treatise was organized on systematic principles, with the choice and the succession of problems reflecting the historically attested interests of the physicists. The material is divided into books

---

165 According to Theophrastus (*Met.* 9a 13–15, 9b 20–10a 4), the subject of physics starts with celestial bodies and ends with animals and plants. Cf. Aët. V, 14.

166 In his writings on plants, Theophrastus repeatedly refers to the Pythagorean Menestor, who wrote on the causes of the falling of leaves, on the difference between warm and cold plants, etc. (32 A 2–7). There were no corresponding divisions in the *Physikōn doxai*, so that Menestor is lacking here.

167 In section 1, Heraclitus is placed in the second group, but is not mentioned subsequently. Anaxagoras is the only true representative of the second group, whereas Alcmaeon, Clidemus, Diogenes, and Democritus do not belong entirely to any of them. It seems that the difficulties of clear-cut systematization prompted Theophrastus to use the simplest, i.e., chronological, principle. Clidemus' position between Anaxagoras and Diogenes is the only chronological indication on this obscure figure, on which basis Diels dated him. In the second part of *De sensibus*, dealing with the objects of the senses, Democritus comes first and Plato after him (sections 59–92). Fritz, K. von. Democritus' theory of vision, *Science, medicine and history. Essays written in honour of Ch. Singer*, ed. by E. Underwood, Vol. 1, Oxford, 1953, 83, considered this work to be critical and historical, presenting theories in chronological order. Cf. Mansfeld, J. Aristote et la structure du *De sensibus* de Théophraste, *Phronesis* 41 (1996) 158–188; Baltussen. *Theophrastus*, 15f.

(parts) and chapters corresponding to the categories of the Peripatetic physics. Within the chapters devoted to specific problems, the main 'units' consisted of the theories of individual philosophers, which were united into groups according to their similarity (if this was relevant) and/or often, but not necessarily, were set in chronological order.[168]

This picture is confirmed by Simplicius' quotations from the chapter *On the Principles*, in which Theophrastus introduces his main characters to the reader. The quotations are given in the commentary to the passage of Aristotle's *Physics* I, 2, where principles of the Presocratics are arranged by the method of division. Simplicius notes that diaeresis of the *Physics* may be developed by dividing the group of the monists on the principle of 'limited – unlimited' and the group of the pluralists on the principle of 'motionless – in motion'.[169] But since the pluralists whose principles would be motionless were unknown to him (as well as to Aristotle), he limits his development of diaeresis to the group of monists alone (*In Phys.*, 22.16–21), characterizing them in accordance with the following scheme (*left to right*):

motionless		neither in motion nor at rest	in motion	
unlimited	limited	neither unlimited nor limited	limited	unlimited
Melissus	Parmenides	Xenophanes	Thales, Hippon, Heraclitus, Hippasus	Anaximander, Anaximenes, Diogenes

Simplicius' explanations preceding his overview of the principles leave no doubt that his order of exposition has hardly anything to do with the sequence of names in the chapter *On the Principles*.[170] In contrast to the schematic pedantry of the late commentator, Theophrastus' fragments feature a historically oriented picture of philosophers' teachings similar in many respects to the historico-doxographical survey in the *Metaphysics*. Simplicius himself pointed

---

168 Cf. Regenbogen's opinion concerning the *Physikōn doxai*: "Der Aufbau scheint nach Sach- und Problemkategorien geordnet gewesen zu sein, innerhalb deren sowohl die zeitliche Folge als auch die angeblichen Schulzusammenhänge bestimmend waren." (Theophrast, 1536).

169 *In Phys.*, 21.34f., 22.20f. Before him, this question was raised by Alexander, who, however, was satisfied with Aristotle's division (ibid., 21.35f.). See von Kienle, *op. cit.*, 59f.

170 ἄμεινον δὲ ἴσως ἐκ τελεωτέρας διαιρέσεως τὰς δόξας πάσας περιλαβόντας οὕτω τοῖς τοῦ Ἀριστοτέλους ἐπελθεῖν (*In Phys.*, 22.20–21). Cf. *Dox.*, 104f.; McDiarmid, *op. cit.*, 88f.; Steinmetz, *op. cit.*, 338ff.; von Kienle, *op. cit.*, 62f.; Wiesner, J. Theophrast und der Beginn des Archereferats von Simplikios' Physikkommentar, *Hermes* 117 (1989) 288–303; Mansfeld. *Studies*, 243ff.

this out at the end of his overview: "This is the summary account of what has been ascertained about the principles, recorded not in chronological arrangement, but according to affinities of doctrines."[171] While κατὰ τὴν τῆς δόξης συγγένειαν characterizes Simplicius' own method of presenting the material, the contrasting οὐ κατὰ χρόνους implies that, in Theophrastus, philosophers followed each other κατὰ χρόνους (cf. *Dox.*, 104 n. 4). To fit the material into his scheme, Simplicius breaks this succession, placing Melissus before Parmenides, Parmenides before Xenophanes, and Thales, the founder of physics, in the middle of the group of monists, which results in a manifest contradiction between the quotations from Theophrastus and the commentary. Simplicius groups the pluralists who admitted a limited number of principles by the number of their principles: two (Parmenides, who had already figured among the monists, and the Stoics), three (Aristotle), four (Empedocles), six (Plato), and ten (the Pythagoreans). They are followed by those who admitted an unlimited number of principles of one kind (Anaxagoras, Archelaus), and, next to them, of principles different in kind (Leucippus, Democritus, Metrodorus). Meanwhile, Theophrastus' words make it clear that in his survey, as well as in Aristotle, Plato, instead of figuring in the middle of the group of pluralists, came after all other physicists.[172] Besides, Theophrastus ascribes to Plato (in physics) only two causes, not six. Simplicius' construction therefore appears obviously artificial.

Theophrastus records the relative chronology of the physicists much more consistently than Aristotle. Thales: the founder of physics who eclipsed his anonymous precursors (fr. 1 Diels); Anaximander: διάδοχος καὶ μαθητής of Thales (cf. *Dox.*, 476n.); Anaximenes: ἑταῖρος of Anaximander (fr. 2); Xenophanes: "is said to have listened to Anaximander" (fr. 6a); Parmenides: a pupil of Xenophanes (fr. 5); Anaxagoras: shared in the philosophy of Anaximenes (fr. 4); Empedocles: "a little younger than Anaxagoras", an admirer of Parmenides (fr. 3); Archelaus: a student of Anaxagoras and an associate of Socrates (fr. 4); Leucippus: shared in the philosophy of Parmenides; Democritus: ἑταῖρος of Leucippus (fr. 8); Diogenes of Apollonia: "almost the youngest of

---

171 *In Phys.*, 28.30–31 = fr. 229 FHSG. Interestingly, in the next sentence (28.32 f.) Simplicius criticizes those whose notions of disagreements between philosophers are based on a superficial knowledge of ἱστορικαὶ ἀναγραφαί. Diels suggested that works like that of Diogenes Laertius are implied here, but Simplicius does not mention Diogenes. More probably, under ἱστορικαὶ ἀναγραφαί, he might have meant the *Physikōn doxai*, which he had just finished quoting (ἡ σύντομος περίληψις τῶν *ἱστορημένων* περὶ ἀρχῶν οὐ κατὰ χρόνους *ἀναγραφεῖσα*) and which was known to him under the title Φυσικὴ ἱστορία (fr. 226b, 228b, 234 FHSG; Sharples. *Commentary*, 12). If this is true, the designation of Theophrastus' doxography as a historical writing will prove revealing, in many respects.

172 "Theophrastus, after giving his account of the other physicists, says: 'After these came Plato, before them in reputation and ability though after them in time.'" (fr. 9 Diels = fr. 230 FHSG).

all the physicists", imitated Anaxagoras and Leucippus (fr. 2); Metrodorus: admitted the same principles as Democritus (fr. 8); Plato: follows chronologically all the rest (fr. 9). Thirteen philosophers have a chronological reference of some kind, the four others (Hippasus, Heraclitus, Hippon, and Melissus) lack it. But it is possible that, though Theophrastus provided it, Simplicius left it out.[173]

References to teacher-student relationships are also more numerous and differentiated than in *Metaphysics* A. In a number of cases, they presuppose personal connection (Thales and Anaximander, Anaximander and Anaximenes, Xenophanes and Parmenides, Leucippus and Democritus, Anaxagoras and Archelaus); the others imply the borrowing of ideas and the influence of the older on the younger (Anaximenes and Anaxagoras, Parmenides and Leucippus, Anaxagoras and Diogenes, Democritus and Metrodorus). The birthplaces of the philosophers are always mentioned. The only exception is Hippon – due, probably, to an omission by Simplicius rather than to Theophrastus' lack of information.[174] Apart from birthplaces, Theophrastus cites the philosophers' patronymics. Simplicius preserved them in five cases (Thales, Anaximander, Anaximenes, Parmenides, Anaxagoras), but there are many reasons to believe that Theophrastus cited the patronymic as a rule, rather than as an exception.[175] Finally, the principle of the *prōtos heuretēs* was also applied: Thales invented natural philosophy (fr. 1 Diels), Anaximander the notion of ἀρχή (fr. 2); Anaxagoras was the first to introduce the efficient cause (fr. 4).[176] Theophrastus did not confine the references to discoverers to the first book of his work: Parmenides' discoveries (fr. 6a, 17) concern astronomy, not the principles. It is no coincidence that all the other references to the discoveries in Aëtius also relate to astronomy:[177] in this science, it was possible to speak about discoveries, i.e.,

---

[173] Theodoretus, who relied on Aëtius, called Melissus the ἑταῖρος of Parmenides (*Dox.*, 286 n. 14).

[174] Aristoxenus thought that Hippon was born on Samos, Meno says his birthplace was Croton, Censorinus Metapontum, Hippolitus and Ps.-Galen Rhegium (38 A 1, 3, 11). What place Theophrastus indicated is unknown. He could have given several variants, as in the case of Leucippus (fr. 8 Diels).

[175] Archelaus' patronymic was certainly given by Theophrastus (*Dox.*, 139, 280.9). For the patronymics, see also: Pythagoras (*Dox.*, 280.17), Xenophanes (284 n. 12), Metrodorus (285.5), Democritus (285 n. 16), Melissus (286 n. 14), Empedocles (286.19), Zeno (289.1), and Plato (289.17). Cf. above, 143 n. 113.

[176] In Aëtius, Pythagoras and Ecphantus are also called πρῶτος (*Dox.*, 280.20, 286.21). Anaximenes is said to be the only one to explain physical processes by condensation and rarefaction (fr. 226b FHSG). Since this obviously contradicts the facts, Usener and Diels corrected μόνος to πρῶτος (*Dox.*, 144 n. 2, 477n.; cf. *DK* 13 A 5n.).

[177] Pythagoras was the first to give the name 'cosmos' to the universe (*Dox.*, 327.8) and to discover the obliquity of the ecliptic (340.21: Oenopides contests his priority), Thales was the first to find the cause of the solar eclipse (353.20) and the source of the moon's light (358.15), Parmenides divided the earth into zones (377.18) and identified the Morning and the Evening Stars with Venus (345b 14), and Anaxagoras was the first to explain the eclipses and phases of the moon (562.26). To be sure, the

about indisputable things and not just about true opinions, as e.g. in meteorology or embryology.

Thus, there is no doubt that the first chapter of the *Physikōn doxai* was organized to unfold for the reader a historical picture of the gradual perfection of philosophy from its first immature ideas to the present state (as Aristotle understands this).[178] Having presented brief biographical data (place of birth, patronymic, name of teacher), Theophrastus focuses his entire attention on the origin and reception of new ideas, on the particular forms of the development of philosophical doctrines, on their succession in time, and on each thinker's individual contribution and his dependence on his predecessors.[179] Theophrastus' striving to give the doxography a historical dimension is obvious, even if we fail to reconstruct safely the original sequence of the names in the chapter *On the Principles*. The relative chronology of the Presocratics was hardly the only method of arranging the material. It is noteworthy that no contemporary history of Greek philosophy is based on chronology alone; affiliation to various schools, such as the Ionians, Pythagoreans, Eleatics, or Atomists, is always taken into account. Pointing out that such an approach to the philosophical past was already widespread in Hellenistic historiography, von Kienle assumed that Theophrastus used an analogous method, arranging the Presocratics in 'successions'. Specifically, he considered it very likely that, after the first series: Thales – Anaximander – Anaximenes – Anaxagoras – Archelaus, Theophrastus turned again to Xenophanes, who was followed by Parmenides – Leucippus – Democritus – Diogenes – Plato.[180] Indeed, the survey begins with Thales and ends with Plato, preceded by Diogenes ("almost the youngest of all the physicists"). Furthermore, each subsequent philosopher is connected with the previous one *in his own series*.[181] At the same time, Xenophanes turns out to be the only philosopher – except for the later eclectic Diogenes (fr. 226a FHSG) – to connect these two series with each other: he is said to have heard Anaximander (fr. 227d FHSG). This reference seems to indicate, quite pertinently in this context, that soon after Archelaus Theophrastus turns back to the sixth century.

information of this kind does not always come from Theophrastus. Cf., e.g., the idea that Thales was the first who called the soul "eternally movable or self-movable" (386.10).

178 *Met.* 993a 15–17; fr. 53 Rose (see above, 121 n. 12).

179 Von Kienle, *op. cit.*, 38ff., 52ff., 58ff.; Mansfeld. Aristotle, 28ff.

180 Von Kienle, *op. cit.*, 61–62.

181 1) Anaximander was a student of Thales, Anaximenes of Anaximander, Anaxagoras followed Anaximander and Anaximenes, and Archelaus his teacher Anaxagoras. 2) Parmenides was a student of Xenophanes, Leucippus developed Parmenides' ideas, Democritus was a student of Leucippus. McDiarmid, *op. cit.*, 89, believed that Theophrastus combined his method of exposition by the four causes (borrowed from *Metaphysics* A) with the chronological one, which resulted in two series: 1) Thales – Anaximander – Anaximenes – Anaxagoras – Archelaus and 2) Xenophanes – Parmenides – Empedocles, and Leucippus – Democritus – Metrodorus.

This is all the more probable because Aristotle, in his doxographical overview, also returns to Xenophanes, Parmenides, and Melissus after the Atomists and the (later) Pythagoreans.[182]

It is also clear enough that Melissus followed Parmenides and that Metrodorus followed Democritus. The situation with others – Hippasus, Heraclitus, Empedocles, and Hippon – is more complicated. The suggestion that Theophrastus included them in a separate group of 'random' philosophers (οἱ σποράδην of the Hellenistic tradition)[183] does not seem convincing. Theophrastus must rather have used the same method as Aristotle, linking Thales and Hippon, Hippasus and Heraclitus by affinity of their principles.[184] Hippon, in this case, is mentioned in violation of chronology with Thales, while Heraclitus and Hippasus are mentioned in accordance with chronology between Anaximenes and Anaxagoras. Empedocles, in contrast, is referred to as both Parmenides' adherent and Anaxagoras' younger contemporary. The latter reference is also found in Aristotle (*Met.* 984a 11), who placed Empedocles after the three pairs of physical monists, though *before* Anaxagoras. Theophrastus could have placed Empedocles either after Parmenides, or before Anaxagoras, but reliable data on this subject is lacking.[185]

Having presented in the first chapter of the *Physikōn doxai* a general genealogical scheme of the development of philosophy, Theophrastus hardly felt compelled to reproduce it in exactly the same way in every chapter. He could have related the opinions of the physicists on particular problems in any order

---

182 See above, 156. It is not clear how this method is connected with the division into different schools. The Pythagorean school was the only one that Aristotle and Theophrastus mentioned directly. Both, however, pointed out the difference between the Eleatics' principles and those of the Ionians, as well as the Atomists' dependence on the Eleatics.

183 Von Kienle, *op. cit.*, 62ff.

184 McDiarmid, *op. cit.*, 89. It follows from Theophrastus' fragments that each of the two pairs (Thales – Hippon and Hippasus – Heraclitus) was mentioned in the same sentence, unlike all the others (fr. 1 Diels = Simpl. *In Phys.*, 23.22f., 23.33f.). Aristotle adopted the same order (*Met.* 984a 2, a 7). Aristotle and Theophrastus provide no data on the chronology of Hippon, Hippasus, and Heraclitus. Though Simplicius, unlike Theophrastus, places Diogenes immediately after Anaximenes, he characterizes their teachings separately and indicates their chronology, doctrinal affiliation, etc. In contrast, Simplicius does not seem to have found in Theophrastus any trace of a specific teaching on principles by Hippon and Hippasus that would have been different from that of Thales and Heraclitus.

185 The order of names in Aëtius' chapter Περὶ ἀρχῶν partly confirms the general reconstruction. Here we find both series: Thales – Anaximander – Anaximenes – Anaxagoras – Archelaus (I,3.1–6), and Xenophanes – <Parmenides> – Leucippus – Democritus – Metrodorus (I,3.12–17), as well as the pair Heraclitus – Hippasus (I,3.11). Anaximenes and Diogenes are named separately (I,3.26), Hippon is absent. On the other hand, we find here Pythagoras (I,3.8), Philolaus (I,3.10), and Ecphantus (I,3.19), who are lacking in Theophrastus' fragments.

that suited him, e.g. according to the degree of their complexity; this order would often, though not necessarily, coincide with the chronological one.[186] Any reader interested in knowing who influenced whom could easily check his hypothesis against the scheme presented at the beginning of the book.

Deciding how the material was organized in Meno's book is still more difficult. First, we have but brief excerpts by a late author from the *Medical Collection*, which are not at all certain to follow the original plan without omissions and compositional changes.[187] Second, the papyrus contains quite a number of lacunae. Third, the chronology of many of the physicians figuring in Meno is highly problematic, since nearly half of them (ten out of twenty-one) are not mentioned anywhere else. The general outline of Meno's scheme is nevertheless clear. All opinions are divided, as in Theophrastus' *De sensibus*, into two groups: those that connected the causes of diseases with the residues of digestion (περιττώματα), and those that explained the diseases by proceeding from the constitutive elements of the body (στοιχεῖα).[188] The doctrines within these groups, to all appearances, were considered in accordance with the degree of their sophistication, which coincides in many cases with the chronological arrangement.[189]

* * *

Going back to the beginnings of Peripatetic historiography, I would like to point out again that its emergence corresponds with the period when Greek science, philosophy, and medicine reached a certain maturity. By that time, Greek poetry and music, which had arrived at their 'perfection' long before, had already become subjects of historical surveys generally organized chronologically and using the *prōtos heuretēs* principle. Early heurematography and doxography, Sophistic theories on the origin of culture, Plato's theory of science, and the expert knowledge of specialists in each of the arts and sciences belong to the most important sources the Lyceum relied on. Yet on the whole, the attempt by Aristotle and his disciples to systematize the entire space of con-

---

186 Diels, H. Über die Excerpte von Menons Iatrika in dem Londoner Papyrus 137, *Hermes* 28 (1893) 415, believed that the *doxai* were arranged by their affinity, yet the systematic order often proved to coincide with the chronological one.

187 Diels (ibid.) was doubtful about Hippon's and Thrasimachus' positions; Grensemann (*op. cit.*, 13f.) suggests a different place for Alcamenes; Manetti (*op. cit.*, 118f.) places Plato after Philistion.

188 Diels. Über die Excerpte, 415f.; Manetti, *op. cit.*, 100f.

189 See e.g. the second group (Manetti, *op. cit.*, 119), containing few unfamiliar names: Philolaus – Polybus (a student of Hippocrates) – Menecrates (a doctor of the fourth century) – Petron – Philistion (a contemporary of Plato) – Plato. The first group runs from Euryphon of Cnidus to Aegimius of Aelis, the younger contemporary of Hippocrates.

temporary culture and to give a historical retrospective of its development was unique in Antiquity and found no analogies until the 18th century.

The key notion of Aristotle's systematics was ἐπιστήμη, embracing theoretical sciences, productive arts (music and poetry), and such practical sciences as he was interested in, like politics and rhetoric. Of course, not every historical outline of any of these fields written in the Lyceum was based on the Aristotelian classification of science, the more so since the latter itself consisted of three different schemes that had emerged at different times: first, the Pythagorean quadrivium, then the division of sciences into three kinds, and finally the later subdivision of theoretical sciences into mathematics, physics, and theology. But in the case of the historiographical project, which inquired into the past of all three theoretical sciences (and into medical theories related to physics, as well), the coincidences between Aristotle's philosophy of science and the history of science written by his disciples are too detailed and numerous to be accidental.

Each of these 'histories' bore individual features, depending upon the nature of the material and the particular task of each treatise. A description of irrefutable discoveries in mathematics and (partly in) astronomy differed, naturally, from that of the contradictory and often erroneous *doxai* of the physicists, which in turn had little in common with a historical overview of 'principles' considered by theologians. Nevertheless, in spite of the predominantly systematic character of the physical and medical doxography, Theophrastus and Meno did their best to build into the very structure of their works the historical perspective shared by all the Peripatetics in their approach to accumulated scientific knowledge. This perspective is quite clearly reflected in Eudemus' works on the history of science. We will turn to these works in the next chapters, drawing parallels from Theophrastus, Meno, and Aristoxenus when necessary.

Chapter 5

The history of geometry

1. Eudemus of Rhodes

We know little about the founder of the historiography of science Eudemus of Rhodes. Ancient sources depict him as a devoted student of Aristotle, who considered Eudemus (along with Theophrastus) a possible scholarch of the Lyceum.[1] We know neither exactly when he was born, nor when he joined Aristotle's Lyceum. Eudemus certainly was younger than Theophrastus (born ca. 370), and after Aristotle's death he returned to Rhodes, where he continued to study and to teach (fr. 88). Eudemus did not lose contact with Theophrastus and corresponded with him on the subject of their teacher's writings (fr. 6).

While Eudemus' *Physics* belongs to his Rhodian period, his works on logic and on the history of science were written while Aristotle was still alive. In practically all of the logical fragments, Eudemus figures together with Theophrastus, which implies a kind of co-authorship. The list of Theophrastus' works contains three writings on the history of science with the same titles as Eudemus' works.[2] Since there are no other traces of such writings in Theophrastus, the editors of his fragments subscribed to Usener's suggestion that these were Eudemus' works, which were later mistakenly added to Theophrastus' list. In the same list we find another work, Τῶν περὶ τὸ θεῖον ἱστορίας α′–ς′, which, contrary to Wehrli's opinion, should be identified with Eudemus' *History of Theology*, known from Damascius.[3] This misunderstanding indirectly confirms that Eudemus' historical works were written before he left Athens, otherwise they would hardly have been in Theophrastus' catalogue. Assuming that these works along with Theophrastus' physical doxography and Meno's medical doxography were a part of Aristotle's historiographical project (4.2), they can be dated between 335/4 (foundation of the Lyceum) and 322/1 (Aristotle's death).

The majority of those who have studied Eudemus' theoretical treatises (*Physics, Analytics*, etc.) agree that in this domain he was not particularly independent. As a rule he followed Aristotle, clarifying the latter's ideas and arranging them more systematically. But though Eudemus, like his colleagues at the

---

1 Fr. 5 = fr. 8 FHSG. To be sure, Aulus Gellius' account does not seem very reliable.

2 D. L. V, 48, 50: Ἀστρολογικῆς ἱστορίας α′–ς′ (137 No. 43 FHSG), Ἀριθμητικῶν ἱστοριῶν (264 No. 2 FHSG, the number of books not indicated), Ἱστορικῶν γεωμετρικῶν α′–δ′ (264 No. 3 FHSG).

3 251 No. 2 FHSG = Eud. fr. 150. See above, 130 n. 51.

Lyceum, did not greatly develop Aristotle's system or create his own philosophical system, this does not mean that he lacked all originality. Several early Peripatetics became prominent not so much in philosophy as in the specific sciences. There is no doubt that ancient Greek botany, geography, and harmonics would appear incomparably inferior without Theophrastus, Dicaearchus, and Aristoxenus. Such an appraisal seems all the more appropriate to the historiography of science, since Eudemus' *History of Geometry*, *History of Arithmetic*, and *History of Astronomy* happened to be not only the first but also the last specimens of that genre in antiquity. Although Eudemus' works were not forgotten (they were still quoted in the sixth century AD) and a special biography was devoted to him,[4] in this particular genre he appeared to have no followers.

This could hardly be explained by Eudemus' failure to found his own school. Even if he had only a few students,[5] Theophrastus had two thousand listeners (D. L. V, 37), and nonetheless his botanical research was not further developed. Meanwhile, in contrast, the Hellenistic writers immediately picked up the biographical genre founded by Aristoxenus and Dicaearchus (about whose students we know nothing), since it corresponded to the interests and the very spirit of their epoch. In spite of the general decay of interest in the exact sciences in the philosophical schools of the Hellenistic age (8.1), one should not think that Eudemus was virtually unknown in this time, especially taking into account that we possess only meager remains of the Hellenistic literature. Eratosthenes and probably Archimedes drew upon his *History of Geometry*; Diogenes Laertius and Clement of Alexandria, known for their extensive use of the Hellenistic sources, cite his *History of Astronomy*.[6] Later, Eudemus' theoretical treatises remained of interest only to Aristotle's commentators, whereas his works on the history of the exact sciences were frequently quoted by those who took these sciences up in one way or another: Theon of Smyrna, Porphyry, Pappus, Proclus, Simplicius, Eutocius. Thus, Eudemus, the expert in the exact sciences and their first and perhaps only historian, was no less important for the classical tradition than Eudemus the true Peripatetic.

---

4 Fr. 1. Only Simplicius refers to this biography, written by a certain Damas. It is not known when Damas lived, but one can guess that he was Eudemus' student, rather than a later biographer (Zeller, *op. cit.* I, 86). Eudemus was not such a popular figure that a late author would be interested in writing his biography; besides, what would a biographer have as possible sources? It should be recalled that Archimedes' biography, written by his student Heraclides, was first quoted by Eutocius (see below, 294f.). Biographies of Eudemus and Theophrastus are mentioned in the Arabic sources (Rosenthal, F. *The classical heritage in Islam*, Berkeley 1975, 36; cf. 4a FHSG).

5 Gottschalk. Eudemus, 25ff. – A fragment of Eudemus' *Physics*, written on Rhodes, depicts a typical picture of a teacher lecturing to a group of students (fr. 88).

6 Fr. 143–144. Philodemus' *De pietate* seems to have used Eudemus' *History of Theology*. See also below, 6.1, 8.1.

To judge from his works, Eudemus received a good mathematical education and was very competent in the problems of contemporary mathematics (which is not always true of Aristotle).[7] This is also manifest in the fact that his histories of the exact sciences are devoted to strictly mathematical problems and methods, rather than to the philosophical interpretation of mathematics that was so characteristic of Plato and his students – Speusippus, Xenocrates, and Philip, as well as Aristotle. Certainly, a professional approach to mathematics was not the only possibility for Eudemus: in his work *On Angle* (fr. 30) he treated an angle as a certain quality, i.e., in the spirit of Aristotle.[8] There is an interesting fragment in Eudemus' *Physics* (fr. 34) that is worth quoting in full to demonstrate one of the possibilities of combining philosophical and mathematical approaches.

> It is difficult to decide whether each science investigates and explains its own principles, or each has some other science about its principles, or there exists a science dealing with all the principles. For mathematicians display their own principles and give its definition to every thing they talk about, so that a person who does not know all this would look ridiculous if he tried to investigate what a line is and every other mathematical object. As for the principles they talk about, mathematicians do not attempt to demonstrate them, they even claim that it is not their business to consider them (ἀλλ' οὐδέ φασιν αὐτῶν εἶναι ταῦτα ἐπισκοπεῖν), but, having reached agreement about them, they prove what follows from them. If there exists some other science about the principles of geometry, as well as those of arithmetic and the principles of every other science, then is it the same for the principles of all the sciences, or for every science in particular? However, whether there exists one general science of the principles or there are different sciences for the principles of each particular science, it will be necessary that these should have their own principles as well. Thus, it will again investigate in the same way whether the principles it uses are its own or otherwise. And if the principles every time prove different, they will go to infinity ... But if they will stop and there will be some sciences or even one specific science of the principles, it will still remain to be investigated and explained why it is a science of its own principles and those subordinate to it, whereas other sciences are not ... This, however, seems more appropriate for another branch of philosophy to examine in details.

Thus, there exists an autonomous complex of mathematical disciplines in which everything happens strictly according to the rules established by specialists.[9] Mathematicians, however, refuse to demonstrate their principles them-

---

7 Aristotle's examples mainly concern elementary mathematics; the mathematical discoveries of his time found little comment in his writings (Heath, T. L. *Mathematics in Aristotle*, Oxford 1949, 1 f.). On Aristotle and Eudoxus' astronomy, see Lloyd, G. E. R. Metaphysics Λ 8, *Aristotle's Metaphysics Lambda*, ed. by M. Frede, D. Charles, Oxford 2000, 245–273.

8 *Cat.* 10 a 11–24, *Phys.* 188 a 25, *Met.* 1020 a 35 – b 8; Heath, T. L. *The thirteen books of Euclid's Elements*, Vol. 1, Oxford 1927, 177 f.; Wehrli, comm. ad loc.

9 Cf. Arist. *Top.* 101 a 5–11, on the premises peculiar to geometry and her 'sister

selves. It is only natural, therefore, that a different science (ἑτέρα φιλοσοφία) deals with them, namely Aristotelian first philosophy.[10] Eudemus' position is thus very close to that of Plato and Aristotle, but in contrast to Plato, who reproaches the mathematicians for lack of interest in proving their principles (*Resp.* 510c–e), Eudemus seems to consider the division of labor between philosophers and mathematicians as quite natural. Indeed, the mathematicians do not want to prove their principles, not because they fail to think things through or are lazy, but rather because such a position corresponds to the general rule that no scientific discipline can prove its own principles. Proceeding from such an understanding of the division of labor, one can suppose that when dealing with subjects that are in the jurisdiction of mathematicians, Eudemus followed their norms and criteria. Indeed, the mathematics found in his works does not constitute any special type of philosophical mathematics; it is exactly what the contemporary scientific community understood by this subject. Thus, Eudemus' history of the exact sciences combines the Peripatetic conceptual framework with the professional approach to the material of geometry, arithmetic, and astronomy. It is this combination that makes him not only a reliable witness to early Greek mathematics and (mathematical) astronomy, but also their first true historian.

2. *The History of Geometry*: on a quest for new evidence

The fragments from the *History of Geometry* where Eudemus' name is mentioned are not numerous. Two of them concern the theorems of Thales (fr. 134–135), two the discoveries of Pythagoreans (fr. 136–137), one concerns Oenopides (fr. 138), another two Antiphon's squaring the circle and Hippocrates' squaring the lunes (fr. 139–140). Yet another fragment deals with Archytas' solution to the problem of duplicating of the cube (fr. 141), and the last one with Theaetetus' theory of irrationals (fr. 141.I). The origin of these fragments is rather accidental, and even taken together they are far from giving us an adequate idea of what the *History of Geometry* originally was like. The five fragments from Proclus (fr. 134–138) deal with the theorems from Euclid's book I, on which Proclus comments. The fragment on the theory of irrationals is preserved in the Arabic version of Pappus' commentary to Euclid's book X (fr. 140.I). Eutocius, commenting Archimedes' book *On Measuring the Circle*, refers to attempts to square the circle (fr. 139); in another commentary he gives (among many others) Archytas' solution of the problem of duplicating the cube (fr. 141). Finally, Simplicius, in his commentary on the passage in Aristotle's *Physics* that touches upon the quadrature of lunes, gives a long quotation from

sciences'. To reason correctly here, one has to follow the accepted definitions and rules of construction.

10 Cf. Arist. *Met.* 995b 4f., 996b 26f., 1005a 19f.

Eudemus on this matter (fr. 140).[11] Thus, five of the nine fragments concern theorems from Euclid's book I, one fragment concerns theorems from book X, and the rest deal with problems absent from the *Elements*. In their entirety, the fragments do not cover even one-tenth of the material that – judging by the *Catalogue of geometers* in Proclus[12] – was presented in the *History of Geometry*. Of the twenty mathematicians mentioned in the *Catalogue*,[13] we find only six in the fragments, including Antiphon, who is omitted from the *Catalogue*.[14]

In reconstructing the original scope of the *History of Geometry*, we can rely on these fragments as solid ground, yet we cannot confine ourselves solely to them. It is well known that, for the late authors, Eudemus was one of the main sources, if not *the* main source of information on pre-Euclidean geometry. This does not mean, of course, that any anonymous evidence concerning early Greek mathematics goes back to Eudemus. Nevertheless, there are many cases in which his authorship seems firmly established. Proclus, for example, informs us about two of Thales' theorems with a reference to Eudemus (fr. 134–135) and about two others without mentioning his name (*In Eucl.*, 157.10f., 250.20f.). It was suggested long ago that the latter two pieces of evidence are also based on Eudemus' authority,[15] which seems to me rather obvious. The same conclusion can be reached about two of Oenopides' discoveries, one of which Proclus mentions with a reference to Eudemus (fr. 138) and the other without it (*In Eucl.*, 283.7f.).[16] It is also very possible that Eutocius, who cites Archytas' solution to the problem of doubling the cube with reference to Eudemus (fr. 141), ultimately owes his information about the solutions of Eudoxus and Menaechmus to the same source.[17]

Here is another example: who was the authority for Proclus' information that the Pythagoreans knew the theorem that only the following polygons can fill up the space around a point: six equilateral triangles, four squares, and three equilateral equiangular hexagons (*In Eucl.*, 304.11f.)? There is no such theorem in Euclid, but his older contemporary Eudemus could have referred to it, since it follows immediately from the theorem on the equality of the angles of the triangle to two right angles (I, 32), which he ascribes to the Pythago-

---

11 See also Eudemus' reference to Hippocrates, omitted by Wehrli: ὥστε καὶ τὸν Εὔδημον ἐν τοῖς παλαιοτέροις αὐτὸν ἀριθμεῖν (Simpl. *In Phys.*, 69.23f.).

12 Procl. *In Eucl.*, 64.16–68.23 = Eud. fr. 133.

13 I hesitantly include in this number Philip of Opus, but not Plato; see above, 3.2 and below, 5.3. Hippias of Elis is mentioned here only as a source, not as a mathematician.

14 The *Catalogue* considers those who contributed to the progress of geometry, whereas Antiphon is only known for his unsuccessful attempt to square the circle.

15 Pesch, J. G. van. *De Procli fontibus* (Diss.), Leiden 1900, 78f.; Heath. *Elements* I, 36.

16 Van Pesch, ibid.; Heath, ibid.

17 Wehrli, com. ad loc.; Knorr *AT*, 21. Probably through Eratosthenes, who derives from Eudemus his knowledge of the solutions of Archytas, Eudoxus, and Menaechmus (3.1).

reans.[18] It seems very likely that two testimonies from the scholia to Euclid can also be attributed to Eudemus: first, that book IV of the *Elements* belongs to the Pythagoreans, and second, that they constructed three of the five regular solids (pyramid, cube and dodecahedron), to which Theaetetus added the octahedron and icosahedron.[19] Eudemus, as we know, wrote both on the Pythagoreans and on Theaetetus; besides, this version contradicts the later erroneous tradition, which ascribed to Pythagoras the construction of all five regular solids (Procl. *In Eucl.*, 65.15f.). The anonymous scholia to book V of the *Elements* twice call Eudoxus the author of this book (280.7f., 282.12f.); here too Eudemus may be the source.

The last proposition of book IV, on the fifteen-angled figure inscribed in the circle, deserves special attention. Proclus, relying on the earlier commentaries on Euclid, notes that the *Elements* contains more than a few theorems and problems useful for astronomy, e.g. problem IV, 16 (269.8f.). The same is said here about problem I, 12, which belongs to Oenopides: "he thought it useful in astronomy" (283.7f.). Since, as we noted, the latter reference goes back to Eudemus, one can suspect that the former remark has the same origin, too. This supposition gains probability if we remember that Oenopides was the very astronomer who first measured the angle of the obliquity of the ecliptic (41 A 7). Proclus explains that by inscribing the side of the fifteen-angled figure in the circle we get the angle between the celestial equator and the zodiacal circle, i.e., 24°. The confirming evidence can be found in a short but regrettably error-ridden summary from Eudemus' *History of Astronomy*: "Oenopides was the first who found the obliquity of the zodiacal circle",[20] whereas "the others found that the angle between the zodiacal circle and the celestial equator is equal to the side of the fifteen-angled figure, or 24°." (fr. 145). As K. von Fritz showed long ago, both statements pertaining to the zodiac originally referred to Oenopides.[21] Therefore, Proclus' evidence on the astronomical significance of the problem IV, 16 also goes back to Eudemus, although Proclus mentions neither his, nor Oenopides' name.[22]

---

18 Fr. 136. See van Pesch, *op. cit.*, 79; Heath. *Elements* I, 36.

19 *Schol. in Eucl.*, 273.3–13, 654.3f. See Burkert. *L & S*, 450; Neuenschwander. VB, 372f.

20 Diels' correction, λόξωσις instead of manuscript διάζωσις (Theon. *Exp.*, 198.15 = 41 A 7), is fully justified by the parallels in Aëtius, Diodorus, and Macrobius (41 A 7): they mention λόξωσις, or oblique circle. Cf. Panchenko, D. Who found the zodiac?, *Antike Naturwissenschaft und ihre Rezeption*, Vol. 9 (1999) 33–44.

21 Fritz, K. von. Oenopides, *RE* 17 (1937) 2260f. They came to be separated when Oenopides' name was taken out of context and placed at the beginning of the chronological list of astronomers: Oenopides, Thales, Anaximander, Anaximenes, others. See also Gundel, H. Zodiakos, *RE* 10 A (1972) 490; Waerden, B. L. van der. *Die Pythagoreer: Religiöse Bruderschaft und Schule der Wissenschaft*, Zurich 1979, 348f., and below, 7.5.

22 The style of IV, 16 differs from the other propositions of book IV, so it seems to be a

In fr. 141.I from the Arabic translation of Pappus, overlooked in Wehrli's first edition, Eudemus speaks about Theaetetus' contribution to the theory of irrationals. This fragment was added to the second edition, but Wehrli, following Burkert, does not include in it the preceding note on the Pythagoreans. It is clear from the text, however, that Eudemus thought that Theaetetus had continued and developed the already existing theory of irrationals. Hence, the preceding note (omitted by Wehrli), "This science had its origin in the school of Pythagoras, but underwent important development at the hands of Theaetetus",[23] perfectly fits in Eudemus' context.[24] Among the predecessors of Theaetetus, at least one person should have been mentioned: his Pythagorean teacher Theodorus, who is named in the *Catalogue* (*In Eucl.*, 66.6). Plato ascribes to him a proof of irrationality of the magnitudes from $\sqrt{3}$ to $\sqrt{17}$.[25] Further, it seems very improbable that Eudemus, who left a detailed account of the authorship of many elementary theorems, should have neglected to say something about the discovery of irrationality as such, which was made *before* Theodorus. In any case, it would be very untypical for Eudemus to refer to Theaetetus (and Theodorus) without saying a word about the *prōtos heuretēs* of irrationality, Hippasus.[26] Although Hippasus' name is not attested in Eudemus' fragments, the probability that he figured in the *History of Geometry* is rather high.

The same passage in Pappus says that Theaetetus classified the irrational lines in accordance with the different means, the geometric, the arithmetic, and the harmonic, whereas in the *Catalogue* we read that Eudoxus added to the three known mean proportionals three new ones (67.2f.).[27] If the latter in-

---

later addition (Neuenschwander. VB, 374). How can we reconcile the Pythagorean origin of book IV with Oenopides' authorship of IV, 16? Since the context of Eudemus' remarks is unknown, we can only suppose that he wrote about the Pythagorean origin of all the theorems of book IV except the last.

23 *The commentary of Pappus on book X of Euclid's Elements*, transl. by G. Junge, W. Thomson, London 1930, 63–64; cf. Burkert. *L & S*, 440 n. 182.

24 Following Pappus, the scholia to book X also call the Pythagoreans the originators of the theory of irrationals (*Schol. In Eucl.*, 415.7, 416.13, 417.12f.). Burkert's position in this question is inconsistent. He states that: 1) Eudemus, quoted by Pappus, does not mention the Pythagoreans; 2) the scholia to book X are mostly from Pappus; 3) the same scholia, ascribing the discovery of the irrationality to one of the Pythagoreans, are based ultimately on Eudemus (*L & S*, 450 n. 13, 457, 458 n. 57, 462 n. 72–73). The contradiction is easily removed by suggesting that Pappus' reference to the Pythagoreans also goes back to Eudemus.

25 *Tht.* 147d = 43 A 4; Papp. *Comm.*, 72–74.

26 Cf. his remark on the discovery of the regular solids by the Pythagoreans and Theaetetus (*Schol. in Eucl.* 654.3f.) and below, 177 n. 47. On Hippasus, see below, 189f.

27 Since the majority of Greek authors used the terms μεσότης (a mean proportional) and ἀναλογία (a proportion) interchangeably (Archytas 47 B 2; Papp. *Coll.* III, 70.16f.; Heath. *History* 2, 292f.; Wolfer, *op. cit.*, 23f.), we shall follow this usage.

formation comes from Eudemus (and there seem to be no grounds to doubt it), we can surmise that he also mentioned the person who discovered the first three means. In this connection, I would like to draw attention to the reports of Nicomachus and Iamblichus on the discovery of the proportionals. According to Nicomachus,

> There are the first proportions that are acknowledged by all the ancients – Pythagoras, Plato, and Aristotle. The very first three are the arithmetic, the geometric, and the harmonic; the other three subcontrary to them have no proper names and are called more generally the fourth, the fifth, and the sixth means. After them the later mathematicians discovered the other four proportions ... (*Intr. arith.*, 122.11f.).

Having considered the first six means, he summarizes:

> These are then the first six means generally known by the ancients: three prototypes that came down to Plato and Aristotle from Pythagoras, and the other three subcontrary to them, which came into use with later writers and followers (ibid., 142.21f.).

Thus, it comes out that the first three proportions were discovered by Pythagoras and the second three by contemporaries of Plato and Aristotle. Nicomachus' evidence is correct, but it lacks details that would allow us to connect it with Eudemus.[28] We find such details, however, in Iamblichus' commentary to Nicomachus:

> Of old there were but three means in the days of Pythagoras and the mathematicians of his times, the arithmetic, the geometric, and the third in order, which once was called the subcontrary, but had its own name changed forthwith to harmonic by Archytas and Hippasus, because it seemed to embrace the ratios that govern the harmonized and tuneful. And it was formerly called subcontrary because its character was somehow subcontrary to the arithmetic... After this name has been changed, those who came later, Eudoxus and his school, invented three more means, and called the fourth properly subcontrary because its properties were subcontrary to the harmonic ... and the other two they named simply from their order, the fifth and the sixth. The ancients and their successors thought that

28 Stated in Nicomachus (II, 22–28), the theory of ten proportions goes back to Eratosthenes' *On Means* (Περὶ μεσοτήτων); it is Eratosthenes who discovered the last four proportions (van der Waerden. *EW*, 385; Wolfer, *op. cit.*, 20ff.). The same theory can be found in Pappus (*Coll.* III, 70.16f., 84.1f.), who mentioned *On Means* several times (ibid., 636.24, 672.5, cf. 662.15). When citing Nicomachus, Pappus repeats his short historical reference, but omitting the names. According to his evidence, the ancients discovered the first three proportions, as well as the second three, while the moderns discovered the last four (84.1f.). Finally, Theon of Smyrna, who also used Eratosthenes, relates the first six (not the first three!) proportions to the Pythagoreans in general (*Exp.*, 116.3f.). It follows that *On Means* was a purely mathematical treatise and contained no (or almost no) historical information, otherwise it is hard to understand why Pappus used Nicomachus' evidence, rather than Eratosthenes'.

> this number, i.e., six, of means could be set up; but the moderns have found four more in addition …[29]

It is clear that we have a fragment of the history of mathematics before us, taken by Iamblichus from some reliable and well-informed source. It contains the names of Hippasus, Archytas, and Eudoxus that were missing in Nicomachus, along with much additional information on the early history of proportions. This information perfectly matches the fragment of Archytas quoted by Porphyry: there are three means in music, the arithmetic, the geometric, and the subcontrary, "which we call harmonic".[30] Since Eudemus ascribed the application of the first three proportions to Theaetetus and the discovery of the three others to Eudoxus, the information provided by Iamblichus likewise has to be related to the Peripatetic. Specifically, he seemed to consider not only Eudoxus but also Pythagoras (in connection with the discovery of the first three means), as well as Hippasus and Archytas (as his followers).[31] Interestingly, Iamblichus turns two more times to the history of proportions, saying again that the first three come from Hippasus and Archytas, whereas the first six were used from the times of Plato till Eratosthenes.[32] His immediate source here was, in all probability, Porphyry's commentary on Euclid's *Elements*, which traced the history of means from Pythagoras to Eratosthenes, relying mainly on Eudemus.[33]

---

29 *In Nicom.*, 100.19–101.9 = 18 A 15, transl. by M. L. D'Ooge.

30 *In Ptol. Harm.*, 93.13 = 47 B 2. Since Philolaus also called this mean harmonic (Nicom. *Intr. arith.*, 135.10f. = 44 A 24), it had to be renamed before Archytas. Tannery (Sur l'arithmétique pythagoricienne, *Mémoires scientifiques*, T. 2, Toulouse 1912, 190) believed that Archytas quoted Hippasus. Even if this was not the case, Archytas surely could have mentioned his name. Cf. Huffman, *Philolaus*, 167 ff.

31 Hippasus made an acoustical experiment with four bronze discs (Aristox. fr. 90), using the so-called musical proportion that includes the arithmetic and harmonic means (12:9 = 8:6). This implies that the first three means were known to Pythagoras, whose experiments Hippasus followed (Zhmud. *Wissenschaft*, 162 ff.). Iamblichus associates the musical proportion with Pythagoras and Philolaus (*In Nicom.*, 118.23f. = 44 A 24). Eudemus' and Aristoxenus' source might have been Glaucus of Rhegium (see below, 195). On the Pythagorean theory of proportions, see Heath. *History* 1, 85f.

32 *In Nicom.*, 113.16f., 116.1f. In the last passage Iamblichus ascribes the last four means not to Eratosthenes, but to completely unknown (and very probably fictitious) Pythagoreans, Myonides and Euphranor (116.5).

33 See below, 186ff. Though Lasserre. *Eudoxos*, 175, also connected Iamblichus' passage with Eudemus, he did not see an intermediary in Porphyry, but in Eratosthenes. Wolfer, *op. cit.*, 24, however, rightly pointed out that the *Platonicus* discusses only the first three proportions known to Plato, while Iamblichus mentions six and even ten proportions. Besides, in Iamblichus' passage Plato is missing, whereas Nicomachus, Theon of Smyrna, and Pappus, who knew the material of the *Platonicus*, do not mention Eudoxus in connection with the discovery of proportions.

The name of the author of the first *Elements*, Hippocrates of Chios, occurs both in the *Catalogue* (Procl. *In Eucl.*, 66.4) and in two of Eudemus' fragments (fr. 139–140). As follows from his detailed account on the quadrature of lunes (fr. 140), Eudemus was well acquainted with Hippocrates' work and estimated his contribution to mathematics highly. Eratosthenes' letter devoted to the problem of the duplication of the cube says:

> Hippocrates of Chios was the first to conceive (πρῶτος ἐπενόησεν) that if, for two given lines, two mean proportionals were found in continued proportion, the cube will be doubled. Whence he turned his puzzle (ἀπόρημα) into another no less puzzling.[34]

This evidence almost certainly goes back to Eudemus, on whose material Eratosthenes relied heavily.[35] In Proclus we find a similar account of Hippocrates. Giving a definition of the ἀπαγωγή, i.e., of the reduction of a complicated problem to another that, if known or constructed, will make the original proposition evident, he adds: so, for example, the problem of doubling the cube was reduced to the finding of two means in continuous proportion between two given straight lines.

> They say that the first to effect reduction (ἀπαγωγή) of difficult constructions (τῶν ἀπορουμένων διαγραμμάτων) was Hippocrates of Chios, who also squared the lune and made many other discoveries in geometry, being a man of genius when it came to construction if there ever was one (*In Eucl.*, 213.7–11, transl. by G. Morrow).

As follows from Proclus' φασι, we have here a reference to a source that seems very close to Eratosthenes' evidence. In both cases, Hippocrates is called the *prōtos heuretēs* of the problem of doubling the cube, and this was one of Eudemus' standard methods of describing mathematical and astronomical discoveries.[36] In both cases, ἀπόρημα, a difficult geometric construction, is mentioned, as well as a problem to which this puzzle was further reduced. It is noteworthy that, for the first time, ἀπαγωγή occurs in Aristotle (*APo* 69a 20f.), who brings as an example of its application the problem of squaring the circle with the help of the lunes (69a 30–34), i.e., the famous problem of Hippocrates.[37] It is only natural that Aristotle's student also applied the term ἀπαγωγή to Hippocrates' method. This is a further proof that Proclus' note derives from Eudemus' *History of Geometry*. The words of admiration for Hippocrates' talent, much more suitable for the classical than for the later author, are probably Eudemus' as well.

---

34 Eutoc. *In Archim. De sphaer.* III, 88.18–23 = 42 A 4, transl. by W. Knorr (cf. above, 85).

35 Eud. Fr. 141, com. ad loc.; Knorr *AT*, 21.

36 See above, 149.

37 Cf. Arist. *SE* 171b 12f. = 42 A 3; see below, 177 n. 45. The same method, which Plato calls ἐξ ὑποθέσεως, is said to be generally accepted in geometry (*Men.* 86e–87c). See Knorr. *AT*, 71f.

In Diogenes Laertius' biography of Archytas, there is an interesting passage where Archytas is called πρῶτος twice:

> He was the first to make mechanics into a system by applying mathematical principles;[38] he also first employed mechanical motion (κίνησις ὀργανική) in a geometrical construction, namely, when he tried, by means of a section of a half-cylinder, to find two mean proportionals in order to duplicate the cube (VIII, 83, transl. by R. Hicks).

Diogenes' immediate source is likely to have been Favorinus, who was very interested in various εὑρήματα and mentioned elsewhere Archytas' studies in mechanics.[39] Favorinus, in turn, could have relied on Eratosthenes, who certainly used Eudemus.[40] The second part of Diogenes' passage closely matches both Eratosthenes' letter that also mentions half-cylinders[41] and Archytas' solution to the problem of doubling the cube, known from Eudemus (fr. 140). Does the first part of the passage that refers to Archytas' pioneering work in mechanics come from Eudemus too? An obvious symmetry between the two parts of the passage indicates that they belong to the same context: Archytas applied mathematics to mechanics and (mechanical) motion to mathematics.[42] Let us recall that Eudemus wrote on Archytas both in the *History of Geometry* and in the *Physics*, where he states that Archytas considered things unequal (τὸ ἄνισον) and uneven (τὸ ἀνώμαλον) to be causes of motion (fr. 60, cf. fr. 65). Krafft connected this idea with the principle of the unequal concentric circles, the main principle of motion in the Aristotelian *Mechanical Problems*, and concluded that it derives from Archytas' mechanics.[43] Hence, Eudemus must have known Archytas' work in mechanics, which indirectly confirms his authority in Diogenes Laertius' passage.

It is Eudemus' evidence again that seems to be the source of the passage from Archimedes' *Quadrature of the Parabola*, where he refers to the geometers who tried to square a circle but, according to most experts' opinions, failed to do this.[44] He clearly means here the attempts of Hippocrates, Anti-

---

38 The translation according to Kühn's conjecture, μαθηματικαῖς ἀρχαῖς.

39 Fr. 66 Mensching = 47 A 10a. The passages in Diogenes Laertius, where he enumerates one εὕρημα after another, usually derive from Favorinus. In his *Manifold History* there was a special book on *prōtoi heuretai*. See Mensching, *op. cit.*, 31f., 161 (index on the word πρῶτος).

40 Cf. fr. 27 Mensching (Eratosthenes as a source).

41 Eutoc. *In Archim. De sphaer.*, 96.6f.

42 κίνησις ὀργανική does not mean, however, that Archytas solved the problem using some mechanical device, as Plutarch suggested (*Quaest. conv.* 718 E: ὀργανικαὶ καὶ μηχανικαὶ κατασκευαί). It might probably refer to the movement generated by the rotation of geometrical figures and bodies.

43 See above, 97 n. 82–83.

44 διόπερ αὐτοῖς ὑπὸ τῶν πλείστων οὐχ εὑρισκόμενα ταῦτα κατεγνωσθέν (II, 263.19–264.1).

phon, and Bryson to square a circle, which Aristotle briefly mentioned[45] and Eudemus treated more extensively (fr. 139–140; but he passes over Bryson in silence). Since Eudemus, in contrast to Aristotle, used to support his judgments by presenting the corresponding geometrical constructions, he was almost surely implied among the experts to whom Archimedes alluded.

The following example is less obvious, but no less interesting. In the introduction to his *Method* Archimedes writes:

> In the case of the theorems, the proof of which Eudoxus was first to discover, namely, that the cone is a third part of the cylinder, and the pyramid of the prism, having the same base and equal height, we should give no small share of credit to Democritus, who was the first to make an assertion with regard to the said figure, though he did not prove it.[46]

Archimedes refers to Democritus only once and hardly knew his works. Eudoxus, it seems, does not mention Democritus either. This follows from the introduction to Archimedes' treatise *On the Sphere and Cylinder*, written before the *Method*, where he refers to the same discoveries of Eudoxus, adding that before him no geometer came to these ideas (I, 4.5f. = fr. 62b Lasserre). Thus, at that time, Archimedes did not know of any of Eudoxus' predecessors in this field. Later, in the *Method*, he corrects his view, but was this correction due to his acquaintance with Democritus' books? The following prompts us rather to suppose that Archimedes took the comparison of Eudoxus and Democritus from Eudemus' *History of geometry*. 1) It was characteristic of Eudemus' style to compare the results of several geometers who worked on the same problem.[47] 2) The expression ἐξηύρηκεν πρῶτος is also typical for him. 3) Eudemus took care to specify whether a strict mathematical proof was given or not.[48] 4) In his *Physics*, Eudemus mentions Democritus (fr. 54a–b); in one place he even addresses him in the vocative (fr. 75). 5) Eratosthenes, to whom the *Method* was addressed, certainly used Eudemus' writings, which reinforces the possibility that Archimedes too was acquainted with the *History of Geometry*. If my assertion is right, we can add Democritus to the list of the mathematicians named by Eudemus. An absence of his name from Proclus' *Catalogue*, which appeared rather strange to many experts,[49] could be due to the Neoplatonic editing of the *History of Geometry*, which entailed considerable reduction of this text.

---

45 *Cat.* 7b 27f., *APo* 69a 30f., 75b 37f., *SE* 171b 12f., b 34f.; *Phys.* 185a 14f.

46 II, 430.1f. = fr. 61c Lasserre. Cf. 68 B 155 *DK*.

47 Fr. 139–140, 146; *Schol. in Eucl.* 654.3; Papp. *Comm.*, 63; on the history of proportions, see above, 173ff.

48 Cf. his remarks on Thales' discoveries: fr. 135; Procl. *In Eucl.* 157.10f., 250.20f.; Heath. *History* 1, 130f.

49 Tannery, P. Sur les fragments d'Eudème de Rhodes relatifs á l'histoire des mathématiques, *Mémoires* I, 172; van Pesch, *op. cit.*, 82; Heath. *Elements* 1, 36; van der Waerden. *EW*, 150.

The wish to attribute to Eudemus the mention of all the significant mathematicians before Euclid is further justified by the fact that the other sources do not tell us of any geometer of that period who do not appear in Eudemus, except for Hippasus and Democritus.[50] Some of them, e.g., Mamercus, Neoclides, Leon, Theudius, Athenaeus, and Hermotimus, are known only from Eudemus. This seems to be a sufficient reason to attempt to connect with his *History of Geometry* even those names that, for various reasons, are omitted from the *Catalogue* (if, indeed, we have independent and reliable evidence on them).

On the other hand, these very facts prompt us to examine the chronology of the geometers *not* mentioned by Eudemus. In Proclus we find several references to a certain Amphinomus, taken from Geminus.[51] In one case Geminus associates Amphinomus with Speusippus, opposing them to the mathematicians of Menaechmus' school (*In Eucl.*, 77.16f.); in the other, in contrast, he writes about the mathematicians from the circle of Menaechmus and Amphinomus (254.3f.). In all the references to Amphinomus, what is in question is not specific mathematical discoveries, but methodological debates and problems of terminology.[52] Thus, Eudemus might have had good reasons to omit Amphinomus in the *History of Geometry.* After all, Speusippus and Xenocrates do not appear in this book either. There is, however, another possibility: perhaps Amphinomus worked after Eudemus had finished his book. This would make him a younger contemporary of Menaechmus, whose generation is the last to appear in Eudemus. According to Geminus, Amphinomus held the view that mathematics does not investigate the causes and that the originator of this view was Aristotle (*In Eucl.*, 201.11). This seems to imply that Amphinomus lived after Aristotle. To be sure, Geminus (or his source) was not interested in chronology and could associate persons according to the similarity of their views, regardless of when they lived. The total silence of the classical sources on Amphinomus seems, rather, to indicate that he lived in the Hellenistic period.[53]

One more reference in Proclus, likewise taken from Geminus, concerns a certain Zenodotus, "who belonged to the succession of Oenopides, although he was a pupil of Andron" (80.15f.). The mathematicians Zenodotus and Andron

---

50 Aristotle mentions several times Bryson's attempt to square the circle (*APo* 75b 37f.; *SE* 171b 15f., 172a 3f.). Later Aristotle's commentators discussed it (Heath. *Mathematics*, 47f.; Mueller. Aristotle, 160ff.; Knorr. *AT*, 76f.), but they, it seems, had no sources apart from Aristotle. That means that Eudemus, while reporting Antiphon's attempts, fails to mention Bryson. Whatever Eudemus' motives could have been, Bryson was certainly not a mathematician.

51 *In Eucl.*, 77.16, 202.11, 220.9, 254.4; cf. van Pesch, *op. cit.*, 112f. On Geminus, see below, 184f., 291f.

52 See Bowen. Menaechmus, 14f.; Knorr. *AT*, 74f.

53 Cf. Tannery. *Géométrie*, 138 n. 1. Knorr. *AT*, 74f. and Tarán. Proclus, 238 n. 37, are indecisive about Amphinomus' chronology; Lasserre included him in the list of the 'Academic mathematicians' (*Léodamas*, 149f., 587f.).

are otherwise unknown; Oenopides of Chios was the older contemporary of Hippocrates. But was he *the* Oenopides that Geminus had in mind? Geminus discusses the distinction between theorems and problems; their definitions given by Zenodotus are very close to those of Posidonius' followers (80.15f.). Thus, in all probability, Oenopides, Zenodotus, and Andron also belong to the Hellenistic period.[54]

The chronology of Aristeas the Elder, who wrote on conic sections and regular solids (Pappus and Eutocius mention him), is still uncertain.[55] In any case, he was younger than Menaechmus, who discovered the conic sections and probably belonged to Euclid's generation.[56]

3. The *Catalogue of geometers*: from Eudemus to Proclus

Moving from the evident cases to the less evident, we come to one of our central problems: who was the author of the *Catalogue of geometers* and how did this document come to Proclus? It was customary since the late 19th century to think of the information in the *Catalogue* as going back, albeit through intermediaries, to Eudemus' *History of Geometry* (fr. 133).[57] Although Proclus does

---

54 Knorr. *AT*, 374 n. 70, cf. Bowen. Menaechmus, 13f. Since Geminus' methodological discussion of theorems and problems is based on Posidonius (*In Eucl.*, 77.7–78.10, 80.15–81.4 = fr. 195 E.-K.), the other references to Amphinomus, Speusippus, Menaechmus, Oenopides, Zenodotus, and Andron might derive from the same source. Aëtius (I,7.17) mentions Oenopides along with the Stoics Diogenes and Cleanthes, and a Stoic idea, the god is the soul of the world, is ascribed to him (see below, 7.5). Von Fritz. Oinopides, 2267f., was wrong to relate the methodological discussion on theorems and problems to Oenopides of Chios: it could not have taken place in the fifth century.

55 Allman, *op. cit.*, 194ff.; Heath, T. L. *Apollonius of Perga*, Cambridge 1896, xxiff; Knorr, W. R. Observations on the early history of the conics, *Centaurus* 26 (1982) 1–24; idem. *AT*, 32f.; *Pappus of Alexandria. Book 7 of the Collection*, transl. by A. Jones, Pt. 2, New York 1986, 404, 577f.

56 More complicated is the case of the mathematician Thymaridas. Iamblichus (*In Nicom.*, 11.2f., 27.4, 62.19, 65.9, 68.3f. = Timpanaro Cardini, M. *Pitagorici – Testimonianze e frammenti*, Pt. 2, Florence 1962, 444f.) quotes his definition of a number and an arithmetic puzzle, the so-called epanthem. Since a certain Thymaridas of Paros is named in the catalogue of the Pythagoreans compiled by Aristoxenus (*DK* I, 447.3), some scholars date him in the fourth century (Heath. *History* 1, 94; Becker, O. *Das mathematische Denken der Antike*, Göttingen 1957, 43f.; cf. Burkert. *L&S*, 442 n. 92). Diels, however, considered it impossible to date Thymaridas' epanthem and definition of number so early (*DK* I, 447.3n.). Federspiel, M. Sur "l'épanthème de Thymaridas", *LEC* 67 (1999) 354, suggests that Thymaridas could be a younger contemporary of Eudoxus. If Thymaridas did live before Eudemus, his absence from the *History of Geometry* could still be explained: his puzzle is purely arithmetical.

57 Spengel, L. *Eudemi Rhodii Peripatetici fragmenta quae supersunt*, Berlin 1865, IX;

not mention Eudemus in connection with the *Catalogue*, he refers (*In Eucl.*, 68.4) to "those who compiled histories of geometry" (οἱ τὰς ἱστορίας ἀναγράψαντες) before Euclid. Besides, Eudemus' fragments, including those quoted by Proclus himself, coincide thematically with the *Catalogue*: they tell us about the development of geometry from Thales to Eudoxus' students (fr. 134–141). However, in the last few decades, this *opinio communis* has been challenged. Lasserre, in particular, on the basis of the similarities between the *Catalogue* and the passage from Philodemus discussed above (3.1), considered Philip, to whom he ascribed this passage, to be the author of the *Catalogue* as well.[58] Indeed, against the background of Eudemus' fragments, the passages of the *Catalogue* concerning Plato and Philip look rather odd. Eudemus could not regard Philip's preoccupation with problems connected with Platonic philosophy as his foremost contribution to mathematics. And was it really relevant for the history of geometry that Plato's writings were "thickly sprinkled with mathematical terms"? Eudemus could have claimed the same for his teacher's works as well. These two passages can hardly belong to a Peripatetic. They are much more likely to come from the Platonist whom Lasserre considered to be the author of the papyrus passage, i.e. Philip. Hence, Lasserre concluded that Philip is the author not only of the second part of the *Catalogue*, which begins with Plato and ends with Philip himself, but of the entire *Catalogue*.

Reasonable as many of Lasserre's observations may seem, I believe that there are no grounds for such a conclusion. First, it is far from evident that Philip was the author of the passage cited by Philodemus (3.2). Second, the *Catalogue* contains too much detailed information on the early Greek geometers that is not related to Plato. It is hard to explain why Philip's book *On Plato* began with the Egyptians and Thales and, even more strangely, ended with Eudoxus' students, who were more than half a century younger than Plato. Third, although Plato could not have been a reference point in the history of the fourth-century geometry written by a Peripatetic, both Academics and Neoplatonists could have considered him one. Thus, it is possible to come up with the following alternatives: either the *Catalogue* was taken from a book by one of Plato's students and does not have any connection with Eudemus, or it was compiled on the basis of Eudemus' work and its Platonic features can be explained by later Neoplatonic redaction. In the latter case, the Neoplatonic redactors could have added to it the material borrowed from the writings of Plato's disciples. The second alternative seems to me preferable, being favored by the following facts.

The traces of Neoplatonic redaction are also discernible in the first part of the *Catalogue*, which is not related to Plato at all, e.g. in the passage where the

---

Tannery. Eudème, 171f.; Allman, *op. cit.*, 2; van Pesch, *op. cit.*, 80: "inter omnes viros doctos constat originem id ab Eudemi historia duxisse".

58 Lasserre. *Léodamas*, 611f. Earlier he attributed the *Catalogue* to Eudemus (Eudox. fr. 22).

discovery of the five regular solids is attributed to Pythagoras (*In Eucl.*, 65.15f.)[59] and typically Neoplatonic terms are encountered (ἀΰλως καὶ νοερῶς). Another such trace is the reference to the late pseudo-Platonic *Anterastai* (66.3), which can belong neither to Philip, nor to Eudemus. Furthermore, if the *Catalogue* has the same origin as Philodemus' passage, he must have known this text. However, in his list of Plato's students, only two of the twelve mathematicians of the fourth century listed in the *Catalogue* are named: Amyclas of Heraclea and Archytas,[60] whereas Philip is inexplicably missing! Both of these names also occur in the source of Diogenes Laertius' list of Plato's students[61] and can be thus traced back to a common tradition that has no connection with the *Catalogue*, where Archytas is *not* called Plato's student.

We know that Eudemus wrote of Archytas and Theaetetus (fr. 141–141.I), the evidence for the duplication of the cube by Eudoxus and Menaechmus also goes back to him, and his *History of Astronomy* mentions Eudoxus and Callippus (fr. 148–149). Thus, it is hardly possible to exclude Eudemus from the *Catalogue*'s sources, i.e., not to number him among those who had been writing the history of geometry before Euclid. Was the author of the quotation in Philodemus also among them? Eudemus might have used the Academics' writings, but it seems unlikely that he should have simply copied the descriptions of Plato and Philip from them. There are many more reasons to relate them to a Neoplatonic redactor. The description of Philip as a faithful disciple of Plato and the fact that Philip is the last mentioned in the *Catalogue* were among Lasserre's major arguments. Lasserre, however, like many others, overlooked the fact that the Platonizing tendency in the *Catalogue* does not end with Philip, it includes Euclid as well. Proclus, once again uniting all the previous mathematicians around Plato, says:

> Euclid was later than the mathematicians around Plato (τῶν περὶ Πλάτωνα), but earlier than Eratosthenes and Archimedes... He belonged to the Platonic school and was at home in this philosophy, and this is why he thought the goal of the *Elements* as a whole to be the construction of the so-called Platonic figures (68.17f., transl. after G. Morrow).

For simple chronological reasons, this phrase could come neither from Eudemus nor from any of Plato's students. It is revealing, however, that it is as similar to the description of Philip as if the two passages were written by the same hand. Both mathematicians worked under Plato's guidance (although in Eu-

---

59 Eudemus, as we remember, ascribed the first three solids to Pythagoreans, the octahedron and the icosahedron to Theaetetus (*Schol. In Eucl.*, 654.3). See above, 174.

60 Gaiser. *Academica*, 110 f., 439 ff.; Dorandi. *Filodemo*, 135 (col. VI). Cf. above, 100 n. 93.

61 Diogenes' list (III, 46) goes back to Plato's biography written by Theon of Smyrna (Gaiser. *Academica*, 439 f., 444). Theon's list, preserved in Arabic, contains both Amyclas and Archytas, whereas Archytas is omitted in Diogenes, since for him Archytas belonged to the Pythagoreans.

clid's case this guidance was not direct), and for both Platonic philosophy was the final goal in mathematics.[62] This proves that the whole historical digression in Proclus' commentary, of which the *Catalogue* is part, was subjected to Platonizing revision *after* the fourth century BC. It is this revision that explains the Platonizing tendency of the *Catalogue* and makes it similar to the quotation in Philodemus.

Let us note again that the *Catalogue* contains chronological indications concerning practically all the mathematicians mentioned in it. The accuracy of these indications differs according to information that was at Eudemus' disposal. Sometimes he only gives the chronological sequence, e.g. Mamercus comes after Thales, Pythagoras after Thales and Mamercus, and Anaxagoras after Pythagoras. Beginning with Oenopides, the information becomes more detailed: the latter was "a bit younger than Anaxagoras", and after him follow Hippocrates and Theodorus (both, as we know, were of the same generation). As with any other historian, the dates in Eudemus become more and more accurate as he approaches his own time. Thus, the indications concerning geometers of the fourth century are more precise: Neoclides and his student Leon were younger than Leodamas, and Eudoxus "was a little younger" than Leon. Then follows the generation of Eudoxus' students: Menaechmus, Dinostratus and others, and Athenaeus, "who lived in the same time". In all cases in which Eudemus' evidence can be verified, it proves correct and serves as a reliable basis for further chronological reconstruction.

What we are dealing with here is not just a collection of the separate dates, but a continuous chronological series, or scale, that connects all mathematicians from Thales till Eudemus' own time. A somewhat analogous genealogical scheme is given in the first chapter of the *Opinions of the Physicists* (4.5), but it is not as consistent as that of Eudemus. Besides, Theophrastus often proceeded not from the real fact of apprenticeship, but from the similarity of doctrines. Eudemus' relative chronology does not depend on doctrinal similarity, nor on the fact of apprenticeship: Anaxagoras was neither a pupil nor a follower of Pythagoras; Oenopides was neither a pupil nor a follower of Anaxagoras; and Eudoxus was neither a pupil nor a follower of Leon. The last case is particularly revealing: Eudemus undoubtedly knew that Eudoxus was Archytas' student, but he preferred to give a more accurate chronological reference rather than mention his teacher: "a bit younger than Leon". Although Eudemus did not use any of the general chronologies that existed by his time (e.g., the chro-

---

62 Note the similarity of the two passages in structure and word usage (67.23f., 68.20f.):

Φίλιππος … καὶ τὰς ζητήσεις ἐποιεῖτο κατὰ τὰς Πλάτωνος ὑφηγήσεις καὶ ταῦτα προύβαλλεν ἑαυτῷ, ὅσα ᾤετο τῇ Πλάτωνος φιλοσοφίᾳ συντελεῖν.

Εὐκλείδης … καὶ τῇ προαιρέσει δὲ Πλατωνικός ἐστι καὶ τῇ φιλοσοφίᾳ ταύτῃ οἰκεῖος, ὅθεν δὴ καὶ τῆς συμπάσης στοιχειώσεως τέλος προεστήσατο τὴν τῶν καλουμένων Πλατωνικῶν σχημάτων σύστασιν.

nology of Olympic games, or of the archons of Athens), his system allowed one to calculate from the dates of one geometer a rather acceptable chronology for almost all his contemporaries.[63]

Revealingly, it is precisely Plato and his faithful disciple Philip who do not fit in this system. Plato is mentioned *earlier* than Leodamas, Archytas, and Theaetetus, who were older than him (3.2); the latter, in turn, are dated in "Plato's time". For a history of geometry, this chronological link seems odd enough and has no parallels in Eudemus. Philip, in contrast, emerges at the end of the *Catalogue*, *after* Eudoxus' students and their younger contemporaries, whereas in fact he belonged to Eudoxus' and Aristotle's generation, if not to an earlier one.[64] This confirms again that the passages on Plato and Philip, as we know them, were inserted in the *Catalogue* later. The 'final' position of Philip is not a mark of his authorship, but a result of Neoplatonic redaction.

In an article on the *Catalogue*, Eggers Lan also points out many passages where Neoplatonic influence is clearly observed, but this leads him to completely different conclusions: that Proclus himself compiled the *Catalogue* and that, except for two or three references, it does not go back to Eudemus, either directly or indirectly.[65] To be sure, in addition to all the aforesaid, the late composition of the entire historical digression in Proclus (*In Eucl.*, 64.16–68.23) is evident from the fact that, besides Euclid, it mentions Archimedes and Eratosthenes. This, however, does not mean that we should exclude Eudemus from its main sources. Even if we did not know about his *History of Geometry*, we could infer the existence of such a work from the *Catalogue*'s detailed information on pre-Euclidean geometry. The very fact that it contains names of geometers from the fourth century, practically unknown to us from other sources, as well

---

63 The only big lacuna in Eudemus' chronology is between Pythagoras (born ca. 570) and Anaxagoras (born ca. 500). It could have emerged due to the disappearance from the *Catalogue* of Pythagoras' student Hippasus (born ca. 530). Other mathematicians are separated from each other by no more than a generation.

64 In the *Suda*, Philip is characterized as Socrates' and Plato's student who lived in the time of Philip of Macedon (Lasserre. *Léodamas*, 20 T 1). His chronology was accordingly considered to be ca. 419–340 (Tarán. *Academica*, 127f.). Lasserre. *Léodamas*, 594, changes the date of his birth to 385/80, making him an exact contemporary of Philip of Macedon (382–336). To this end, he had to assume that the Socrates mentioned in the *Suda* was Socrates the Younger! But even then, contradictions remain unresolved. To date somebody at the time of Philip of Macedon means to relate this person to the date of Philip's death (336), not of his birth. A person who died 15–20 years after Philip of Macedon would rather be related to Alexander's time. But if Philip of Opus was the *Catalogue*'s author, he must have lived at least till the 320s to be able to describe achievements of Eudoxus' students. Well, Philip *could* have been born ca. 385/80 and *could* have lived till the end of the century, but how can one reconcile this chronology with the fact that in the *Catalogue* he is named *after* Eudoxus' students? Obviously, it is impossible to date Philip on the basis of the *Catalogue*.

65 Eggers Lan, *op. cit.*, 154f.

as precise chronological indications on them, suggests its early origin. Where, if not in Eudemus, could a later author get the information that Neoclides was younger than Leodamas (66.18) and that his student Leon was a little older than Eudoxus (67.2), if Neoclides and Leon are not mentioned elsewhere?

Considering Eudemus the main source of the *Catalogue*, the scholars replied differently to the questions whether his *History of Geometry* was available to Proclus and whether Proclus himself could have been the compiler of the *Catalogue*. Whereas Tannery's answer was negative in both cases,[66] Heiberg, van Pesch (who studied Proclus' sources), and Heath were inclined to the conclusion that, although Proclus probably had access to Eudemus' book, he did not compile the *Catalogue*, but took it from another source.[67] While agreeing with this opinion, let us emphasize that there are no grounds to believe that Eudemus' writings had been lost in the fire of the Alexandrian library and so were not available after 389 AD.[68] One should recall that Simplicius gives a long verbatim quotation from the *History of Geometry* on the quadrature of the lunes (fr. 140), cites almost a hundred passages from Eudemus' *Physics*, and transmits three of the seven preserved fragments from the *History of Astronomy* (fr. 146, 148–149), praising Eudemus' concise and clear style (fr. 149). Eutocius' words also imply that the *History of Geometry* was available to him (fr. 139),[69] whereas Damascius seemed to use Eudemus' *History of Theology* (fr. 150) without any intermediaries. All this simply does not fit with the disappearance of Eudemus' writings by the time of Proclus, who had at his disposal the same library of the Academy in Athens as did Simplicius after him.[70]

We also have to admit that Geminus, whom many, after Tannery,[71] regarded as the *Catalogue*'s compiler and the intermediary between Eudemus and Proclus, is hardly fit for this role either.[72] All we know about Geminus[73] and his mathematical encyclopedia[74] agrees neither with Neoplatonic influence in the

---

66 Tannery. Eudème, 171.

67 Heiberg. Jahresberichte, 345; van Pesch, *op. cit.*, 84; Heath. *Elements* 1, 37f. Nobody, except for Eggers Lan, as far as I know, argued in detail for Proclus' authorship.

68 That was Tannery's view (Eudème, 171).

69 Cf. Knorr. *TS*, 126 n. 124 and 128 n. 146.

70 See Heath. *History* 2, 530f.; idem. *Elements* 1, 35.

71 Tannery. Eudème, 172f.; idem. *Géométrie*, 71f.

72 See van Pesch, *op. cit.*, 80f.; Heath. *Elements* 1, 37; Eggers Lan, *op. cit.*, 140f.

73 Schmidt, M. Philologische Beiträge zu den griechischen Mathematikern, *Philologus* 45 (1886) 63–81, 278–320; Tannery. *Géométrie*, 18f.; van Pesch, *op. cit.*, 87f., 95f.; Tittel, K. Geminos, *RE* 7 (1912) 1026–1050; Heath. *Elements* 1, 38f. For Geminus' chronology (ca. 70 BC), see Jones, A. Geminus and the Isia, *HSCPh* 99 (1999) 255–267.

74 Its title is given as Περὶ τῆς τῶν μαθημάτων τάξεως (Papp. *Coll.* VIII, 1026.9) and as Μαθημάτων θεωρία (Eutoc. *In Apol. Con.* II, 168.17f.). Schmidt. Philologische Beiträge, 71; Tannery. *Géométrie*, 18f.; Tittel, *op. cit.*, 1040f., and Heath.

Catalogue, nor with its special interest in the predecessors of Euclid who compiled the *Elements*. Geminus' encyclopedia was devoted to the foundations of the exact sciences and to their classification; his main goal was to show the logical consistency of Euclid's mathematics and to refute its critique by the Sceptics and Epicureans.[75] There is no evidence that he was particularly interested in the history of mathematics before Euclid or that he knew Eudemus' book.[76] The only pre-Euclidean geometer mentioned in those passages of Proclus' commentary that can be safely attributed to Geminus[77] is Menaechmus. In two of the three cases, the foundations of mathematics and methodological debates are in question (*In Eucl.*, 72.24f., 78.9f.),[78] and once only Geminus refers to Menaechmus' discovery of the conic sections (111.21f.). But even here, his authority is Eratosthenes' epigram, and not Eudemus. One gets the same impression from Geminus' *Introduction to Phaenomena*: Eudemus' *History of Astronomy* clearly was not used there.[79]

Generally, one has to admit that Tannery's few arguments in favor of Geminus are so unconvincing that it seems that only his authority promoted this idea.[80] Meanwhile, apart from Geminus' encyclopedia, there were three com-

---

History 2, 223, favored Eutocius' version. Mansfeld. *Prolegomena*, 24 n. 21, opts for Pappus' version.

75 Tittel, *op. cit.*, 1040ff. The same was characteristic of his teacher Posidonius (fr. 46–47, 195–199 E.-K.). See above, 179 n. 54. Lasserre believed Posidonius to be the intermediary between Philip and Geminus (*Léodamas*, 20 F 17–23, 614f.). On Geminus, cf. below, 291f.

76 Tittel, *op. cit.*, 1048, suggested that Geminus knew Eudemus only through Eratosthenes. In this case, Eratosthenes becomes the *Catalogue*'s author, which is impossible (see above, 173 n. 28, 174 n. 33).

77 See van Pesch, *op. cit.*, 112f.

78 This material comes from Posidonius (see above, 179 n. 54) and not from Eudemus.

79 Geminus briefly mentions the Pythagoreans (10.5), Eudoxus' calendar (108.5. 17) and parapegma (210.17–18), and in more detail the 19-year cycle of Euctemon, Philip, and Callippus (120.6ff.) and the 76-year cycle of Callippus (122.16ff.). Heraclides Ponticus' name is attested in Geminus' summary of Posidonius' *Meteorologica* (Simpl. *In Phys.*, 291.21f. = Her. Pont. fr. 10 = Posid. fr. 18 E.-K.).

80 The passage that precedes the *Catalogue* says that according to Aristotle, sciences emerged and perished many times (64.9–15). Tannery (*Géométrie*, 71, cf. van Pesch, *op. cit.*, 81) relates this idea to the Stoics rather than to the Peripatetics and, therefore, ascribes it to Geminus. Meanwhile, the Aristotelian provenance of this idea is obvious (*Cael.* 270b 19, *Pol.* 1329b 19), so Proclus rightly refers to Aristotle, and not to the Stoics. In the passage that follows the *Catalogue*, Proclus discusses Euclid's *Pseudaria*, which is now lost. Though Proclus' praise suggests that this work was available to him, Tannery excludes this possibility and considers the whole passage to have been derived from an earlier author. "Who could have been this author, if not Geminus?" (*Géométrie*, 72). Even if Tannery was right in his first supposition (that Proclus knew *Pseudaria* through some intermediary), there is nothing to confirm the second one (van Pesch, *op. cit.*, 81, 83 n. 1). Believing his remarks to provide

mentaries to the *Elements* among Proclus' sources: by Hero, Porphyry, and Pappus. There is no point in discussing Hero as a Platonizing redactor of Eudemus. Pappus, although he suits this role better than Hero,[81] does not mention Eudemus in his vast *Collectio* and only once in his commentary on book X of Euclid. Plato's name occurs only twice in the *Collectio*: first, in connection with the "nature of proportion" and second, concerning the so-called Platonic bodies.[82] Although Plato is mentioned more often in the commentary on book X, Pappus was clearly too 'technical' an author to be enthusiastic about Plato's contribution to geometry.[83] As for the Neoplatonist Porphyry, he is the most appropriate option for the role of the compiler of the *Catalogue*. Actually, Tannery had already considered this option, but rejected it in favor of Geminus,[84] who, unlike Porphyry, did not comment on the theorems of Euclid's book I; his attention was focused on the foundations of mathematics, including definitions, axioms, and postulates. Hence Tannery postulates, in a rather mechanical way, the prevailing influence of Geminus in the first part of Proclus' commentary, which comprises the *Catalogue*, and of Porphyry in the second one. Yet this picture is far from being correct: both introductions to Proclus' commentary contain material from both Geminus and Porphyry.[85] What makes Porphyry the most preferable candidate?

1) In his *Commentary*, Proclus refers five times to Porphyry's work[86] that was, in all probability, a commentary to Euclid's book I.[87] In any case, the last of Proclus' quotations covers five pages and could have been taken only from a special work on the *Elements*.[88] Such a commentary, written by the Neoplaton-

---

a sufficient proof that the *Catalogue* is preceded and followed by Geminus' material, Tannery does not bring any specific arguments in favor of his authorship.

81 On Neoplatonic influence on Pappus, see Mansfeld. *Prolegomena*, 99ff.

82 Papp. *Coll.* III, 86.19f.; V, 352.10f. Both times he is named ὁ θειότατος Πλάτων.

83 On Pappus' attitude toward philosophers, see Cuomo, S. *Pappus of Alexandria and the mathematics of late Antiquity*, Cambridge 2000.

84 Tannery. Eudème, 171f.

85 Proclus refers to Porphyry's Σύμμικτα ζητήματα in the same second introduction, where the *Catalogue* is located (*In Eucl.*, 56.24). The direct references, however, do not give an adequate picture. In the first introduction, Proclus borrows extensively from Iamblichus, whose name never occurs in the entire *Commentary*. See Mueller, I. Iamblichus and Proclus' Euclid commentary, *Hermes* 115 (1987) 334–348.

86 255.12–14, 297.1–298.3, 315.11–316.13, 323.5–326.5, 347.20–352.14 = fr. 482–486 Smith.

87 Tannery. Eudème, 170f.; van Pesch, *op. cit.*, 127f.; Heath. *History* 2, 529; idem. *Elements* 1, 24; Mueller, I. Mathematics and philosophy in Proclus' commentary on book I of Euclid's Elements, *Proclus*, 311f. Mansfeld's suggestion (*Prolegomena*, 24) that this work was a part of Σύμμικτα ζητήματα is not convincing. According to Porphyry's biographer Eunapius, he studied all branches of knowledge, including arithmetic, geometry, and music (*Vit. Soph.*, p. 457 Wright).

88 Porphyry's *Elements* are mentioned in Arabic sources (422 T Smith), according to which they had been translated into Syriac and consisted of one book (I am grateful

ist Porphyry, could explain both the Platonism of the second part of the *Catalogue* and its interest in the predecessors of Euclid. Characteristically enough, it is in the passages devoted to the two figures of particular importance to Porphyry – Pythagoras and Plato – that his influence is the most easily traced. 2) Proclus twice refers to Eudemus immediately after Porphyry, the first time in the space of one page (297.4–298.10 and 299.3, on prop. XIV and XV), and the second time in the very same line (352.14, on prop. XXVI). This reinforces the probability that Porphyry's commentary included references to Eudemus' book (in both cases theorems attributed to Thales are in question).[89]

3) Porphyry was the author of a commentary on Ptolemy's *Harmonics* containing the only extant fragment from Eudemus' *History of Arithmetic* (fr. 142). 4) The passage from the *Catalogue* relating to Pythagoras (65.15f.) contains Neoplatonic terms and partly coincides with the passage from the work by Porphyry's student Iamblichus (*De comm. math. sc.*, 70.1f.). This work makes a brief mention of Theodorus and Hippocrates (77.24f.), which is missing in the parallel passage of his biography of Pythagoras (*VP* 89). This remark is very similar to the place in the *Catalogue* where both mathematicians are also mentioned in one sentence (66.4f.). It looks as though Iamblichus used the same source as Proclus,[90] i.e., Porphyry. The same source must be the origin of Iamblichus' information on the development of the theory of means from Pythagoras to Eratosthenes.[91]

5) Porphyry wrote a history of philosophy that starts from its Oriental precursors, moves up to Thales and other Presocratics, and ends with Plato, i.e., it embraces practically the same period as the *Catalogue*.[92] Porphyry's attitude toward Plato's system as the consummation of the whole of earlier philosophy

---

to Maroun Aouad for his assistance on this point). The book could hardly have been a commentary on Euclid.

89 This does not necessarily mean (*pace* Tannery. Eudème, 170f.) that in these two cases Proclus used Eudemus through Porphyry. It is equally possible that Porphyry's references prompted Proclus to look in Eudemus' book. Simplicius also used Eudemus' *History of Geometry* and *Physics* both directly and indirectly, through Alexander of Aphrodisias' commentary on Aristotle's *Physics* (e.g., fr. 43, 82b, 140). See Knorr. *AT*, 29ff.; Sharples, R.W. Eudemus' *Physics*: Change, place and time, *Eudemus of Rhodes*, 114f.

90 Björnbo, A. A. Hippokrates von Chios, *RE* 8 (1913) 1782; Burkert. *L & S*, 458 n. 59. We find in this remark two 'progressive' terms, ἐπιδιδόναι and προαγαγεῖν, so characteristic of the *Catalogue*. See below, 212 n. 222.

91 See above, 173f.

92 φιλόσοφος ἱστορία ἐν βιβλίοις δʹ = fr. 193–224 Smith. Apart from the fragments, only one part of book I is preserved, containing the *Life of Pythagoras*. The *Catalogue* shares with the *Life of Pythagoras* the idea that geometry comes from Egypt and arithmetic from Phoenicia (*In Eucl.*, 65.3f.; Porph. *VP* 6), but this can be found already in Plato (*Leg.* 747a–c); cf. Strab. 16,2.24, 17,1.3. See also Panchenko, D. Ὅμοιος and ὁμοιότης in Thales and Anaximander, *Hyperboreus* 1.1 (1994) 40 n. 36.

is very close to the 'Platocentrism' of the second part of the *Catalogue*, where the construction of the five Platonic bodies is called the final goal (τέλος) of the whole *Elements* (68.18f.). This provides additional support to the hypothesis that Porphyry was the principal intermediary between Eudemus and Proclus.[93]

Thus, we find in Porphyry exactly what is missing in Geminus that would have enabled us to consider him the compiler of the *Catalogue*: a predisposition to Platonism, obvious knowledge of Eudemus' works, interest in the intellectual history of the classical period in general and in Euclid's predecessors in particular. It seems that we owe to Porphyry the present form of the *Catalogue*, i.e., a very abridged and tendentiously revised version of the *History of Geometry*. Following Eudemus' chronological exposition, he compiled a short outline of geometry from Thales to the immediate forerunners of Euclid and used it as a historical introduction to his commentary on the *Elements*. It is conspicuous that, as soon as Porphyry leaves Eudemus' text and turns to Euclid, he begins to lack reliable historical data. The absence of any followers of Eudemus' in the Hellenistic period made it very problematic for Porphyry to figure out even a rough date for Euclid: he had to rely on casual remarks of Archimedes and Eratosthenes and on historical anecdotes (68.10). What a contrast with Eudemus' chronologies, which are accurate almost to a decade!

The last circumstance proves that Eratosthenes' *Platonicus* was historical only to the degree that it relied on Eudemus' work. The chosen genre did not allow Eratosthenes to continue the *History of Geometry*; nor is it known whether he wanted to continue it. The fragment concerning the history of the discovery of means, which Iamblichus probably borrowed from Porphyry,[94] can scarcely derive from Eratosthenes. Although Eratosthenes found the last four means and, therefore, must have known of the first six, there is no evidence that he described the history of their discovery in the *Platonicus* or in other works.[95] In this case, Porphyry must have relied directly on Eudemus' book.

It would be a mistake, however, to think that the final version of the *Catalogue* is due to Porphyry. Specifically, the passage that ascribes to Pythagoras the discovery of irrationality and of the five regular solids (65.15f.) can probably be attributed to Proclus. At any rate, we do not find this information in the parallel passage in Iamblichus (*De comm. math. sc.*, 70.1f.) nor in his other works on the Pythagoreans. Meanwhile, Pan-Pythagoreanism is much more characteristic of Iamblichus than of Proclus, and if he had known this tradition,

---

93 Porphyry could have derived additional information from the same work as Philodemus (3.1), which would explain certain similarities between the *Catalogue* and column Y. It is noteworthy that Porphyry knew Hermodorus' book *On Plato*: he cited a long passage from it, which he had found in Dercyllides (fr. 7–8 Isnardi Parente = *FGrHist* 1008 F 2a–b = fr. 146 Smith).

94 Porphyry's interest in the early theory of means is confirmed by his quotation from Archytas (*In Ptol. harm.*, 92 = 47 B 2).

95 See above, 173 n. 28, 174 n. 33.

he would have certainly mentioned it.[96] While the version referring the five regular solids to Pythagoras was known before Proclus,[97] the idea that Pythagoras was the author of the theory of irrationals (τῶν ἀλόγων πραγματεία) is not attested in any other ancient source. In this connection it has been long proposed to change the reading τῶν ἀλόγων πραγματεία into τῶν ἀνὰ λόγων πραγματεία, i.e., into the theory of proportions.[98] To judge from Iamblichus' passages that connect Pythagoras with the discovery of proportions,[99] Porphyry and, correspondingly, Eudemus wrote about τῶν ἀνὰ λόγων πραγματεία, not τῶν ἀλόγων πραγματεία. This is confirmed both by Nicomachus and by the obvious fact that Pythagoras did, indeed, study proportions.[100] What made Proclus change the 'theory of proportions' into the 'theory of irrationals'?

I suppose he did it for the same reason that made him remain silent about the Pythagorean Hippasus, one of the important mathematicians of the early fifth century. The late tradition connects with Hippasus two significant discoveries: the construction of the dodecahedron inscribed in the sphere and the discovery of irrationals. One part of the evidence (Clement of Alexandria, Iamblichus) mentions Hippasus in this connection, the other speaks about some anonymous Pythagorean.[101] At least two of the last group's testimonies concerning the discovery of irrationality are based on Eudemus: Pappus' commentary on book X of Euclid and the scholia to the same book.[102] To them we should add the first scholium to book XIII that relates to the Pythagoreans the discovery of the first three regular solids, including the dodecahedron. Does it follow from this evidence that Eudemus referred to Pythagoreans in general, without mentioning any names, while the late tradition filled in Hippasus' name? If so, we are not in a position to determine who in fact was the author of these discoveries. Besides, it means that Hippasus totally disappears from the history of mathematics, since no other discoveries are ascribed to him. In other

---

96 This means that Porphyry most probably shared Eudemus' view on the discovery of the regular solids (*Schol. In Eucl.*, 654.3). Cf. below, n. 97.

97 See e.g. Aët. II,6.5 = 44 A 15. For other evidence, see Sachs, E. *Die fünf platonischen Körper*, Berlin 1917, 8 ff. The five regular solids are mentioned in Speusippus' *On Pythagorean Numbers* (fr. 28 Tarán), which was probably the original source of this version (Burkert. *L & S*, 71; cf. Tarán. *Speusippus*, 256 f.). Later it can be found in doxography (Sachs, *op. cit.*, 65 f.). If Achilles' version (*Dox.*, 334 not.), which is parallel to Aëtius, goes back to Posidonius (Sachs, *op. cit.*, 10, 51 f.; Burkert. *L & S*, 70 n. 113), then the latter, unlike Eudemus, ascribed all the five regular solids to the Pythagoreans.

98 *DK* I, 98.23; Heath. *History* 1, 84 f. This reading is attested only in one manuscript of Proclus, all the others have τῶν ἀλόγων πραγματεία: Stamatis, E. S. Die Entdeckung der Inkommensurabilität durch Pythagoras, *Platon* 29 (1977) 188. On this, see Zhmud. *Wissenschaft*, 158 f.

99 See above, 173 f.

100 See above, 173 f., 174 n. 31.

101 For the evidence and its analysis, see Zhmud. *Wissenschaft*, 170 f.

102 Papp. *Comm.*, 63 f.; *Schol. in Eucl.*, 415.7, 416.13, 417.12 f. See above, 172 n. 24.

words, if Eudemus and his contemporaries did not know the mathematician Hippasus, he did not exist.

Meanwhile, Aristotle and Theophrastus knew Hippasus as a philosopher;[103] Aristoxenus referred to his acoustical experiment based on mathematical proportions;[104] and Iamblichus, relying on the tradition that most likely derives from Eudemus, likewise connected his name with the first three proportions.[105] Thus, the Pythagorean Hippasus who took up philosophy, harmonics, and mathematics did not merely exist but was known in the late fourth century. On the other hand, Eudemus mentioned the discovery of the irrationals and the construction of the dodecahedron by the Pythagoreans. I therefore believe that the late tradition assigning these discoveries to Hippasus contains a historical core and might go back to Eudemus. But if Eudemus named Hippasus, then why is his name missing from the *Catalogue* and not mentioned in Proclus at all? There is at least one important reason for such an omission. Proclus attributed to Pythagoras the very discoveries that the other authors associate with Hippasus: the theory of irrationals and the construction of all five regular solids, including the dodecahedron. Therefore, there was no place left in the *Catalogue* for the mathematician Hippasus. One might surmise that Proclus chose to trust the tradition that persistently connected Hippasus with plagiarism and with divulging the Pythagorean secrets[106] and to sacrifice this figure by 'returning' his discoveries to Pythagoras.

The omission of Hippasus' name in Pappus, who referred to an anonymous Pythagorean discoverer of the irrationals,[107] can be explained in a simpler way. Pappus was not particularly concerned with naming the authors of the mathematical discoveries he presented in his work. While Nicomachus says that the first three means "came down to Plato and Aristotle from Pythagoras", Pappus, quoting him, omits all the names.[108] This is not the only example: in book IV of the *Collectio*, Pappus presents three methods of angle trisection and two methods of dividing an angle in a given proportion, without any attribution.[109]

103 Arist. *Met.* 984a 7; Theophr. fr. 225 FHSG, cf. Aët. I,3.11, IV,3.4 = 18 A 9.

104 Fr. 90. See above, 174 n. 31.

105 *In Nicom.*, 100.23, 113.17, 116.4.

106 This tradition is widely presented in Iamblichus, including the work used by Proclus (*De comm. math. sc.*, 77.18 f., cf. *VP* 88, 246–247).

107 See above, 189 n. 102. The scholia repeat Pappus.

108 *Coll.* III, 84.1 f.; see above, 173 n. 28.

109 *Coll.* IV, 272.15 f. On Pappus' method of working with sources, see Knorr. *TS*, 227 ff. We should not exclude the possibility that Hippasus' name was already omitted from Pappus' source. He mentions Eudemus only once and most likely used his works at second hand.

4. Early Greek geometry according to Eudemus

We will now try to bring together the data contained in Eudemus' fragments, in the *Catalogue*, and in the evidence we relate to the *History of Geometry*. This does not imply the reconstruction of early Greek geometry on the basis of Eudemus' writings. Such a reconstruction, involving a detailed analysis of all available sources on the subject, does not belong to the historiography of Greek mathematics. Our goal is more modest: we seek to get a general picture of *what, whom, and how* Eudemus' *History of Geometry* is written about. Different parts of this picture can be reconstructed more or less reliably, depending in each case on the character of evidence at our disposal. Probably the most difficult question concerns the origin of Eudemus' information on the geometers of the sixth and the early fifth century. Although we know that he used the works of Oenopides, Hippocrates, Archytas, Theaetetus, Eudoxus, and Eudoxus' students, it is very hard to say definitively which of them contained any information on the earliest mathematicians.

Like most of his predecessors, Eudemus considered Egypt a birthplace of geometry and explained the discipline's appearance there with the practical needs of land surveying (*In Eucl.*, 64.17f. = fr. 133). In its development, geometry passes through three stages: αἴσθησις, λογισμός, and νοῦς (65.1f.). These can be related to the tripartite scheme of *Metaphysics* A (ἐμπειρία, τέχνη, and ἐπιστήμη) and interpreted respectively as the acquisition of practical skill in land surveying, the emergence of a practically oriented applied discipline, and its further transformation into a theoretical science. Eudemus appeared to attribute the first two stages to Egypt and the third to Greek mathematics.

He interprets the passing of knowledge from one culture to another within the framework of two traditional formulas: πρῶτος εὑρετής and μάθησις (μίμησις) – εὕρεσις (2.3). Thales, having visited Egypt, was the first to bring geometry to Greece and discovered many things in it himself. Regrettably, the stereotypes Eudemus used are still popular in the historiography of science. It is very probable that the Greeks did in fact borrow from Egypt a lot of knowledge needed for land surveying and building, the more so since early Greek architecture and sculpture bear obvious traces of Egyptian influence. It is hard to believe, however, that the Greeks would have waited for Thales to get from Egypt the practical knowledge they needed. Stone building in Greece was resumed in the eighth century, Naucratis was founded in the mid-seventh century, and the famous architects Theodorus, Chersiphron, and Metagenes were Thales' contemporaries. The passage from empirical to theoretical geometry is indeed connected with Thales, but the practical knowledge he relied on must have been of a Greek, rather than Egyptian origin.

According to Eudemus, Thales 1) was the first to prove that the diameter divides the circle into two equal parts (Eucl. I, def. 17); 2) was the first to learn and state (ἐπιστῆσαι καὶ εἰπεῖν) that the angles at the base of any isosceles triangle are equal (I, 5), calling them, in the archaic manner, *similar,* not equal; 3) was the first to discover that if two straight lines cut one another, the vertical angles are equal (I, 15), whereas the scientific proof for this theorem was given later by Euclid, as Proclus adds; and 4) knew the theorem about the equality of the triangles that have one side and two angles equal (I, 26), which he must have used to determine the distances of ships from the shore.[110] Did Thales really need to go Egypt to learn there that the diameter divides the circle in half? On the other hand, his theorems concerning angles and triangles are not in the least related to Egyptian mathematics, which was never preoccupied with comparing angles and establishing the similarity of triangles.[111] In fact, Egyptian geometry lacked the notion of an angle as a measurable quantity – it was, in this respect, 'linear', unlike the 'angle' geometry of the Greeks, in which angles first became objects of measurement.[112]

The statement that the diameter divides the circle in half is not proved in Euclid, but accepted as a definition. Thales, in his demonstration, must have resorted to the method of superposition,[113] which in early Greek geometry was employed much more often than in the time of Eudemus and Euclid. As von Fritz observed,

> All theorems ascribed to Thales are either directly related to the problems of symmetry and can be 'demonstrated' by the method of superposition, or such that the first step of the demonstration is evidently based on considerations of symmetry while the second, which brings the argument to conclusion, is simply an addition or subtraction.[114]

In the fourth century, mathematicians tried to avoid this visual and overly empirical method. It is still to be found, however, in some theorems from Euclid's book I (4, 8). When saying that Thales treated certain things in geometry more empirically (αἰσθητικώτερον), Eudemus could well have meant the method of superposition. It does not follow, however, that Thales appealed in his demonstrations to nothing but the visualizability of the geometrical drawing. In the case of I, 5 we are able to verify this claim: there is a proof of this theorem in Aristotle that is different from the one found in Euclid and might well go back to Thales.[115] It is based on the equality of mixed angles – the angles of a semicircle and the angles of a circular segment, in particular – which, in turn, could only be demonstrated by superposition or follow from the definition of such angles. The proof proceeds in the following way:

---

110 Fr. 134–135, Procl. *In Eucl.*, 157.10f., 250.20f.

111 Vogel, K. *Vorgriechische Mathematik*, Hannover 1958–1959, Pt. 1, 72; Pt. 2, 23 n. 2, 39 n. 4.

112 Gands, S. The origin of angle-geometry, *Isis* 12 (1929) 452–482.

113 Heath. *Elements* 1, 225; von Fritz. *Grundprobleme*, 401 ff., 477 f.

114 Von Fritz. *Grundprobleme*, 568 n. 79.

115 *APr* 41 b 13–22. See Heath. *Elements* 1, 252 f.; von Fritz. *Grundprobleme*, 475 f.; Neuenschwander. VB, 358 f.

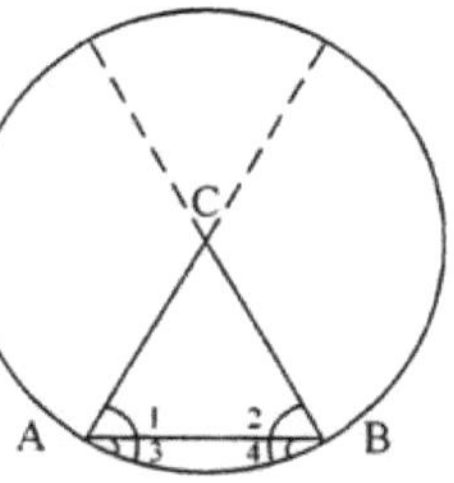

ABC is an isosceles triangle with the apex in the center of a circle. Prove that the angles at the base are equal. ∠ 1 = ∠ 2 since both of them are angles of the semicircle. ∠ 3 = ∠ 4 since two angles of any segment are equal to each other. Subtracting equal angles 3 and 4 from equal angles 1 and 2, we obtain that angles CAB and CBA are equal to each other.

The proof in Aristotle derives most probably from the *Elements* by Leon or Theudius, whereas Eudemus' source must have retained Thales' archaic terminology (ὅμοιοι instead of ἴσοι).

The theorem on the equality of vertical angles (I, 15) could also have been proved by the method of superposition. Euclid gives a different demonstration of this theorem, based on I, 13. According to Neuenschwander's reconstruction, I, 13–15 in their present form entered the *Elements* either in the time of Euclid or shortly before him.[116] That fully agrees with Proclus' remark that the scientific proof of I, 15 belongs to Euclid. Proclus (or his source) could have come to this conclusion by comparing the proof of Thales cited by Eudemus with Euclid's proof, which was, naturally, more rigorous.

Unlike the first three, Thales' fourth theorem is ascribed to him by Eudemus on the force of an indirect argument: the method Thales used to determine the distances of ships from the shore presupposes the use of this theorem. Without going into the details of different reconstructions of this method,[117] we note that Eudemus used his sources with discrimination and was perfectly able to distinguish between what was passed on directly by the tradition, on the one hand, and his personal conjectures and hypotheses, on the other. The fact that here we might be dealing with a mere reconstruction, possibly a fallacious one,[118] does not cast doubt on other things we learn from Eudemus about Thales – no history of mathematics can dispense with reconstructions. Numerous attempts to cast doubt on the facts Eudemus reports, and, along with them, Thales' place in the history of Greek geometry have so far all proved futile.[119] The information re-

116 Neuenschwander. VB, 361f.

117 The most convincing one is suggested by Heath. *History* 1, 131f., cf. van der Waerden. *EW*, 144f.

118 Gigon. *Ursprung*, 55, supposed that the method of calculating the distance to a ship at sea derives from the *Nautical Astronomy*, ascribed to Thales (11 B 1). But this work was written in verse and so hardly suitable for the exposition of geometrical proofs. The practical value of this method for navigation is doubtful as well.

119 See von Fritz's detailed answer (*Grundprobleme*, 337ff.) on doubts expressed by K. Reidemeister (*Das exakte Denken der Griechen*, Hamburg 1949, 18ff.) and Neugebauer (*ES*, 142). The hypercritical position of D. R. Dicks (Thales, *CQ* 9 [1959] 294–309, esp. 301f.) is not convincing; cf. Zaicev. *Griechisches Wunder*, 210f.

ported by Eudemus is rich and accurate and includes details that could not have been invented; the theorems he attributes to Thales show close mutual interconnection.[120] Even if we cannot identify the source of Eudemus' information on Thales' geometry, this only means that he had at his disposal certain texts that are inaccessible to us.[121]

About Mamercus, the next geometer after Thales, Eudemus must have only known the little he learned from Hippias (*In Eucl.*, 64.11f.). Judging by the general context in Hippias' *Collection*, investigated by A. Patzer, this reference was hardly isolated. Most likely, Mamercus was named along with the other famous geometers of his time, Thales and Pythagoras. This suggestion seems the more probable since, in the early doxography, Hippias played an important role as a transmitter of evidence about Thales. Though Patzer himself believed that geometry also belongs to the circle of themes examined by Hippias,[122] his own reconstruction of the *Collection* leaves little room to suppose that it contained particular mathematical material. It is important to make clear what is meant by 'geometry' here. The section of the *Collection* on 'famous men' could have said that Thales, Mamercus, and Pythagoras became famous for their studies of geometry, and that among the 'barbarians' the Egyptians excelled in geometrical knowledge. It could even have contained references to some particular problems studied by these mathematicians. What seems to me highly improbable is that Hippias quoted geometrical propositions at length, including their proofs, e.g., Thales' demonstration of the equality of the angles at the base of the isosceles triangle. Eudemus clearly knew substantially more about the proofs offered by Thales and by the other early geometers than could be included in Hippias' book, focused on looking for similarities between the ideas of Greek and 'barbarian' poets and sages.[123]

Eudemus' information on the Pythagoreans is especially rich, which corresponds to the role of this school in the development of early Greek geometry. Eudemus ascribes the "transformation of geometry into the form of a liberal education" and the discovery of the first three mean proportionals to Pythagoras himself, the study of proportions and (probably) the discovery of irrationals to his student Hippasus.[124] The following discoveries are related to the Pythagorean school as a whole: 1) the theorem that the sum of the interior angles of a triangle is equal to two right angles (I, 32); 2) the theorem, omitted from the *Elements,* that the space around a point can only be filled up with six triangles, four squares or three hexagons; 3) the theory of the application of the areas set forth mainly in books I and II of the *Elements*; 4) the entire book

---

120 See Becker. *Denken*, 37f.

121 Panchenko. Ὅμοιος and ὁμοιότης, 42f., believes that Hippocrates of Chios could have mentioned Thales, but this remains a conjecture.

122 Patzer, *op. cit.*, 106f.

123 Eudemus certainly used the *Collection* in his *History of Theology* (fr. 150).

124 Procl. *In Eucl.*, 65.15f.; Iambl. *In Nicom.*, 100.19–101.9, 113.16f., 116.1f., 118.23f.; Papp. *Comm.*, 63f.; *Schol. In Eucl.*, 415.7, 416.13, 417.12f.

IV of the *Elements*; 5) the construction of three regular solids (cube, pyramid, and dodecahedron); and 6) the beginnings of the theory of irrationals.[125] Besides, the *Catalogue* mentions two other Pythagorean mathematicians, Theodorus and Archytas, who should be discussed separately. The abundance and variety of this information shows that Eudemus used various sources. Since Archytas approvingly refers to his Pythagorean predecessors in mathematics (οἱ περὶ μαθήματα), retelling their views (47 B 1 and A 17) and, elsewhere, a theory by Philolaus' student Eurytus (A 13), his writings might have contained evidence on earlier geometers. Eudemus' other possible source was Glaucus of Rhegium, the author of the book on the history of music and poetry.[126] In this field, Glaucus was a predecessor of Eudemus, and his book, used by Heraclides Ponticus and Aristoxenus,[127] must have been known to Eudemus as well. It is revealing that Glaucus and Eudemus organize their material in a similar way: both of them proceed chronologically, from one *prōtos heuretēs* to another. Glaucus' origin in Rhegium in Southern Italy might point to his Pythagorean connections, so it does not seem odd that in his work he mentions Empedocles and the Pythagorean teachers of Democritus (fr. 5–6 Lanata). Glaucus' name is also attested in Aristoxenus' description of Hippasus' acoustical experiment:

> Hippasus made four bronze discs in such a way that, while their diameters were equal, the first disc was one-third as thick as the second (4:3), a half as thick as the third (3:2), and twice as thick as the fourth (2:1). When struck, the discs sounded in a certain consonance. It is also said that when Glaucus heard the notes produced by the discs, he was the first to master the art of playing on them (fr. 90).

Hippasus, as we noted,[128] made the discs in accordance with the musical proportion (12:9 = 8:6), which was probably known to Pythagoras, and arrived at the same intervals as the latter: the octave, the fifth, and the fourth. It is very likely that Glaucus is the source of other two testimonies: 1) on Hippasus' contemporary Lasus of Hermione, musician and theoretician of music; and 2) on acoustical experiments carried out by Lasus and Hippasus.[129] Thus, Glaucus' book could have provided Eudemus with information not only on Hippasus, but also on the early theory of proportions, which sprang from harmonics and remained closely connected with it for quite a long time.[130]

It seems that Eudemus' main source on the early Pythagorean mathematics was a mathematical compendium that preceded Hippocrates' *Elements* and

---

125 Fr. 136–137; Procl. *In Eucl.*, 301.11f.; *Schol. In Eucl.*, 273.3–13, 654.3; Papp. *Comm.*, 63f.

126 See above, 49 n. 18.

127 Aristox. fr. 90–91. See also Wehrli's commentary on Heracl. Pont. fr. 157–163.

128 See above, 174 n. 31.

129 1) Ps.-Plut. *De mus.* 1141 C (through Heraclides Ponticus or Aristoxenus); 2) Theon. *Exp.*, 59.4f. = 18 A 13 (through Aristoxenus?).

130 See above, 174 n. 30–31, cf. 47 B 2 and Eud. fr. 142.

contained the basis of the first four books of Euclid.[131] In this Pythagorean compendium, Eudemus could have found the information about specific proofs as well as about entire theories or books, such as the theory of the application of areas or book IV of the (future) *Elements*. Apart from planimetrical propositions, the compendium must have included a number of stereometrical ones, at least those that concerned the three regular solids discovered by the Pythagoreans. And, still more important, the compendium contained the first explicitly formulated definitions and axioms of geometry, laying the basis for geometrical demonstration. Van der Waerden, in particular, ascribes to the Pythagoreans the formulation of axioms 1–3 and 7–8.[132]

Like the treatises of the Hippocratic corpus, the compendium, representing the achievements of the school as a whole, did not contain the names of its authors. That is why most of the information we find in Eudemus on the Pythagoreans does not concern the individual representatives of the school, but the Pythagoreans as a whole, by which we should understand the mathematicians of the late sixth and the first half of the fifth century. For the same reason, Eudemus' information on Pythagoras is very general, exactly like our notions of Hippocrates of Cos as a doctor. Apart from the theory of proportions, closely linked with Pythagoras' acoustical experiments,[133] the historian mentions only one, though very important, achievement of his: the transformation of geometry into the form of a liberal education (σχῆμα παιδείας ἐλευθέρου), which aimed at acquiring knowledge, rather than serving practical needs.[134] The last testimony implies that Eudemus knew about the role of Pythagoras' school in the formation of the mathematical quadrivium.[135] The tradition on Pythagoras as an advocate of the *vita contemplativa* is familiar to us through Aristotle's *Protrepticus*;[136] elsewhere he testifies that "Pythagoras devoted himself to the study of mathematics, in particular of numbers".[137] According to Aristoxenus, Pythagoras was the first to turn arithmetic into a theoretical science

---

131 For its convincing reconstruction, see Waerden, B. L. van der. Die Postulate und Konstruktionen in der frühgriechischen Geometrie, *AHES* 18 (1978) 354 ff. Van der Waerden relied on Neuenschwander's historical analysis of the first four books of the *Elements* (Neuenschwander. VB).

132 Van der Waerden. *Pythagoreer*, 360 f. According to Favorinus, Pythagoras was the first to give definitions in geometry (D. L. VIII, 48). On the axiomatico-deductive character of Pythagorean arithmetic, see below, 221 ff.

133 See Zhmud. *Wissenschaft*, 187 ff. Xenocrates credits Pythagoras with the discovery of the numerical expression of musical intervals (fr. 87 Isnardi Parente).

134 Cf. Aristotle's characterization of philosophy, μόνην ἐλευθέραν τῶν ἐπιστημῶν (*Met*. 982 b 27): philosophy like a free person exists for its own sake.

135 See above, 63 f. Already Isocrates had associated Pythagoras' name with geometry, arithmetic, and astronomy (*Bus*. 23, 28).

136 The following fragment is particularly revealing: καλῶς ἄρα κατά γε τοῦτον τὸν λόγον Πυθαγόρας εἴρηκεν ὡς ἐπὶ τὸ γνῶναί τε καὶ θεωρῆσαι πᾶς ἄνθρωπος ὑπὸ τοῦ θεοῦ συνέστηκεν (fr. 11 Ross = fr. 20 Düring, cf. fr. 18 Düring).

137 Πυθαγόρας… διεπονεῖτο περὶ τὰ μαθήματα καὶ τοὺς ἀριθμούς (fr. 191 Rose).

(fr. 23), which gives us a further proof that both Eudemus and Aristotle relied on a common tradition.[138]

Ascribing to the Pythagoreans the discovery of the theorem about the sum of the interior angles of the triangle, Eudemus quotes their proof (fr. 136), which is different from that given in the *Elements*. In Euclid the theorem I, 32 combines two propositions: 1) for any triangle, the exterior angle formed by the continuation of any of its sides is equal to the two inner alternate angles; 2) the three interior angles of a triangle are equal to two right angles. The Pythagorean proof, more simple and elegant, concerns the second proposition only and uses a different geometrical construction. It is noteworthy that its only premise is the equality of the alternate angles (I, 29), one of the most evident corollaries of the properties of parallels:

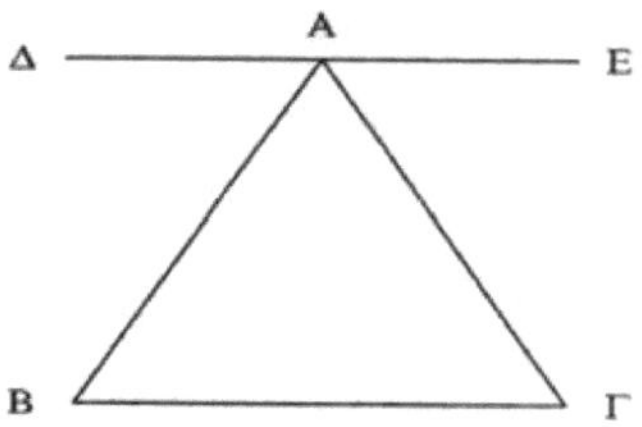

Let ABΓ be any triangle and let us draw through A a line ΔE parallel to BΓ. Since BΓ and ΔE are parallel, the alternate angles are equal. Then, the angle ΔAB is equal to the angle ABΓ and the angle EAΓ to the angle AΓB. Let us add to each sum the common angle BAΓ. Therefore, the angles ΔAB, BAΓ, ΓAE, i.e., the angles ΔAB, BAE, i.e., two right angles, are equal to the three angles of the triangle ABΓ. Therefore, the three angles of a triangle are equal to two right angles.

Eudemus, who was acquainted with several different versions of the *Elements*, realized, naturally, that I, 32 could be proved in various ways. If he nevertheless says "the Pythagoreans demonstrated this theorem as follows", it means that the text he cites must go back to the early Pythagorean compendium, rather than to the *Elements* by Leon or Theudius.[139] This is confirmed by the fact that one of Aristotle's references to this theorem implies the proof given in Euclid, and not the Pythagorean one.[140] As Neuenschwander has shown, the theorem I, 32 antedated Hippocrates' *Elements*: it is quoted verbatim in the fourth book, which comes from the Pythagoreans (IV, 15), as well as in III, 22, used by Hippocrates.[141] The theorem that the angles of the triangle equal two right angles is

---

138 See below, 219 f.

139 Cf. van der Waerden. *Pythagoreer*, 337 f.

140 *Met.* 1051 a 24. See Heiberg, I. L. *Mathematisches zu Aristoteles*, Leipzig 1904, 19; Heath. *Elements* 1, 320.

141 Neuenschwander. VB, 333, 375 f.; van der Waerden. Postulate, 353 f. Mueller, I. Remarks on Euclid's *Elements* I, 32 and the parallel postulate, *Science in Context* 16 (2003) 292, is ready to date the Pythagorean proof to the middle of the fifth century. – Proposition III, 31 (an angle in a semicircle is a right angle), containing a reference to I, 32, was also known to the early Pythagoreans. Pamphila (first century AD) ascribed it to Thales, while others, including Apollodorus the Calculator, related it to Pythagoras (D. L. I, 24–25, cf. VIII, 12). This Apollodorus may be identified with Apollodorus of Cyzicus, an author of the later half of the fourth century BC (Burkert. *L & S*, 428). Since it is III, 31 that follows from I, 32, and not vice versa, its attribu-

one of Aristotle's favorite examples: he refers to it about ten times in the *Analytics* alone.[142] That is probably why Eudemus paid particular attention to it and cited its earliest demonstration.

Aristotle's reference to I, 32 gave rise to a curious episode in the historiography of Greek geometry. Geminus alleged that the ancients had investigated this theorem for each individual type of triangle: first, for the equilateral, then for the isosceles, and finally for the scalene, whereas the later geometers had demonstrated the general theorem.[143] Since the Pythagorean proof quoted by Eudemus refers to the general case, the earlier stages were associated with Thales and even with his Egyptian teachers. Geminus' evidence might seem to undermine our conclusion that he did not use Eudemus and showed little interest to the history of pre-Euclidean geometry.[144] In fact, three stages of the proof of I, 32 pointed out by Geminus are wholly fictitious, since he relied not on the real but on the hypothetical example, given by Aristotle (*APo* 74a 25f.).[145]

The incommensurability of the side of a square with its diagonal, along with theorem I, 32, was one of Aristotle's favorite mathematical examples.[146] The indirect proof of this theorem, to which he several times alludes (*APr* 41a 24f., 50a 37), relies on the Pythagorean theory of the odd and even numbers.[147] In Aristotle's time, this proof was still a part of the *Elements*; later it was excluded by Euclid, who employed Theaetetus' general theory of the irrational magnitudes. Since Eudemus obviously took account of the mathematical examples Aristotle often referred to, we can assume that in considering the Pythagorean origin of the theory of irrationals he cited as an example the very indirect proof that might be related to the early Pythagorean mathematical compendium.

Another proposition going back to this compendium is the theorem that only three regular polygons – the triangle, the square, and the hexagon – can fill the space around a point. It was not included in the *Elements*, but Aristotle refers to it (*Cael.* 306b 5f.), adding that the same is true only for two regular solids, the cube and the tetrahedron. The Pythagorean theorem Eudemus refers to is linked

tion to Thales remains highly doubtful (Heath. *History* 1, 133f.; idem. *Elements* 1, 319). In any case, theorem III, 31, featuring mixed angles, belongs to the oldest stratum of Greek geometry. Aristotle refers to it several times (*APo* 94a 24f., *Met.* 1051a 26f.), implying a proof different from that of Euclid (Heath. *Mathematics*, 71f.).

142 *APr* 66a 13, 67a 13–20; *APo* 71a 17, 71a 27, 73b 31, 74a 16–b4, 85b 5f., b 11f., 86b 25. For further references, see Heiberg. *Mathematisches zu Aristoteles*, 19f.

143 Eutoc. *In Apollon. Con. comm.* II, 170.4–8.

144 See above, 184f.

145 Heiberg. *Mathematisches zu Aristoteles*, 20; Heath. *History* 1, 135f.; idem. *Elements* 1, 319f.; idem. *Mathematics*, 43ff.; Kullmann, W. Die Funktion der mathematischen Beispiele in Aristoteles' *Analytica Posteriora*, *Aristotle on science*, 255f. Cf. Becker. *Denken*, 39; Neuenschwander, E. Beiträge zur Frühgeschichte der griechischen Geometrie I, *AHES* 11 (1973) 127–133.

146 Heiberg. *Mathematisches zu Aristoteles*, 24, gives more than 15 references to this theorem.

147 Ibid., 24. See below, 223 n. 40.

with both I, 32 and Thales' theorem on the equality of the alternate angles (I, 15). The latter has a corollary that the space around a point is divided into angles whose sum equals four right angles. It follows from I, 32 that the angle of an equilateral triangle is equal to ⅔ of the right angle, which means that the sum of six such triangles is equal to four right angles. Accordingly, the angle of a hexagon equals 1⅓ of the right angle, and the angle of a square is right, so that the area around a point can be filled by either three hexagons or four squares. All other regular polygons, as Proclus notes (*In Eucl.*, 304.11f.), give in sum either more or less than four right angles.

The entire book IV of the *Elements* on the relations between the regular polygons and the circle, which Eudemus ascribes to the Pythagoreans, must be a part of the early Pythagorean compendium, as well. It is evident, at any rate, that it was written before Hippocrates, who used it in his attempt to square the lunes, and underwent only insignificant revision later.[148]

The theory of the application of areas with excess and deficiency (ἥ τε παραβολὴ τῶν χωρίων καὶ ἡ ὑπερβολὴ καὶ ἡ ἔλλειψις), which Eudemus, pointing out its antiquity (ἔστι μὲν ἀρχαῖα), attributed to the 'Pythagorean muse' (fr. 137), relates to the transformation of areas into equivalent areas of different shape.[149] The propositions of this theory, comprising theorems I, 44–45, the entire book II of the *Elements*, and several theorems of book VI (27–29), can be reformulated into algebraic identities and quadratic equations and, for this reason, were often termed 'geometrical algebra'.[150] For example, the application of areas with a deficiency means the construction on a given line a of the rectangle ax, so that by subtracting from it the square x^2, the given square b^2 is obtained ($ax-x^2=b^2$). This does not mean, however, that the application of areas really stemmed from the solution of quadratic equations, let alone that they are of Babylonian origin.[151] The Pythagorean character of this theory is obvious: the area of a rectilinear figure (II, 14) is determined by finding the geometric mean x between lines a and b; i.e., a square with the side x equals a rectangle ab ($x^2=ab$). Since Hippocrates of Chios was familiar with this theory and developed it, the application of areas can be dated in the first half of the fifth century.

It seems to follow from Eudemus' words that book II, related entirely to the application of areas, along with book IV, was created by the Pythagoreans.

---

148 Neuenschwander. VB, 374f.; van der Waerden. *Pythagoreer*, 341ff.

149 In general form this problem is formulated as the application to a given straight line a rectangle equal to a given rectilinear figure and exceeding or falling short by a square (Heath. *History* 1, 151).

150 E.g. proposition II, 2 can be reformulated as the identity $(a+b)c=ac+bc$, and II, 14 as the equation $x^2=ab$. See Heath. *Elements* 1, 343f.; idem. *History* 1, 150ff.; Bekker. *Denken*, 60f.; van der Waerden. *Pythagoreer*, 341ff.

151 The algebraic and Babylonian origin of book II, which used to be almost unanimously accepted, was subjected to shattering criticism in recent decades. For references, see Zhmud. *Wissenschaft*, 149 n. 37.

Why, then, did Eudemus refer in this case to the application of areas, rather than to book II directly? First, this theory is expounded not only in book II, but also in books I and VI. Proclus, in particular, quotes Eudemus in his commentary to I, 44, whereas the terms ὑπερβολή and ἔλλειψις, which (unlike παραβολή) are missing in book II, occur only in VI, 27–29. Second, not long before Euclid, book II underwent some mainly stylistic changes, and a new notion, the parallelogram, was introduced into it.[152] That was why Eudemus could have preferred to speak of the theory whose antiquity was warranted by its authentic form, which was different from the *Elements* of his own time, as well as by its provenance from the Pythagorean compendium.

Eudemus tells much less about Ionian geometers before Hippocrates than about the Pythagoreans. This is easily accounted for by the fact that, till the mid-fifth century, the Pythagorean school was foremost in geometry. But one cannot ignore the selective approach of Eudemus himself, let alone those who subsequently used his *History of Geometry.* Anaxagoras is mentioned in the *Catalogue* as a man who touched upon many geometrical problems. What his discoveries in geometry actually were, remains unclear; nor can we be sure that Eudemus knew anything about them.[153]

Eudemus regarded Oenopides, the founder of the Chian school of mathematicians, as having been "a little younger" than Anaxagoras. The traditional date of Anaxagoras' birth is ca. 500; ὀλίγῳ νεώτερος refers here, as in the other places in the *Catalogue*, to a time span smaller than one generation, i.e., 10–15 years. We can, accordingly, date Oenopides' birth at ca. 490–85 and his *floruit* in the mid-fifth century.[154] Eudemus attributes to Oenopides two elementary geometrical constructions that later entered Euclid's book I,[155] as well as the last proposition of book IV concerning a regular fifteen-angled figure inscribed in the circle.[156] According to Eudemus, propositions I, 12 and IV, 16 were important not only for geometry, but also for mathematical astronomy; in the first case, moreover, he refers to the opinion of Oenopides himself.[157]

152 Neuenschwander. VB, 371 f.

153 Plutarch says that Anaxagoras, while he was in prison, worked on the problem of squaring the circle (59 A 38, cf. A 39). This evidence does not seem to come from Eudemus.

154 The date of Anaxagoras' birth must have been known to Aristotle (*Met.* 984 a 11 f.) and served him as a kind of starting point. Following Aristotle, Theophrastus noted that Empedocles οὐ πολὺ κατόπιν τοῦ Ἀναξαγόρου γεγονώς (fr. 227 a FHSG). Here, as in Eudemus, "a little later" implies the difference of 10–15 years. Empedocles' birth is usually dated to 490/85.

155 To draw a perpendicular to a given straight line from a point outside it (I, 12); at a point on a given straight line, to construct a rectilinear angle equal to a given rectilinear angle (I, 23).

156 Fr. 138; Procl. *In Eucl.*, 283.7 f., 269.8 f. This also shows that book IV was written before the middle of the fifth century.

157 On Oenopides' mathematical astronomy, see below, 7.5.

Eudemus' familiarity with Oenopides' work is further confirmed by the fact that he points out an archaism in the formulation of I, 12: Oenopides called the perpendicular κατὰ γνώμονα, since the gnomon also stands at right angles to the horizon.

The attention Eudemus pays to problems of terminology, which we already noted when discussing a theorem of Thales (I, 5), is an additional proof of the historical character of his work on mathematical sciences. Like a modern historian of mathematics, Eudemus was interested not only in the discovery itself, but also in details of the proof and its correspondence with demonstration current in his own day, in peculiarities of terminology, connections with other sciences, etc. This aspect of Eudemus' works, testifying to his conscientious approach to sources, is one of the guarantees that he avoided introducing arbitrary changes into his material unless he had to.

Oenopides was obviously not the first to have drawn a perpendicular to a line or to have constructed a rectilinear angle. His propositions were consequently regarded as the first attempt at a formal geometrical approach to these constructions, deliberately limited to the use of compasses and ruler alone.[158] Disputing the latter opinion, Knorr denied the formal geometrical character of Oenopides' constructions, believing them to come from Oenopides' astronomical treatise that considered the construction of astronomical instruments as well.[159] Knorr is right in maintaining that Oenopides is unlikely to have written a special treatise on geometry. As an astronomer, he could use a number of instruments, apart from ruler and compasses. But in cases where he turned to geometrical constructions important for astronomy, Oenopides followed the already existing formal requirements known to him, in particular, from the Pythagorean compendium.[160] Obviously, Oenopides determined the angle of the obliquity of the ecliptic empirically. But to give it an accurate geometrical measure, he constructs a regular fifteen-angled figure inscribed in the circle, in full accordance with the rules for the construction of polygons laid out in book IV.[161]

Another Ionian mathematician, Democritus, was a younger contemporary of Oenopides, whom Democritus mentioned in one of his works (41 A 3). Like Oenopides, Democritus was also associated with Pythagorean mathematics: according to Glaucus of Rhegium, he had Pythagorean teachers (68 A 1.38 =

---

158 Heath. *History* 1, 175f.

159 Knorr. *AT*, 15f.

160 He shared also some astronomical views of the Pythagoreans (cf. 41 A 10 and 58 B 37c).

161 Szabó, Á., Maula, E. ΕΓΚΛΙΜΑ. *Untersuchungen zur Frühgeschichte der antiken griechischen Astronomie, Geographie und Sehnentafeln*, Athens 1982, 118f. It is worth noting that IV, 16 differs somewhat from other propositions of this book. In all of them, the construction of a polygon is followed by a proof that the constructed figure does possess the required qualities. Such proof is lacking in the case of IV, 16 (Neuenschwander. VB, 374).

fr. 6 Lanata), to whom he probably owed his vast knowledge in mathematics.[162] Democritus, who asserted that "nobody excelled him in drawing lines with proofs", was the author of about ten mathematical works.[163] It is difficult, then, to find any reason why Eudemus should have omitted him from his *History of Geometry.* If my hypothesis is true, he considered Democritus to be the author of propositions that the cone is equal to one-third of the cylinder and the pyramid to one-third of the prism with the same base and height (Eucl. XII, 7 and 10), while noting at the same time that the scientific demonstration of these theorems was given by Eudoxus.[164]

The level of geometry in ca. 440–430 is better appreciated, not from the scant information about Oenopides and Democritus, but rather from the non-trivial problems of duplicating the cube and squaring the circle studied by Hippocrates of Chios, a younger contemporary and probably a student of Oenopides. These problems found responses outside the circle of geometers as well. In the *Republic* we find some hints at duplicating the cube, which later gave rise to a legend about Plato's part in solving this problem (3.4). Whereas Aristotle remained indifferent to it, Eudemus attests the achievements of Hippocrates and cites the solutions offered by Archytas, Eudoxus, and his students. The problem of squaring the circle aroused still greater interest in wide circles: Aristophanes mentioned it (*Av.* 1004–1010), it preoccupied Antiphon and Bryson, and Aristotle often referred to it.[165] The latter circumstance attracted to the problem the attention of Aristotle's commentators Alexander and Simplicius, who brought to us Eudemus' evidence on Antiphon and Hippocrates. In the first case, Eudemus follows Aristotle's judgment: Antiphon does not admit the basic principles of geometry, in particular, that geometrical magnitudes are infinitely divisible.[166] Unlike Aristotle, however, who accused Hippocrates of having committed a logical mistake by squaring the circle with the help of lunes (*SE* 171b 12f.; *Phys.* 185a 14f.), Eudemus found no fault with him.[167]

---

162 Zhmud. *Wissenschaft*, 40 n. 69. One of Democritus' works is related to irrational lines – the problem that only the Pythagoreans had treated before him. Democritus wrote a book on Pythagoras (14 A 6 = fr. 154 Luria).

163 68 B 299 = fr. 137 Luria. The list of mathematical writings by Democritus includes works on geometry, arithmetic, astronomy, and the theory of perspective (D. L. IX, 47–48).

164 See above, 177, and 68 B 155 on the cutting of a cone by parallel planes. Cf. fr. 125 Luria with commentary; Waschkies, *op. cit.*, 267 ff.

165 *Cat.* 7b 31, *APr* 69a 30–34, *APo* 75b 40f., *SE* 171b 13–172a 8, *Phys.* 185a 16f., *EE* 1226a 29.

166 Fr. 140, p. 59, 9–12 Wehrli. For a modern appraisal of Antiphon's approach, see Heath. *History* 1, 221f.; Mueller, I. Aristotle and the quadrature of the circle, *Infinity and continuity*, 146–164; Knorr. *AT*, 26f.

167 Lloyd, G. E. R. The alleged fallacy of Hippocrates of Chios, *Apeiron* 20 (1987) 103–128.

According to Eudemus, Hippocrates, being an expert in geometrical constructions, 1) was the first to apply the method of reduction (ἀπαγωγή, one of the methods foreshadowing analysis) to complex constructions; 2) in particular, was the first to reduce the problem of doubling the cube to finding two mean proportionals in continuous proportion between two given magnitudes; 3) discovered the quadrature of lunes; and 4) was the author of the first *Elements*.[168] Proclus defines ἀπαγωγή as a "transition from a problem or a theorem to another that, if known or constructed, will make the original proposition evident" and identifies it with the method Hippocrates employed to solve the problem of doubling the cube (*In Eucl.*, 212.24–213.11). The problem itself was connected, of course, not with the demand of the Delphic oracle to double the altar on Delos, but with Pythagorean mathematics.[169] The Pythagoreans solved the problem of squaring a rectilinear figure by finding the mean proportional x between two lines ($x = \sqrt{ab}$). This problem had to be followed, naturally, by that of finding two mean proportionals between two lines. It is to this latter problem that Hippocrates reduced the duplication of the cube using the method of ἀπαγωγή.[170]

The attempt at squaring a circle followed, in turn, from the squaring of the rectangular figure considered in book II of the *Elements*.[171] Of course, Hippocrates could not solve the problem of squaring a circle. He succeeded, however, in squaring three lunes – figures limited by two circular arcs.[172] According to Simplicius,

> In book II of the *History of Geometry* Eudemus says the following: "The quadratures of lunes, which were considered to belong to an uncommon class of propositions on account of the close relation (of lunes) to the circle, were first investigated (ἐγράφησαν) by Hippocrates, and his exposition was thought to be in correct form.[173]

It is worth noting that Eudemus refers here to an opinion of specialists, and not to that of Aristotle, who erroneously believed that Hippocrates pretended to have solved the problem of squaring a circle. Eudemus also points out that the solution offered by Hippocrates was of a general character:

> Eudemus, however, in his *History of Geometry* says that Hippocrates did not demonstrate the quadrature of the lune on the side of a square (only), but generally, as one might say. For every lune has an outer circumference equal to a

---

[168] 1) Eutoc. *In Archim. De sphaer.*, 88.18–23 (from Eratosthenes); 2) fr. 133, 140; 3) Procl. *In Eucl.*, 213.7–11; 4) ibid., 66.4f. = fr. 133.

[169] Heath. *History* 1, 200f.; Knorr. *AT*, 23f.

[170] See Saito K. Doubling the cube: A new interpretation of its significance for early Greek geometry, *HM* 22 (1995) 119–137.

[171] Neuenschwander. Beiträge, 127; cf. above, 199 n. 150.

[172] The fact that, using compasses and a ruler, one can square only five types of closed circular lunes was demonstrated only in the last century.

[173] Fr. 140, p. 59.28–60.2 Wehrli, transl. by T. Heath.

semicircle or greater or less, and if Hippocrates squared the lune having an outer circumference equal to a semicircle and greater and less, the quadrature would appear to be proved generally.[174]

Simplicius further adds that he is going to give a literal quotation from Eudemus on the squaring of lunes, expanding, for clarity's sake, his "brief proofs in the ancient manner" (ibid., 59.26). Interestingly, it is precisely where Eudemus promised "to deal with the quadratures of lunes at length and to go through them" (ἐπὶ πλέον ἁψώμεθα τε καὶ διέλθωμεν, ibid., 60.1) that Simplicius characterizes his account as brief. In fact, Eudemus gave a concise description of some of Hippocrates' demonstrations and quoted some others verbatim, which, from Simplicius' point of view, was still not complete enough.[175] The reconstruction of Hippocrates' solution reported by Eudemus has already given rise to a vast literature, and there is no point in considering it here in detail.[176] I would like to note only that the considerable length of the text that Eudemus devoted to the squaring of lunes (about 4.5 pages of a Loeb format) allows us to estimate the length of his *History of Geometry*. Eudemus, as we know, numbered Hippocrates among the earliest geometers[177] and treated the squaring of lunes in book II of the *History of Geometry*. Most of the mathematicians mentioned in this work belong to the period after Hippocrates, so that if their discoveries were treated at comparable length, one may suppose that the *History of Geometry* comprised at least four books.[178] The first book might have con-

---

174 Ibid., 59.19–24, transl. by I. Bulmer-Thomas.

175 "That in Eudemus' text not everything said is proved, corresponds to the character of Eudemus as a historian." (Becker. Zur Textgestaltung, 415). At any rate, Simplicius' interventions in Eudemus' text are more or less equally distributed among all four of Hippocrates' demonstrations, no matter how fully they were reported by Eudemus.

176 Rudio, F. *Der Bericht des Simplicius über die Quadraturen des Antiphon und des Hippokrates*, Leipzig 1907 (summarizes the earlier works of Allman, Tannery, and Heiberg); Björnbo. Hippokrates; Heath. *History* 1, 183ff.; Becker. Zur Textgestaltung, 411–419; ibid. *Denken*, 58f.; Böker, R. Würfelverdoppelung, *RE* 9 A (1961) 1198f.; Bulmer-Thomas, I. Hippokrates of Chios, *DSB* 6 (1972) 410–418; Knorr. *AT*, 26ff. Recently, Reviel Netz (Eudemus of Rhodes, Hippocrates of Chios and the earliest form of a Greek mathematical text, *Centaurus* 46 [2004] 243–286) offered a very ingenious but admittedly speculative account of how Eudemus' report relates to Hippocrates' text. Cf. Federspiel, M. Sur la locution ἐφ' οὗ / ἐφ' ᾧ servant à designer des êtres géometriques par des lettres, *Mathématiques dans l'Antiquité*, ed. by J.-Y. Guillaumin, Saint-Étienne 1992, 9–25.

177 ὥστε καὶ τὸν Εὔδημον ἐν τοῖς παλαιοτέροις αὐτὸν ἀριθμεῖν (Simpl. *In Phys.*, 69.23f.). Netz (Eudemus) seemed to overlook this evidence, though it could reinforce his case, viz. that Hippocrates was in a sense *the* first Greek geometer. Now, Eudemus might have considered Hippocrates to be one of the earliest mathematical *writers* whose works were available in the late fourth century, but hardly the founder of Greek geometry as it was known to him.

178 This corresponds to the number of books given in the list of Theophrastus' works (fr. 264 No. 3 FHSG). See above, 166 n. 2–3.

sidered discoveries by Thales and the Pythagoreans, the second geometers of the second part of the fifth century, and the last two geometry of the fourth century.

Hippocrates' *Elements* is the first of a number of mathematical treatises under the same title that ultimately led to Euclid's collection. The title itself, Στοιχεῖα, i.e., the basic, fundamental elements, indicated its task: to organize interrelated mathematical propositions in their logical sequence, starting from the very basic ones.[179] Apart from the axioms and definitions, Hippocrates' *Elements* might have contained the first three construction postulates,[180] although the last point is disputable.[181] At any rate, by the last decades of the fifth century, geometry acquired features of a truly scientific axiomatico-deductive system.

Theodorus of Cyrene, a contemporary of Hippocrates, appears in the *Catalogue* probably owing to his contribution to the theory of irrationals. According to Plato (*Tht.* 147d), Theodorus had proved the irrationality of magnitudes from $\sqrt{3}$ to $\sqrt{17}$. This is practically all we know about his discoveries in geometry. Even if Eudemus provided further information on Theodorus, it did not attract the attention of the later commentators. The historian attributes the creation of the general theory of irrationals (fr. 141.I), set forth in book X of the *Elements*, to Theaetetus, a student of Theodorus. It was also Theaetetus who, adding the icosahedron and the octahedron to the three regular solids known to the Pythagoreans (*Schol. in Eucl.*, 654.3), developed a general theory of regular solids (book XIII).

The discoveries of Theaetetus' contemporary Leodamas remain unknown to us. The stories in which Leodamas figures as the receiver of the method of analysis developed by Plato are unlikely to be true and can hardly go back to Eudemus.[182] The *Catalogue* does not mention the discovery of analysis. Its application is associated with Eudoxus, not Leodamas, though the former is not named as its inventor. This implies that Eudemus related analysis to an earlier period, but it remains unclear whether he associated its discovery with Leodamas or Hippocrates. The author of the Academic work quoted by Philodemus, as we remember, dated the discovery of analysis to Plato's time (3.1).

Archytas, who proceeded from the results of Hippocrates' research, was the first to solve the problem of doubling the cube. His remarkable stereometrical construction, which for the first time introduces movement into geometry, employed the intersection of the cone, the torus, and the half-cylinder, which pro-

---

179 Burkert, W. ΣΤΟΙΧΕΙΟΝ. Eine semasiologische Studie, *Philologus* 103 (1959) 167–197.

180 Van der Waerden. *Pythagoreer*, 361f.

181 See Mueller. Remarks, 293f.: Aristotle does not discuss any of Euclid's postulates.

182 D. L. III, 24 = Favor. fr. 25 Mensching; Procl. *In Eucl.* 211.18f. See Mensching, *op. cit.*, 103f. On the application of analysis in the fifth century, see Allman, *op. cit.*, 41 n. 62, 97f.; Heath. *History* 1, 291; Cherniss. Plato as mathematician, 418f. See above, 92f.

duced the necessary curve.[183] Eutocius cites Archytas' solution with reference to Eudemus (fr. 141); this text, however, unlike the verbatim quotation in Simplicius, bears obvious traces of a later revision.[184] Diogenes Laertius' evidence that Archytas was "the first to employ mechanical motion (κίνησις ὀργανική) in a geometrical construction, when he tried, by means of a section of a half-cylinder, to find two mean proportionals in order to duplicate the cube" (VIII, 83) might also go back, through Favorinus and Eratosthenes, to Eudemus, as might the related reference to Archytas as the founder of mechanics.[185] It is also very probable that Archytas figured in the Eudemian outline of the history of proportions, which is preserved in Iamblichus.[186]

We know nothing about the discoveries of Neoclides, who follows Archytas in the *Catalogue*. Eudemus ascribes to Leon, his disciple: 1) the authorship of new *Elements* exceeding the older ones "both in number and in the utility of propositions proved in it", and 2) the discovery of the method of diorism, allowing one "to determine when a problem under investigation is capable of solution and when it is not" (*In Eucl.*, 66.22f.). A particular interest in compilers of *Elements* manifested in the *Catalogue* can only partly be explained by the fact that this text formed part of Porphyry's commentary on the *Elements*. Eudemus obviously regards the appearance of new and improved *Elements* as progress in geometry. Some historians of mathematics believe that such basic principles of mathematics as axiom, postulate, hypothesis had already been acknowledged and defined in Leon's *Elements*.[187] As for the method of diorism, though the Academic work relates its discovery to the mid-fourth century, we have good reason to suppose that it had been used earlier.[188] It is possible that Leon formulated the method clearly or improved it, rather than invented it.

Much more detailed is the information Eudemus provides on Eudoxus. According to the *Catalogue*, Eudoxus 1) "was the first to increase the number of the so-called general theorems", 2) added three new proportionals to the three already known, and 3) "multiplied the theorems concerning the section ...,

183 See Heath. *History* 1, 246ff.; Becker. *Denken*, 76f.; van der Waerden. *EW*, 150f.; Böker, *op. cit.*, 1203f.; Knorr. *AT*, 50f.

184 Neuenschwander, E. Zur Überlieferung der Archytas-Lösung des delischen Problems, *Centaurus* 18 (1974) 1–5; Knorr. *AT*, 50f.; idem. *TS*, 100ff. It does not mean that the *History of Geometry* was inaccessible to Eutocius: he refers to it in another work (*In Archim. De dimens. circ.*, 228.20 = Eud. fr. 139). Like Simplicius, he could have quoted Eudemus at second hand as well, if it was convenient for him. Knorr. *TS*, 100ff., suggested Geminus or Sporus as possible intermediaries between them, but there is no evidence that Geminus used Eudemus (see above, 184f.).

185 See above, 176f.

186 See above, 173ff.

187 Lasserre. *Birth*, 18.

188 Heath. *History* 1, 319f.; Lasserre. *Léodamas*, 516f. This method can be clearly traced, in particular, in Plato's *Meno*, written ca. 385. See Knorr. *AT*, 73f.; Menn, S. Plato and the method of analysis, *Phronesis* 47 (2002) 193–223.

employing the method of analysis for their solution".[189] To this one should add: 4) the information from the scholia that Eudoxus was the author of book V of the *Elements*, containing the general theory of proportions, 5) Eratosthenes' and Eutocius' evidence on Eudoxus' solution to the problem of doubling the cube, and, finally, 6) the words of Archimedes that Eudoxus proved the theorem that each cone equals one-third of the cylinder and every pyramid equals one-third of the prism with the same base and height.[190] Nos. 1 and 4 refer, it seems, to one and the same discovery, namely, to the general theory of proportions. This, in any case, is how "the so-called general theorems" (τὰ καλούμενα καθόλου θεωρήματα) are usually understood.[191] Similar expressions occurring in Aristotle (ἡ καθόλου μαθηματική and τὰ καθόλου ἐν ταῖς μαθήμασιν), are normally interpreted as a reference to Eudoxus' theory of proportions applied equally to all magnitudes (numbers, lines, figures, etc.).[192] By contrast, the "theorems concerning the section" (τὰ περὶ τὴν τομὴν), whose number Eudoxus multiplied, applying to them the method of analysis, remain obscure. Lasserre, following Bretschneider, was inclined to relate them to the section of line in extreme and mean ratio (golden section).[193] More probable, however, is the interpretation preferred by Tannery and Heath that 'section' means the section of solids,[194] which might correspond to no. 6 on our list.

Also unclear remains the question of Eudoxus' solution to the problem of doubling the cube. Eratosthenes' epigram and his letter to King Ptolemy III mention 'curved lines' (καμπύλαι γραμμαί); Eutocius, however, who quotes both these texts, leaves Eudoxus' solution out. As he explains, the 'curved lines' Eudoxus mentions in the 'introduction' are not found in the text of the proof and a discrete proportion is used here as if it were continuous (*In Archim. De sphaer.* III, 56.4f.). It seems that Eutocius either was using a distorted version of Eudoxus' solution or simply failed to understand it. Leaving this problem to the historians of mathematics,[195] we merely point out that Euto-

---

189 *In Eucl.*, 67.2f. = fr. 22 Lasserre. Information on the discovery of proportions found in Iamblichus (*In Nicom.*, 100.19–101.9) confirms the evidence of the *Catalogue*, adding to it that Eudoxus called the fourth mean 'subcontrary'. See above, 173ff.

190 4) *Schol. in Eucl.*, 280.7f., 282.12f. = fr. 32–33 Lasserre; 5) Eutoc. *In Archim. De sphaer.*, 56.4f., 90.7, 96.18 = fr. 24–25, 29 Lasserre; 6) Archim. *Meth.*, 430.1f. = fr. 61c Lasserre.

191 Heath. *History* 1, 323; Lasserre. *Eudoxos*, 162.

192 *Met.* 1026a 25–27, 1064b 8–9; 1077a 9–10, 1077b 17–18, cf. *APr* 85a 37; *APo* 74a 17–25. See Ross, *op. cit.*, 356f.; Heath. *Mathematics*, 222f.; Lasserre. *Eudoxos*, 166f.; Fiedler, *op. cit.*, 52ff.; Kouremenos, T. Aristotle's mathematical matter and Eudoxus' proportion theory, *WS* 109 (1996) 61ff.

193 Bretschneider, *op. cit.*, 168; Lasserre. *Eudoxos*, 176f.

194 Tannery. *Géométrie*, 76; Heath. *History* 1, 325.

195 See Tannery, P. Sur les solutions du problème de Délos par Archytas et Eudoxe, *Mémoires scientifiques* I, 53–61; Heath. *History* 1, 249f.; Böker, *op. cit.*, 1207ff.;

cius' immediate source here was not Eudemus, but probably Eratosthenes' *Platonicus*.[196]

While no evidence has been left of the discoveries of Amyclas of Heraclea, two of Eudoxus' students, Menaechmus and Dinostratus, are known for their research relating to the curves. Menaechmus solved the problem of doubling the cube by finding two mean proportionals through the intersection of a hyperbola with a parabola,[197] while Dinostratus constructed the so-called quadratissa, a curve used to square a circle.[198] It was traditionally thought that Menaechmus constructed the hyperbola and parabola by sectioning a cone.[199] Knorr, however, doubts whether he might have developed even a rudimentary theory of conic curves, believing that such a theory was hardly conceivable nearly half a century before Euclid.[200] To this it ought to be objected that Menaechmus, being a disciple of Eudoxus (born ca. 390), could hardly have been born before 375/70, and consequently is only a generation older than Euclid. Menaechmus' solution was significantly revised by Eutocius as well as probably by his source.[201] The latter could have been Eratosthenes, who relied on Eudemus and considered the history of doubling the cube in detail.[202]

By contrast, there is no reason to associate Eudemus with the evidence on Menaechmus' theoretical views in mathematics that we find in Geminus.[203]

Lasserre. *Eudoxos*, 163f.; Riddel, *op. cit.*; Knorr. *AT*, 52f.; idem. *TS*, 77ff. Knorr believed that the next solution, ascribed by Eutocius to Plato, could belong to Eudoxus. Cf. Netz, R. Plato's mathematical construction, *CQ* 53 (2003) 500–509.

196 Wolfer, *op. cit.*, 51, believed that *Platonicus* did not reach Eutocius, however Eutocius did have two other texts of Eratosthenes at his disposal, which makes his familiarity with *Platonicus* rather probable.

197 Eutoc. *In Archim. De sphaer.*, 78.13–80.24. An anonymous solution via the intersection of two parabolas that Eutocius also cites does not belong to Menaechmus. See *Diocles. On burning mirrors*, ed. and transl. by G. J. Toomer, New York 1976, 169f.; Lasserre. *Léodamas*, 552; Knorr. *TS*, 94f, 98.

198 Papp. *Coll.* IV, 252.26ff. See Allman, *op. cit.*, 180ff., Becker. *Denken*, 95; cf. Knorr. *AT*, 80, 84f. Pappus took this construction from Sporus, whose ultimate source must be Eudemus (Lasserre. *Léodamas*, 561f.).

199 Geminus, while referring to Eratosthenes' epigram (μηδὲ Μεναιχμείους κωνοτομεῖν τριάδας), called Menaechmus the author of the theory of conic sections (Procl. *In Eucl.*, 111.21f.; cf. Eutoc. *In Archim. De sphaer.*, 96.17); the 'triads' are usually taken to mean parabola, hyperbola, and ellipsis. Proclus too says that Menaechmus had solved the problem of doubling the cube by means of κωνικαὶ γραμμαί (*In Plat. Tim.*, 34.1f.). See Schmidt. Fragmente; Allman, *op. cit.*, 166f.; Heath. *Apollonius*, xviiff; idem. *History* 1, 251f.; 2, 110; Becker. *Denken*, 82f. In the time between Menaechmus and Euclid, Aristeas the Elder developed the theory of conic sections (see above, 179). The names of the three curves (παραβολή, ὑπερβολή, ἔλλειψις) go back to the Pythagorean application of areas (Eud. fr. 137).

200 Knorr. *AT*, 61ff. See also Böker, *op. cit.*, 1211f.

201 Knorr. *TS*, 94ff.

202 Lasserre. *Léodamas*, 550.

203 Procl. *In Eucl.*, 72.23–73.12 (on two meanings of the word 'element', with examples

Nor, actually, is it certain that this Menaechmus is the same person as Eudoxus' student. The *History of Geometry* was concerned with mathematical discoveries, not with discussions of differences between theorems and problems, or the meanings of the word 'element'. Besides, it is unlikely that Geminus used this work of Eudemus at all.[204] The main source of the discussion related by Geminus and figuring Amphinomus, Speusippus, Menaechmus, and Zenodotus as its protagonists was undoubtedly Posidonius,[205] who had made use of both classical and Hellenistic sources. The topics of this discussion were unlike those treated in the treatises of professional mathematicians. Most likely, Posidonius relied here on the works of philosophers who took an interest in mathematics (Speusippus, Posidonius, Geminus, and Proclus belonged to this category) or on some Hellenistic introductory courses to mathematics. Though we cannot rule out the possibility that Menaechmus wrote a popular treatise on mathematics that reached Posidonius, the Stoic could well have meant a different Menaechmus.[206]

The group of Eudoxus' students seems to have included the last author of pre-Euclidean *Elements*, Theudius of Magnesia, and his contemporary Athenaeus of Cyzicus.[207] All that is known about Theudius' *Elements* is that they were better arranged and gave a more general character to many partial theorems (*In Eucl.*, 67.12f.).[208] Athenaeus of Cyzicus was "famous in mathematical sciences, geometry in particular" (67.16f.), but we know nothing about his specific discoveries. Hermotimus of Colophon extended the investigations of Eudoxus and Theaetetus and discovered many new theorems of the *Elements* (67.20f.). Besides, he composed a writing on the so-called loci (τόποι), i.e., on the theory of geometrical places,[209] developed later by Aristeas the Elder and Apollonius.

---

from Euclid!); 77.7–78.10 (all mathematical propositions are problems, the latter being of two types), 253.16–244.5 (reversibility of theorems).

204 See above, 184f.

205 See above, 179 n. 54. This was admitted by Lasserre (*Léodamas*, 552f.), who nevertheless saw the ultimate source in Philip's book *On Plato* (cf. above, 89f.).

206 I shall treat this subject in a forthcoming paper "On two Menaechmi". As a preliminary, it is worth pointing out that we find in Geminus quite a few other 'doubles'. 1) The Theodorus attested by Geminus (Procl. *In Eucl.*, 118.7) is definitely not Theodorus of Cyrene, but a Hellenistic mathematician; 2) the Oenopides to whose 'succession' Zenodotus belonged (ibid., 85.15f.) is not Oenopides of Chios, but most likely a Stoic philosopher (see above, 179 n. 54, and below, 260 n. 134); 3) the Hippias who found the quadratrix (ibid., 272.8f., 356.6f.) is not the famous Sophist Hippias, but a Hellenistic mathematician (Knorr. *AT*, 82f.).

207 See above, 98f.

208 Translation according to Grynaeus' conjecture μερικῶν instead of the manuscript ὁρικῶν.

209 Knorr. *AT*, 142 n. 30, 371 n. 25.

5. Teleological progressivism

Generally, Eudemus paid attention not only to the results and the form of their exposition, but also to the method of the proof and its correspondence to what was customary in his own time. In the *Catalogue*, the improvement of geometrical methods is discussed in much detail. Thales' method is characterized as in some cases more empirical, and in others more general, i.e., scientific, whereas Pythagoras is credited with transforming geometry into an abstract science and including it in the canon of education of a free man. Owing to the efforts of Leodamas, Archytas, and Theaetetus, geometry became "more scientific and systematic", Leon discovered the method of diorism, and Eudoxus used the method of analysis, while Amyclas, Menaechmus, and Dinostratus "made the whole of geometry even more perfect" (ἔτι τελεωτέραν ἐποίησαν τὴν ὅλην γεωμετρίαν, 67.11). It is significant that the *Catalogue* starts with the words about progressing "from the imperfect to the perfect" (ἀπὸ τοῦ ἀτελοῦς εἰς τὸ τέλειον, 64.25) and ends with the notice that "those who compiled histories of geometry bring the perfecting (τελείωσις) of geometry up to this point" (68.4f.). Of course, no author who lived after Eudemus could have the idea of the 'completion' of geometry at the end of the fourth century, i.e., before Euclid, Archimedes, and Apollonius. This confirms again that the beginning of the *Catalogue*, as well, goes back to Eudemus.

I have suggested above that, in his histories of the exact sciences, Eudemus employed the professional approach – of course, to the extent that this was possible for a Peripatetic philosopher. This does not contradict the fact that Eudemus' views on development in general, and on the development of science in particular, rely on the Aristotelian doctrine that we call teleological progressivism. According to this doctrine, everything in nature, society, and culture develops from a primitive state to a perfect one,[210] and for many things this state of perfection was thought to have already been achieved or almost achieved. Tragedy, for example, had already attained its perfection, society had found its best and final state in the *polis*, and philosophy, whose early ideas were as immature as childish speech, would be soon completed.[211] Against this background, Eudemus' idea that geometry had achieved or almost achieved perfection seems quite natural.[212] To be sure, such progressivism (not necessarily a teleological one) was characteristic not only of Aristotle, but of many other

210 "There is evolution towards a state of excellence all over in the design of nature; the goal is the end, *telos*." (Burkert, W. Impact and limits of the idea of progress in Antiquity, *The idea of progress*, ed. by A. Burgen et al., Berlin 1997, 31).

211 *Poet.* 1449a 15; *Pol.* 1252a 26–1253a 9; *Met.* 993a 15f. "Since in few years great progress has been achieved, philosophy will be finished and perfected in a short time." (fr. 53 Rose). Cf. similar remarks on the arts (*EN* 1098a 23f.) and political systems (*Pol.* 1264a 3).

212 Theophrastus expressed similar views, though more cautiously (fr. 34a FHSG). See Edelstein, *op. cit.*, 148 n. 31.

authors of the period of classical rationalism: Hippocratic physicians, scientists, rhetoricians, and poets.[213] Without casting doubt upon Eudemus' Aristotelianism, we should emphasize that in this case (as well as in many others) he did not follow his teacher's ideas slavishly, but rather shared with him the views common for that period.

The same common attitude is seen in the thesis that all the sciences appeared due to practical necessity. Geometry, in particular, was discovered by the Egyptians, and arithmetic by the Phoenicians, who were employed in trade (*In Eucl.*, 64.17). The author of *Ancient Medicine* also believed that medicine had been discovered due to necessity and need. We find analogous notions of the origin of τέχνη in many classical writers, including Aristotle.[214] The idea that geometry originated with the Egyptians has even more predecessors.[215] But whereas Herodotus noticed the practical origin of Egyptian geometry, Plato the practical character of Egyptian geometry and Phoenician arithmetic, and Philip wrote that the Greeks bring to perfection the knowledge they receive from the barbarians, Aristotle points out that μαθηματικαὶ τέχναι were first discovered in Egypt because the local priests had leisure.[216] This remark clarifies the preceding passage, where he states that in every civilization practical crafts are born first, then appear the fine arts, and after them theoretical sciences whose end is knowledge. For the latter, adds Aristotle, leisure time is needed.

As we have already noticed, the same historical scheme was known before Aristotle; some of his predecessors noticed the role of leisure, whereas others did not.[217] It is therefore possible that in this case Eudemus followed the simpler version that also occurs in Aristotle.[218] But does this mean that Aristotle denied the practical origin of Egyptian geometry? He would hardly argue against the fact that the Egyptians' practical geometry had preceded the scientific geometry. The general line of cognitive development known from the *Catalogue* has quite obvious parallels in Aristotle.[219] In accordance with this

---

213 *VM* 2; *De locis in hom.* 46; *De arte* 1; Isoc. *Nic.* 32, *Euag.* 7, *Antid.* 82, 185, *Paneg.* 10; Chairem. *TrFG* 71 F 21; Alex. fr. 31 K.–A. See above, 2.2, 2.4.

214 Democritus (68 B 144); Isocrates (*Bus.* 12–15, 21–23); Philip (*Epin.* 974d 8f., 975c 9f.); Aristotle (fr. 53 Rose = *Protr.* fr. 8 Ross; *Pol.* 1329b 25f.; *Met.* 981b 12–22, 982b 22f.). See above, 2.2–3.

215 Herodotus (II, 109); Isocrates (*Bus.* 28, cf. 23); Plato (*Phaedr.* 274d 1f.; *Leg.* 747a–c); Philip (*Epin.* 986d 8f., 987d 9); Aristoxenus (fr. 23). See also Democr. test. XIV Luria.

216 *Met.* 981b 23. Further he adds that the difference between τέχνη and ἐπιστήμη was explained in detail in the *Ethics* (981b 26f.).

217 Democritus (68 B 144) says nothing about leisure. The first to mention leisure was Isocrates (*Bus.* 21–23), followed by Plato (*Crit.* 110a). But leisure plays no role in the *Laws* (677a–683b), where Plato's theory on the origin of culture is stated in great detail, or in the *Epinomis* (974d 3–977b 8).

218 See e.g. fr. 53 Rose = *Protr.* fr. 8 Ross; *Pol.* 1329b 25f., where leisure is not mentioned.

219 *Met.* 981a 12 f., 981b 10 f.; *EN* 1139a 17f. See also Wehrli's commentary on fr. 133.

scheme, Eudemus emphasized the practical *origin* of Egyptian geometry, since even Thales, who borrowed this science from Egypt, still proved some things in a more particular and some in a more general way (fr. 133).

The idea of the progressive growth of knowledge is older than both Aristotle and Eudemus. Already Isocrates employs ἐπίδοσις and ἐπιδιδόναι as notions designating qualitative development and advancement (*Paneg.* 10, 103, 189). The verbs from the same semantic group, such as αὐξάνειν, προαγαγεῖν, and προέρχεσθαι, acquire a similar meaning with Isocrates.[220] These notions are often found where he discusses the discovery and development of various τέχναι, in particular the progress of rhetoric and the constant quest for novelties.[221] Aristotle employs the verb προαγαγεῖν when referring to the advancement of mathematics by the Pythagoreans (*Met.* 985b 23f.) and ἐπίδοσις for expressing the idea of rapid progress in contemporary mathematics and in all the τέχναι in general (*EN* 1098a 24–25). In Philodemus' quotation from the Academic treatise, ἐπίδοσις is used twice: first, with regard to all mathematical sciences, and second, in connection with geometry. In Eudemus, ἐπίδοσις appears only once, and conspicuously in the passage of the *Catalogue* praising Plato's role in geometry, but αὐξάνειν, προαγαγεῖν, προέρχεσθαι occur in this text unusually often.[222]

One of the criteria of progress in mathematics, for Eudemus, was the degree of generality of mathematical propositions.[223] Thales was the first to teach geometry "more generally" (*In Eucl.*, 65.10), Eudoxus augmented the number of "general theorems" (67.4) by developing a new theory of proportions, and Theudius gave to many partial propositions a more general character (67.15). This also coincides with Aristotle's notion of the development of sciences from the particular to the general. Another notion shared by the Peripatetics was the idea, going back to Plato, of the cyclical character of the historical process.[224] Humankind, being eternal, periodically goes through a number of regional catastrophes in which most arts and sciences perish, so that later generations are compelled to discover everything (or nearly everything) anew.[225] Proclus refers to Aristotle's opinion on the periodical emergence of sciences before the very beginning of the *Catalogue* (64.8f.), which, in Wehrli's edition, opens with the

---

220 See above, 77 n. 140.

221 See e.g. *Nic.* 32, *Antid.* 81–83, 185, *Paneg.* 10.

222 ἐπαυξάνειν (66.16), αὐξάνειν (67.5), προαγαγεῖν (67.7, 67.22), προέρχεσθαι (66.17). See Edelstein, *op. cit.*, 92; Thraede. Fortschritt, 141f., 154.

223 Lasserre. *Eudoxos*, 161f.

224 See above, 109 n. 133.

225 Arist. fr. 13 Rose (= fr. 463 Gigon), fr. 53 (= *Protr.* fr. 8 Ross = fr. 74.1 Gigon); *Cael.* 270b 16–24; *Mete.* 339b 25–30; *Met.* 1074b 10–13; *Pol.* 1269a 5f., 1329b 25–33; Theophr. fr. 184 FHSG; Dicaearch. fr. 24. Cf. οὐ γὰρ μόνον πόλεις τε καὶ ἔθνη, φησίν, ἀρχὰς καὶ τέλη λαμβάνουσιν, ὡς εἰς παντελῆ λήθην ἐκπεσεῖν, ἀλλὰ καὶ δόξαι καὶ τέχναι καὶ ἐπιστῆμαι τοῦτο πάσχουσιν (Philop. *In Arist. Mete.*, 17.26f.). See Festugière. *Révélation*, 219f.; Palmer, *op. cit.*, 192ff., 196 n. 26.

following words: "Since we have to examine the beginnings of arts and sciences in the course of the present period ...".[226] Uncertain as we are whether these words go back to Eudemus himself, the idea formulated here is quite appropriate for an introduction to the *History of Geometry.* In his *Physics,* Eudemus criticized the Pythagorean idea of recurrence of all phenomena and events in exactly the same way (ἀριθμῷ, fr. 88), but this is obviously not what the Aristotelian theory implies. Defending this theory, Theophrastus stated that the discoverers of sciences had lived about a thousand years earlier (fr. 184.145f. FHSG), which brings us back to the period immediately preceding the Trojan War, the period with which the Greeks used to start the history of their culture.[227] There is no reason to suppose that Eudemus rejected the idea of the cyclical development of sciences, especially since in all other cases the coinciding or at least the proximity of the basic historical notions of the teacher and the student are obvious.

Hence, it is Aristotle's historico-philosophical notions of the development of humankind in general and of arts and sciences in particular that formed the conceptual basis of the *History of Geometry.* That is what distinguished Eudemus' history from the ordinary histories that, according to Aristotle, describe a great number of events coinciding in time but having no connection with each other (*Poet.* 1459a 16–29).[228] Unlike them, Eudemus' ἱστορία contains, in compliance with Aristotle's requirements, both ἀρχή and τέλος, and possesses an intrinsic unity as a result. The events of this history, i.e., the discoveries of Greek mathematicians, were connected by causal relationship, which Eudemus never failed to emphasize, and demonstrated, in total, the general regularities in the progressive growth of knowledge: from the simple to the complex, from the particular to the general, from the humble empirical origins in Egypt, which were born of practical necessity, to the perfection of theoretical geometry in Eudemus' own times.

[226] ἐπεὶ δὲ χρὴ τὰς ἀρχὰς καὶ τῶν τεχνῶν καὶ τῶν ἐπιστημῶν πρὸς τὴν παροῦσαν περίοδον σκοπεῖν, λέγομεν ... (64.16f.). Heath. *History* 1, 120, attributed these words to Proclus.

[227] See above, 48ff.

[228] Weil. Aristotle's view, 203.

Chapter 6

The history of arithmetic and the origin of number

1. The fragment of Eudemus' *History of Arithmetic*

The only surviving fragment of Eudemus' *History of Arithmetic* comes from Porphyry's commentary to Ptolemy's *Harmonics*. Its content is related, accordingly, to the mathematical theory of music, rather than to arithmetic. The broad context of Porphyry's commentary is the following. Discussing the Pythagorean mathematical theory of concordant intervals, Ptolemy on the whole agrees with it, while criticizing some of its propositions.[1] For his starting point he takes the Pythagorean method of associating equal numbers with tones of equal pitch and unequal numbers with tones of unequal pitch (*Harm.*, 11.8ff.). Further, the Pythagoreans divide tones of unequal pitch into two classes, concordant and discordant intervals, and associate the first class with epimoric and multiple ratios, and the second with epimeric ratios. The reason for this is that as concordant intervals are 'finer' than discordant, so epimoric and multiple ratios are 'better' than epimeric, because of the simplicity of the comparison.[2] Ptolemy acknowledges that concordant intervals differ from each other by the degree of their proximity to absolute equality: the less the difference between the terms of their ratios, the better. Thus, the octave is the finest of the concordant intervals, since its ratio (2:1) "alone makes an excess equal to that which is exceeded" (i.e., the difference between its terms is equal to the smaller term of the ratio); it is immediately followed by the fifth (3:2) and the fourth (4:3).

The principle of comparing the tones of unequal pitch by their closeness to equality is further elaborated in Ptol. *Harm.* I, 7 (15.18f.). In his commentary to this passage, Porphyry emphasizes that the method comes from the Pythagoreans and adds: many of them took their starting point not in equality alone, but in the so-called *pythmenes*, or the 'first numbers', as well, i.e., the ratios of the concordant intervals expressed in their lowest terms.[3] Earlier, Porphyry defined

---

1 *Harm.* I, 5–7. See Barker, A. *Scientific method in Ptolemy's* Harmonics, Cambridge 2000, 54ff.

2 For details, see Düring, I. *Ptolemaios und Porphyrios über die Musik*, Göteborg 1934, 176f.; Barker. *GMW* II, 284f.; idem. *Scientific method*, 61f.

3 *In Ptol. Harm.*, 114.23ff. πυθμήν, 'base', is the first in a series of equal ratios that is expressed in the lowest terms. Thus, the ratio 2:3 is a *pythmēn* in the series 2:3, 4:6, 8:12, etc. With this meaning, *pythmēn* is found already in Plato (*Res.* 546c 1); cf. Speus. fr. 28 Tarán. See Nicomachus of Gerasa. *Introduction to arithmetic*, transl. by M. L. D'Ooge, New York, 1926, 216.

the first numbers or *pythmenes* as the smallest numbers in which concordant intervals are produced.[4] Now he says:

> For Eudemus makes clear in the first book of the *History of Arithmetic* that they demonstrated the ratios of concordant intervals through the *pythmenes*, saying about the Pythagoreans the following word for word: "(They said) moreover that it turned out that the ratios of the three concords, of the fourth, the fifth and the octave, taken in the first numbers (ἐν πρώτοις), belong to the number nine. For 2 and 3 and 4 are nine." (fr. 142)

The ratios 2:1, 3:2, and 4:3 are expressed in the 'first numbers' indeed. It is not clear, however, what precisely Eudemus had in mind and why it was important for the Pythagoreans that the numerators of these three ratios make up nine. Unfortunately, the quotation is too fragmentary to conclusively suggest any immediate context. Besides, it is different from other fragments of Eudemus' works on the history of science, which usually deal with particular discoveries. Let us nevertheless make the most of the scant information we find in Porphyry.

It follows from the quoted passage that Eudemus' *History of Arithmetic* comprised at least two books, i.e., was of about the same length as his *History of Astronomy*. If Eudemus treated the history of arithmetic in as much detail as he did the history of astronomy, he must have had enough material for it at hand. Euclid's arithmetical books (VII–IX) are known to have preserved only part of his predecessors' theories, the rest remained outside the *Elements*[5] and in part was subsequently included in compendia by Nicomachus, Theon of Smyrna, Iamblichus, and other later authors. In the early fourth century, arithmetic already enjoyed the status of an exact science; some believed its demonstrations to be more conclusive than even geometrical ones.[6] No wonder then that Eudemus considered arithmetic to be worth a special historical treatise.

The fact that the only quotation from it relates to the mathematical harmonics of the Pythagoreans can hardly be fully explained by its coming from commentary to Ptolemy's *Harmonics*. Theoretical arithmetic, created by the early Pythagoreans, was closely connected to harmonics,[7] especially as con-

---

4 πρώτους λαβόντες ἀριθμούς, οὓς ἐκάλουν πυθμένας … τουτέστιν ἐν οἷς ἐλαχίστοις ἀριθμοῖς συμφωνίαι ἀποτελοῦνται (*In Ptol. Harm.*, 107.18f.). Cf. Theon of Smyrna's definition: "Of all the ratios … those that are expressed in the smallest numbers and prime to each other are called firsts among those having the same ratio or *pythmenes* of the same species." (*Exp.*, 80.15).

5 E.g., the arithmetical part of the theory of irrationality is not retained in book X (Knorr. *Evolution*, 311). Speusippus analyzed linear and polygonal numbers (fr. 28 Tarán), about which Euclid remains silent. See also Philip of Opus' *On the Polygonal Numbers* (Lasserre. *Léodamas*, 20 T 1).

6 Archytas (47 B 4), Aristotle (*APo* 87a 34–7, *Met.* 982a 26f.). On the role of arithmetic in Archytas' and Plato's classification of sciences, see Knorr. *Evolution*, 58 n. 71, 90f., 311.

7 Van der Waerden. *Pythagoreer*, 364ff., 406ff.

cerns the theory of means. It is highly possible that the development of arithmetic in its application to music could also have been reflected in the *History of Arithmetic*, for example, in problems that occupied Archytas (47 A 17, 19, B 2).[8] Further considerations lead in the same direction. Of the quadrivium of mathematical sciences, harmonics was the only one to which Eudemus did not devote a special treatise. Assuming that the Peripatetic project supposed the description of all fields of knowledge that had a long enough history behind them, the absence of the history of harmonics may, indeed, appear strange. That Eudemus could have treated the problems of mathematical harmonics in his *History of Arithmetic* provides a possible explanation for this.[9] This hypothesis seems the more probable in that the Aristotelian classification of sciences links arithmetic and harmonics with each other as a main and a subordinate discipline: first, they are based on common principles and, second, harmonics depends on arithmetic inasmuch as it uses the latter's methods of demonstration.[10]

Let us return to the text of the fragment. To judge from the verbatim quotation and the reference to the first book of the *History of Arithmetic*, Porphyry must have cited this work first-hand. Since 'the Pythagoreans' are not mentioned in the quotation, they must have been dealt with outside the quoted passage, too. On the other hand, Porphyry never refers to the *History of Arithmetic* elsewhere. Hence, he could have borrowed this passage from an intermediary source that quoted Eudemus *verbatim* and more amply. This source may well have been the book *On the Differences Between the Theories of Aristoxenians and the Pythagoreans* by Didymus of Alexandria, a musicologist of the early first century AD.[11] Both Porphyry and Ptolemy used this book, particularly as a source on Pythagorean harmonics.[12] According to Porphyry, Ptolemy borrowed much from Didymus, though not always referring to him.[13] Indeed, in

---

8 See above, 173f. In his treatise on harmonics, Archytas discussed, in particular, the mean proportionals (47 B 2). His approach to harmonics is taken up in Euclid's *Sectio canonis*, where a number of arithmetical theorems are proved: van der Waerden. *Pythagoreer*, 364ff., 406ff.; Barker. *GMW* II, 42f., 48f. Further, Theaetetus and Eudoxus also worked on the theory of means (see above, 173f.).

9 Cf. above, 129.

10 *APo* 75a 38–75b 17, 76a 9–15. Geometry and optics relate to each other in the same way.

11 See Barker. *GMW* II, 230, 241f.; idem. Greek musicologists in the Roman empire, *Apeiron* 27 (1994) 53–74, esp. 64ff.; West, M.L. *Ancient Greek music*, Oxford 1992, 169f., 239f.

12 Didymus was Porphyry's source for Xenocrates' fragment: "Pythagoras discovered also that musical intervals do not come into being apart from numbers, for they are an interrelation of one quantity with another." (*In Ptol. Harm.*, 30.1f. = fr. 87 Isnardi Parente). See Düring. *Ptolemaios*, 155f.

13 *In Ptol. Harm.*, 5.12f. Ptolemy mentions Didymus only in book II of *Harmonics*, doing so, as a rule, in connection with Archytas and other early theoreticians. But he used Didymus' material in book I as well (Düring. *Ptolemaios*, 139f.). Porphyry, in

Harmonics I, 6, where Ptolemy criticizes another Pythagorean method and mentions "the first numbers that make up the ratios" (i.e., the *pythmenes*), Didymus does not figure among his sources, whereas Porphyry adduces the same arithmetical method in a fuller version, referring to both Didymus and Archytas.[14]

According to Didymus, to determine the most concordant intervals, the Pythagoreans proceeded as follows. Taking the 'first numbers', which they called *pythmenes* (i.e., 2:1, 3:2, 4:3), and assigning them to concordant intervals, they subtracted a unit from each of the terms of the ratio and compared the remainders. Thus, subtracting a unit from both terms of the octave (2:1) they obtained one, from the fourth (4:3) five and from the fifth (3:2) three. The smaller the remainder, the more concordant was the given interval considered. The whole procedure hardly seems convincing, from either a mathematical or musical point of view.[15] No wonder Ptolemy calls it 'utterly ludicrous' (*Harm.*, 14.6). He notes, in particular, that a ratio remains the same whether it is expressed by *pythmenes* or not, while the Pythagorean method is valid for *pythmenes* only. Porphyry on the whole agrees with him (*In Ptol. Harm.*, 109.1ff.), repeatedly pointing out, however, that it is precisely on the lowest terms that the Pythagorean theory was based. This was precisely his reason for quoting Eudemus' passage dealing with ratios taken ἐν πρώτοις. It remains unclear whether the number nine, being the sum of the numerators of the three *pythmenes*, had a particular mathematical or 'numerological' sense.

Though Didymus must have referred to Archytas,[16] the theory he sets forth cannot belong to the latter; it relates, most likely, to the earlier stage of the Pythagorean harmonics.[17] One can suggest that Eudemus' Pythagoreans belong to the same period, the more so because his histories of geometry and astronomy generally place 'the Pythagoreans' in the first part of the fifth century. Archytas, too, constituted ratios of the tetrachords in their lowest terms (ἐν πρώτοις ἀριθμοῖς, 47 A 16), but his theory has nothing to do with calculations described above. In the arithmetical books of the *Elements*, the term πρῶτοι (ἀριθμοί) was consistently replaced by the more technical one, οἱ ἐλάχιστοι ἀριθμοὶ τῶν αὐτὸν λόγον ἐχόντων, the least numbers of those that have the same ratio – probably to avoid confusion with the prime numbers. But Euclidean 'numbers prime to one another' (πρῶτοι πρὸς ἀλλήλους ἀριθμοί) are the same as the Pythagorean 'first numbers' (*Elem.* VII, 21). Revealingly, the

---

his turn, did not fail to mention his sources. The fact that he refers to Eudemus only once means either that he found nothing worthy of notice in him, or that he quotes him second-hand.

14 Porph. *In Ptol. Harm.*, 107.15ff. = Archytas A 17.

15 For interpretations of this method, see Barker. *GMW* II, 35 n. 29; idem. *Scientific method*, 71f.; Huffman, C. A. *Archytas of Tarentum: Pythagorean, philosopher and mathematician king*, Cambridge 2005, 428ff.

16 Düring. *Ptolemaios*, 157; Barker. *GMW* II, 34f.

17 Barker. *GMW* II, 34 n. 25; Huffman. *Archytas*, 432f.

ancient Pythagorean demonstration of the incommensurability of the square's diagonal with its side uses the notion of the lowest numbers having the same ratio (ἐλάχιστοι τῶν τὸν αὐτὸν λόγον ἐχόντων), i.e., of *pythmenes.*[18] The 'first numbers' obviously proved to be quite a useful tool in the Pythagorean arithmetic.

2. Aristoxenus: *On Arithmetic*

Unfortunately, no other references to the *History of Arithmetic* have survived in the ancient tradition. For later authors, this subject was obviously less attractive than the history of geometry and astronomy. It is revealing that we also have only one fragment (cited in Stobaeus) from Aristoxenus' Περὶ ἀριθμητικῆς (fr. 23), which is luckily more informative.

From Aristoxenus' books *On Arithmetic.*[19]

§ 1. Pythagoras seems to have valued the science of numbers most of all and to have advanced it, separating it from the merchants' business and likening all things to numbers. For number contains all else as well, and there is a ratio between all the numbers to each other <...>[20]

§ 2. The Egyptians, for their part, believe numbers to be the invention of Hermes, whom they call Thoth. And others derived numbers from the circular paths of the divine luminaries.

§ 3. A unit is a beginning of number, and a number is a multitude consisting of units. Of numbers, the even are those that are divisible into equal parts, and the odd are those that are divisible into unequal parts and have a middle.

§ 4. It is considered, therefore, that crises and changes in illnesses relating to their beginning, culmination, and end occur on odd days, since an odd number has a beginning, a middle and an end.

Before we turn to the contents of the fragment, let us consider whether Aristoxenus is likely to have written a special treatise *On Arithmetic* at all. In his commentary, Wehrli wrote that fr. 23 "in its present form of an elementary introduction to numerical notions" does not stem from Aristoxenus. On this basis, Wehrli denied the existence of Aristoxenus' treatise *On Arithmetic* and considered the part of this fragment that belongs to Aristoxenus to come from one of his three treatises on the Pythagoreans. It is not clear, however, why Aristoxenus' *On Arithmetic* could not contain definitions he borrowed most probably from an arithmetical treatise of Pythagorean origin. Wehrli does not

---

18 *Elem.* III, 408.12, 410.4. See below, 223 n. 40.

19 For greater convenience, I have broken the fragment into four paragraphs.

20 Τὴν δὲ περὶ τοὺς ἀριθμοὺς πραγματείαν μάλιστα πάντων τιμῆσαι δοκεῖ Πυθαγόρας καὶ προαγαγεῖν εἰς τὸ πρόσθεν, ἀπαγαγὼν ἀπὸ τῆς τῶν ἐμπόρων χρείας, πάντα τὰ πράγματα ἀπεικάζων τοῖς ἀριθμοῖς. τά τε γὰρ ἄλλα ἀριθμὸς ἔχει καὶ λόγος ἐστὶ πάντων τῶν ἀριθμῶν πρὸς ἀλλήλους... (Wehrli, following Diels and Meineke, marked a lacuna here).

seem to have real grounds for denying the existence of Aristoxenus' work *On Arithmetic* and for relating fr. 23 to Aristoxenus' book *On Pythagoras and His Disciples* (fr. 11–25). Stobaeus never refers to this book; all of Aristoxenus' fragments that he cites (except fr. 23) come from *Pythagorean Precepts* (fr. 34–37, 39–41), which he dutifully mentions quotation by quotation. Hence, in citing this fragment, Stobaeus is all the more likely to indicate the right source.[21] Since Pythagoreans are mentioned in many of Aristoxenus' writings (see, e.g. fr. 43, 90, 131), there is no particular reason to relate this fragment to any of his three special works on this school.

Unlike Eudemus' *History of Arithmetic*, Aristoxenus' book bore the title *On Arithmetic*, which fully accords with the wealth of historical, scientific, and philosophical material we find in the fragment cited by Stobaeus. Having mentioned the founder of arithmetical science (§ 1), Aristoxenus adduces several versions of the origin of number as such (i.e., of the art of calculation), which go back to the Academy (§ 2), quotes the definitions of unit, number, and even and odd numbers (§ 3), and proceeds to characterize the role of numbers in nature (§ 4). Of course, we cannot be wholly certain that Stobaeus' quotation presents the text in its continuity rather than a set of separate fragments. The passage, however, seems quite coherent and is more likely to come from a book on arithmetic than from a work on the Pythagoreans.[22] Hence, the fact that fr. 23 goes back to Aristoxenus' work *On Arithmetic* does not seem to raise any serious doubts. The onus, at any rate, lies with those who state the contrary. Rash as it is to base conjectures on the subject of the book on a single fragment, the work seems to have been of a popular philosophical, rather than historical character. It is also clear that its material did not repeat that of Eudemus' *History of Arithmetic*, though certain themes treated in both of them, as well as the positions of their authors, could well coincide.

In § 1, relying on a certain tradition (δοκεῖ), Aristoxenus tells us that Pythagoras highly valued the science of numbers (ἡ περὶ τοὺς ἀριθμοὺς πραγματεία)[23] and advanced it; he turned arithmetic into a theoretical science by separating it from practical need (ἡ τῶν ἐμπόρων χρεία). Obvious similarities between this passage and Eudemus' *Catalogue* seem to reflect some common notions of the development of exact sciences (5.5). First of all, familiar from the *Catalogue* (*In Eucl.*, 64.17f.) and characteristic of the Peripatetics in general is the contrast between χρεία as the primary impulse towards acquiring knowledge and πραγματεία, the scientific discipline created by Pythagoras from the study of numbers. Furthermore, the verb προαγαγεῖν, which belongs, as we remember, to the semantic group meaning 'progress', occurs twice in

21 On Stobaeus' working method, see Mansfeld, Runia. *Aëtiana*, 196ff.

22 Aristoxenus' treatise *On Music* contained, apart from theoretical material, much information on various musical discoveries and their authors (fr. 78–81, 83).

23 πραγματεία, in the sense of 'scientific discipline' or 'branch of science', is repeatedly found in Aristoxenus (*Elem. harm.*, 5.6, 6.1, 7.5 etc.).

Eudemus (ibid., 67.7, 67.22). And, last but not least, Aristoxenus associates with Pythagoras the same transformation of arithmetic into a theoretical science as Eudemus does with geometry (εἰς σχῆμα παιδείας ἐλευθέρου μετέστησεν). We should recall that Aristotle attributed to Pythagoras the study of mathematical sciences, in particular numbers; in *Metaphysics*, he says that the Pythagoreans were the first to advance mathematics.[24]

While Eudemus limited his *History of Geometry* to purely scientific problems, Aristoxenus, following Aristotle, thought it important to note the connection between the Pythagorean study of numbers and the philosophical and scientific ideas of this school: according to him, Pythagoras likened all things to numbers, "for number contains all else as well, and there is a λόγος between all the numbers".[25] The similarity between Aristoxenus' words and Aristotle's *Metaphysics* 985b 23f. was noted in particular by Frank and Burkert, who interpreted it to the effect that Aristoxenus, lacking any independent evidence on Pythagoras as a mathematician, simply quoted *Metaphysics*, while substituting 'Pythagoras' for 'the Pythagoreans'.[26] However, the parallels with Eudemus' *Catalogue* and Aristotle's fr. 191, quoted above, give this similarity a different meaning: all the three testimonies reflect the notion, common to the Lyceum, that Pythagoras made a decisive contribution to transformation of mathematics into a theoretical science.

Besides, one cannot fail to note the essential differences between the Pythagorean 'number philosophy' in Aristotle's interpretation, on the one hand, and Aristoxenus' understanding of the resemblance between things and numbers, on the other. What Aristoxenus meant by τά τε ἄλλα that number has (ἔχει) is not wholly clear, but his reference to λόγος existing between all numbers seems to indicate that he understands the Pythagorean tradition of relationship between things and numbers in the same epistemological sense as Philolaus, for whom "all the things that are known have number, for without it is impossible to understand or to know anything" (44 B 4).[27] Aristotle, who indeed mentioned certain ὁμοιώματα between numbers and things perceived by the Pythagoreans (*Met.* 985b 28f.), meanwhile inclined, rather, to an ontological interpretation, according to which mathematical principles are, at the same time, the principles of all being (985b 25, 986a 16, *etc.*). The connection of an odd number, which has a beginning, a middle, and an end, with medical prognostics (§ 4), as mentioned by Aristoxenus, differs substantially from the numerical metaphysics Aristotle imposed on the Pythagoreans.[28] It is revealing that while

---

24 Πυθαγόρας … διεπονεῖτο περὶ τὰ μαθήματα καὶ τοὺς ἀριθμούς (fr. 191 Rose); τῶν μαθημάτων ἁψάμενοι πρῶτοι ταῦτά τε προήγαγον (*Met.* 985b 23f.).

25 Cf. "In numbers they seemed to see many resemblances to the things that exist and come into being – more than in fire and earth and water" (Arist. *Met.* 985b 27f.).

26 Frank, E. *Plato und die sogenannten Pythagoreer*, Halle a. S. 1923, 260 n. 1; Burkert. *L & S*, 414f.

27 See Huffman. *Philolaus*, 172ff.; Zhmud. Philolaus, 255ff.

28 Zhmud. *Wissenschaft*, 320f.

Aristoxenus speaks precisely of medicine, treated at length in his works on the Pythagoreans,[29] Aristotle keeps complete silence upon this subject. The definitions cited by Aristoxenus are of a purely mathematical character (§ 3), whereas Aristotle explicitly states that, for the Pythagoreans, even and odd are στοιχεῖα τοῦ ἀριθμοῦ (986a 16). Empty as it is of mathematical meaning, this statement is of crucial importance for the numerical ontology he attributes to the Pythagoreans. As for Aristoxenus, he did not at all mean that a disease *is* a number, nor that it *consists* of numbers, nor that the *principles* of number are the same as the principles of disease. His example has to be understood as pointing to a likeness between the odd numbers and the periods of an illness: since of both them have a beginning, a culmination, and an end, it is on odd days that changes in the course of illness occur.[30] Therefore, illness comes to resemble number for the sake of prognosis (one needs to define the day of a possible crisis), not for metaphysical *identification* of things and numbers, nor the principles of both. This is confirmed by the popularity the doctrine of critical days won in an empirically oriented Hippocratic medicine.[31]

Hence, Aristoxenus associates the birth of arithmetic with Pythagoras and then, having mentioned several versions of the origin of number, turns to the 'principles' of theoretical arithmetic (§ 3). Three of the four definitions he cites (those of unit, odd number, and even number) differ from those given in Euclid's book VII:

Aristoxenus	Euclid
A unit is a beginning (ἀρχή) of number.	A unit is that by virtue of which each of the things that exists is called one (def. 1).
A number is a multitude composed of units.	A number is a multitude composed of units (def. 2).

[29] Fr. 21–22, 26–27; Iambl. *VP* 163–164 = *DK* 58 D 1.6–16. This fact additionally supports the authenticity of the second part of Aristox. fr. 23.

[30] See similar ideas in the Hippocratic corpus: "The odd days must be especially observed, since on them patients tend to incline in one direction or the other." (*De victu in acutis* [Appendix] 9 Littré; cf. *Epid.* I, 12; *De sept. partu*, 9). On the possible Pythagorean origin of the doctrine of critical days, see Zhmud. *Wissenschaft*, 237 f.

[31] See e.g. Jouanna, J. *Hippocrate*, Paris 1992, 475 f. It is revealing that, in his natural-scientific works, Aristotle himself was not averse to the 'Pythagorean' likening of things and numbers. See *Mete.* 372a 1 ff., 374b 31 f. and especially *Cael.* 268a 10 f.: "For, as the Pythagoreans say, the world and all that is in it is determined by the number three, since beginning and middle and end give the number of an 'all', and the number they give is the triad. And so, having taken these three from nature as (so to speak) laws of it, we make further use of the number three in the worships of the gods ... And in this, as we have said, we do but follow the lead which nature gives." See also πάντα τρία καὶ οὐδὲν πλέον ἢ ἔλασσον τούτων τῶν τριῶν (Ion of Chios, 36 B 1); Theon. *Exp.*, 46.14 f.: they say that three is the perfect number, for it is the first to have beginning, middle, and end.

Even numbers are those that are divisible into equal parts (εἰς ἴσα).	An even number is that which is divisible into two parts (δίχα) (def. 6).
Odd numbers are those that are divisible into unequal parts (εἰς ἄνισα) and have a middle.	An odd number is that which is not divisible into two parts (μὴ διαιρούμενος δίχα), or that which differs by a unit from an even number (def. 7).

Since the definitions Aristoxenus gives also occur before him, for example in Aristotle, they can go back either to one of the pre-Euclidean versions of *Elements* or, still more likely, to an elementary 'introduction' to arithmetic of a Pythagorean origin. In Nicomachus, Theon, and Iamblichus, who have preserved some Pythagorean material not included in the arithmetical books of the *Elements*, we find similar definitions of odd and even numbers: they also speak of the division of even numbers into *equal* parts (instead of by two, as in Euclid) and of odd numbers as having a *middle*.[32] Meanwhile, one can speak of a number as having a middle only if it is represented by *psēphoi*, counting stones, as the early Pythagoreans used to do.[33] In Euclid, by contrast, where numbers are represented by line segments, a 'middle' does not figure in the definition, since the middle of a segment is a point, not another segment.

Let us recall again that pre-Euclidean arithmetic was not limited to the material of books VII–IX of the *Elements*, it contained various, even competing, traditions. Thus, e.g., Euclid's definition of unit bears some archaic features and obviously shares its origin with other propositions of book VII, whereas Aristoxenus' much clearer definition comes from a different source.[34] The existence of different traditions in arithmetic is also attested by the fact that both definitions of an odd number cited above are found in Aristotle;[35] obviously, they cannot come from the same arithmetical treatise.

In connection with the definitions of even and odd numbers given by Aristoxenus, it is worth noting that Philolaus also mentions the division of numbers into even and odd,[36] whereas Plato repeatedly calls arithmetic the science of

---

32 Theon. *Exp.*, 21.22f.; Nicom. *Intr. arith.* I,7.2–3; Iambl. *In Nicom.*, 12.11f.: even numbers are divisible into equal parts, odd numbers are not. See Knorr. *Evolution*, 53 n. 18. Cf. also a fragment from the pseudo-Pythagorean treatise *On Numbers* by Butheros: "The odd is more perfect than the even, for it has a beginning, a middle and an end, while the even lacks a middle." (Thesleff, H. *The Pythagorean texts of the Hellenistic period*, Åbo 1965, 59.10f.).

33 Arist. *Met.* 1092b 10f.; Theophr. *Met.* 6a 15f. = 45 A 2–3.

34 Cf. Nicom. *Intr. arith.* I,8.2: unit is the beginning of all numbers; Theon. *Exp.*, 19.21: unit is the beginning of numbers. There was another definition of unit, "a point without position" (στιγμὴ ἄθετος, Arist. *Met.* 1084b 26), the opposite of the definition of point as a unit having position (*Met.* 1016b 24f.). Both these definitions go back to the Academy, not to the Pythagoreans (Burkert. *L & S*, 66f.).

35 *Top.* 142b 6f., 149a 29f.; *SE* 173b 8.

36 "Number, indeed, has two proper kinds, odd and even, and a third from both mixed together, the even-odd." (44 B 5), transl. by C. Huffman.

even and odd.[37] Such a view of arithmetic undoubtedly has an *early* Pythagorean origin, since arithmetic of the fourth century, as we find it in Euclid, is not the science of even and odd in the least. In all the arithmetical books of the *Elements*, the definitions of even and odd (VII, def. 6–7) occur only once – namely, in the archaic theory of even and odd numbers that, as Becker has shown, belongs to the earliest stratum of Pythagorean mathematics.[38] This theory, consisting of propositions IX, 21–34 based on definitions VII, 6–11,[39] is of an elementary character and stands in no logical connection with the material of other arithmetical books of the *Elements*.[40] Becker himself dated this theory to the first half of the fifth century, and van der Waerden to around 500.[41] Though neither of them attributes it directly to Pythagoras, Aristoxenus' fragment, which they did not take into account, seems to be pointing in this direction.[42]

The four definitions given by Aristoxenus in § 3 are likely to have opened a Pythagorean arithmetical treatise, a sort of introduction to arithmetic setting forth, in particular, a theory of even and odd numbers.[43] In Aristoxenus, this theory appears as a specimen, an example of the 'science of numbers' (ἡ περὶ τοὺς ἀριθμοὺς πραγματεία) practiced by Pythagoras, especially as being directly related to his notions of the role of numbers in nature. To judge from the evidence of Epicharmus, Philolaus, and, in particular, Plato (for whom arithmetic, as we have noted already, was a science of even and odd), this theory remained quite popular through the whole of the fifth century, even outside the narrow circle of specialists. In the mathematics contemporary with Plato and Aristotle, it remained in the background, still functioning as an elementary in-

---

37 *Prot.* 357a 3; *Gorg.* 451 b1, 451c 2; *Res.* 510c 4; *Charm.* 166a 5–10; *Tht.* 198a 6.

38 Becker, O. Die Lehre von Geraden und Ungeraden im IX. Buch der Euklidischen Elemente, *Q & St* B 3 (1934) 533–553 (= *Zur Geschichte der griechischen Mathematik*, 125–145).

39 Definitions 8–11 relate to the so-called even-odd numbers (cf. Philolaus 44 B 5). The Pythagorean origin of definitions 6–11 is pointed out in the scholia to Euclid (*Elem.* V, 364.6).

40 The only exception is the ancient demonstration of the incommensurability of the square's diagonal with its side (cf. Arist. *APr* 41a 26, 50a 37), which figures in several manuscript copies of Euclid's book X as appendix 27 (*Elem.* III, 408–410). It also points to the Pythagorean school, namely, to Hippasus' discovery of irrationality (Becker. Lehre, 544f., 547; cf. Knorr. *Evolution*, 22ff.).

41 Its antiquity is confirmed, in particular, by the quotation from Epicharmus' comedy (23 B 2), which seems to reflect the Pythagorean studies in even and odd numbers.

42 Becker, O. *Grundlagen der Mathematik in geschichtlicher Entwicklung*, Freiburg 1954, 38; van der Waerden. *Pythagoreer*, 392. If Hippasus used the theory of odd and even to demonstrate the incommensurability of the square's diagonal with its side, this theory must indeed go back to Pythagoras' time.

43 It is worth noting that the later introductions to arithmetic by Nicomachus, Theon, and Iamblichus arrange the material according to the same pattern: having defined a unit and a number they proceed to even and odd numbers and their derivatives.

troduction to arithmetic for laymen. Otherwise it is hard to explain why Plato,[44] Speusippus (fr. 28 Tarán), and Aristotle[45] turn so often to odd and even numbers, as well as to their derivatives.[46] By the end of the fourth century, after Eudoxus' studies in particular, this theory had become a sort of mathematical rarity that lacked any intrinsic connection with the main body of arithmetical science. No wonder Euclid placed it at the end of the last of his arithmetical books (IX, 21–34). From Aristoxenus' point of view, however, the antiquity and primitive character of this theory could be an argument for associating it with Pythagoras as the founder of theoretical arithmetic.

3. The origin of number

Let us turn now to the two versions of the origin of number, or the art of calculation, mentioned by Aristoxenus (§ 2). The reference to the Egyptian god Thoth points to Plato's dialogue *Phaedrus* (274c–d), where, for the first time in the Greek tradition, this god is called the *prōtos heuretēs* of numbers and counting (i.e., arithmetic), geometry, astronomy, writing, and even the games of draughts and dice.[47] Interestingly, Aristoxenus refers not to Plato, but directly to the Egyptians, probably taking Socrates' story of Thoth's inventions as historical evidence. Meanwhile, Phaedrus himself seems to regard this story as Socrates' fabrication (275b 3–4), as he does the story of cicadas preceding it. Hence, Plato gives the reader to understand that the story should not be taken too seriously.[48] Which does not mean, however, that we ought not to look for possible sources (both Greek and Egyptian) of the tradition in which Thoth figures as the inventor of sciences.

Though Thoth had been known to the Greeks under the name of Hermes at least since the founding of Naucratis (seventh century BC),[49] Plato was the first to mention him under his own name.[50] Relying on the Egyptian tradition and,

---

44 See above, 223 n. 37.

45 See e.g. *Cat.* 12a 6; *APr* 41a 26, 50a 37; *APo* 71a 32f., 73b 20f., 76b 7; *Top.* 120b 3, 142b 7, 149a 30; *SE* 166a 33, 173b 8; *Met.* 986a 18, 990a 9; *Pol.* 1261b 29, 1264b 20; *Rhet.* 1407b 3. Revealingly, Aristotle criticizes the definitions of odd and even contemporary to him (*Top.* 142b 6f.; 149a 30; *SE* 173b 8).

46 Under the derivatives I mean even times odd, odd times even and other similar kinds of numbers considered in propositions IX, 32–34 of the *Elements*. See Philolaus (44 B 5); Pl. *Parm.* 143e–144a; Arist. fr. 199 Rose.

47 Θεύθ … πρῶτον ἀριθμόν τε καὶ λογισμὸν εὑρεῖν … ἔτι δὲ πεττείας τε καὶ κυβείας, καὶ δὴ καὶ γράμματα (274c–d); γράμματα … μνήμης τε γὰρ καὶ σοφίας φάρμακον (274e). Cf. λήθης φάρμακ' (Eur. fr. 578 Nauck).

48 Hackforth, R. *Plato's Phaedrus*, Cambridge 1952, 157 n. 2; Heitsch, E. *Platon. Phaidros*, Göttingen 1993, 188f. To be sure, in *Philebus* (18b–d), Thoth figures again as the inventor of writing.

49 See already Hdt. II, 68, 138.

50 The next was Aristoxenus, which points again to Plato as the source of this version.

still more, on the identification of Thoth with the wise and ingenious Hermes, classical literature after Plato makes Thoth the inventor of writing, sciences, law, and even language itself.[51] Meanwhile, Egyptian notions of Thoth were somewhat different. He was supposed to be the sage counselor of Re, the lawgiver, the keeper of all sorts of wisdom. He was also the patron of scribes, associated as such with the calendar and the art of counting. On the other hand, Egyptian literature makes practically no mention of Thoth as the *inventor* of what Greek and Roman authors attributed to him.[52] Egyptians, as well as the Greeks of the Homeric age, regarded their gods as patrons, protectors, and givers of wise and useful things, hardly as their inventors (1.1). At least, this aspect of the gods' activities was never brought to the foreground.[53] Plato probably did know Thoth as the patron of writing and scribes, but it was all he really needed to know; the other elements of his story going back to Greek, not Egyptian, tradition.

Most of the authors of the sixth and fifth centuries ascribe the discovery of writing, arithmetic, and astronomy to Palamedes, who was already recognized as a sage in the archaic epoch,[54] whereas in Aeschylus' *Prometheus* (500–504) the inventor of all these things turns out to be a philanthropic titan. The picture drawn by Plato reflects two tendencies already described (1.3–4): the 'secondary sacralization' of the inventors, on the one hand, and the growing role of the Orient as the homeland of all the arts and sciences, on the other. While the earlier literature spoke of the Egyptians as the first discoverers of geometry and cal-

---

51 Rusch, A. Thoth, *RE* 7 A (1936) 351–362, esp. 356 f. Plutarch (*De Iside*, 3) attributed to him even the invention of music, probably following the tradition that ascribed the invention of the lyre to Hermes.

52 Roeder, G. *Urkunden zur Religion des Alten Ägyptens*, Jena 1915; idem. Thoth, *Roscher's Lexikon der Mythologie* 5 (1924) 825–863, esp. 849ff.; Rusch, *op. cit.*, 356 ff.; Bonnet, H. *Reallexikon der Ägyptischen Religionsgeschichte*, Berlin 1971, 805–812; Bleeker, C. J. *Hathor and Thoth*, Leiden 1973, 140 f.; Kurth, D. Thoth, *Lexikon der Ägyptologie* 6 (1986) 497–523, esp. 503 ff.

53 Modern literature has a pronounced tendency to represent Thoth as a Greek god-inventor: everything that lies within the scope of his knowledge and skill is automatically regarded as his discovery. See e.g. "Das Epitheton 'der alle Dinge berechnet' kennzeichnet Thoth als Erfinder der Rechenkunst" (Kurth, *op. cit.*, 506); in a similar way "Zähler der Zeiten, Monate, Jahre" becomes the first discoverer of the calendar and astronomy, "Schützer der Schrift und der Bücher" the inventor of writing, etc. Though references to Thoth as the god who *gave* people the gift of language and created different languages are indeed found in Egyptian literature (e.g. in the Ebers papyrus I, 8; Cerny, J. J. Thoth as creator of languages, *JEA* [1948] 12–22; Bleeker, *op. cit.*, 140), reliable sources on his *discoveries* are not easy to detect; see, however: "das Schreiben begann am Anfang", "Sprache erfand" (Kurth, *op. cit.*, 503).

54 Kleingünther, *op. cit.*, 78f. See Stesichorus (fr. 213 Page), Aeschylus (fr. 182 Nauck), Sophocles (fr. 399 Nauck), Euripides (fr. 578 Nauck), Gorgias (76 B 11a, c. 30), and Alcidamas (*Od.* 22).

endar astronomy (Hdt. II, 108; II, 4),[55] in Plato it is an Egyptian deity who becomes the *prōtos heuretēs* of these sciences. Interestingly, he attributes to Thoth even the invention of draughts and dice, ascribed in the Greek tradition almost unanimously to Palamedes[56] – probably in order to subsequently contrast these 'unserious' things (including writing) to the serious ones, such as dialectic.[57]

As a result, Palamedes appears to be deprived of all the discoveries traditionally associated with him, though his name remains a synonym for inventor: in the same *Phaedrus*, Zeno of Elea, the inventor of dialectic, figures as "Elean Palamedes" (cf. Arist. fr. 65 Rose). Plato outlined this transformation earlier, in the *Republic*, where he ridicules the notion, passing from one tragedy to another, of Palamedes as an inventor of the art of counting (ἀριθμὸς καὶ λογισμός) – as if Agamemnon before him could not count his own legs properly (522c–d)! In later dialogues, in particular in *Philebus* (18b–d), the invention of writing is once more related to Thoth, while in the *Laws* (677c–d) Plato mentions Palamedes and other traditional *prōtoi heuretai* (Daedalus, Orpheus, Marsyas, Olympus), only to give the readers to understand that all important things were invented many thousand years before them, in the 'antediluvian' epoch, and have only recently been 'revealed' to the Greeks.[58]

Among Plato's precursors in the 'orientalization' of the traditional Greek εὑρήματα was Isocrates, whose influence is clearly traceable in *Phaedrus*.[59] Though *Busiris* does not belong to Isocrates' serious works, it is in this epideictic speech, laying no claim on trustworthiness (*Bus.* 9), that the rhetorician expands on the subject, so important for later classical thought, of Greek customs, laws, philosophy, and exact sciences as having been borrowed from Egypt.[60] Many elements of Isocrates' story are close not only to the passage in *Phaedrus* that attributes the invention of astronomy, arithmetic and geometry to Thoth, but to other Platonic dialogues as well, particularly to the *Republic*, with its vast program of mathematical education for future guardian-philosophers, the main of the three classes of the Platonic polity. In spite of Isocrates' light tone, subsequent philosophical and historical thought took most of the things he describes quite seriously,[61] and the idea of Pythagoras' traveling to

55 Hecataeus makes Danaus the inventor of the alphabet (*FGrHist* 1 F 20), which also points to Egypt.

56 ἐφηῦρε … πεσσοὺς κύβους τε (Soph. fr. 438 Nauck); πεσσούς τε σχολῆς ἄλυπον διατριβήν (Gorg. *Palam.* 30 = *DK* II, 302.2); Herodotus, referring to the Lydians, attributes to them the invention of dice (I, 94), leaving the invention of draughts to the Greeks, however.

57 Heitsch, *op. cit.*, 197 n. 436.

58 τὰ μὲν Δαιδάλῳ *καταφανῆ γέγονεν*, τὰ δὲ Ὀρφεῖ, τὰ δὲ Παλαμήδει (677d).

59 The commentators of *Phaedrus* point, in particular, to Plato's polemic against two of Isocrates' speeches, *Against the Sophists* and *Helen* (Heitsch, *op. cit.*, 257 ff.).

60 See above, 52 f.

61 Aristotle (*Met.* 981b 20 f.), Eudemus (fr. 133), Aristoxenus (fr. 23). Aristotle, in particular, mentions priests' leisure (cf. Isoc. *Bus.* 21). See also σχολή in Plato (*Crit.* 109d–110a).

Egypt, first put forward explicitly in *Busiris*, became a commonplace in biographical tradition.

Going back to Aristoxenus, let us note again that he, like Eudemus, discerned two distinct stages in the development of mathematics: first, the birth of practical arithmetic, probably in the Orient, and second, Pythagoras' transforming it into a theoretical science. A similar variant of a theory on the origin of culture is found in Philip' *Epinomis,* to which the second version of the origin of numbers related by Aristoxenus refers: "and others derived numbers from the circular paths of the divine luminaries".[62] According to the *Epinomis*, the necessary τέχναι, which appeared first, were followed by those that serve pleasures, then by the 'defensive' ones, and finally by the ἐπιστήμη, based on the notion of number.[63] Following Plato, Philip considered the knowledge of number to be a gift of the deity, whom he identified with the visible universe (978b 7f.). The inhabitants of Egypt and Syria were the first to observe the movements of heavenly bodies, while the Greeks turned astronomy into real wisdom, owing to their ability to bring to perfection everything they borrowed from others (987d 3f.).

Our digression into the sources of notions, popular in the fourth century, of the Oriental origin of sciences, in particular arithmetic, once again demonstrates that the Peripatetics' approach to this problem, serious as it is, does not rule out the use of information that had figured previously in genres and contexts that were far from historically reliable. To be sure, Aristoxenus, while mentioning Thoth, refers to Egyptians, thereby distancing himself from this version (to immediately offer another one), while Eudemus does not mention the divine discoverers at all. His version of the origin of arithmetic in Phoenicia (fr. 133), however, is hardly original either: it seems to be suggested in Herodotus,[64] while Plato makes a direct mention of the Phoenicians (along with the ubiquitous Egyptians) in the passage that relates to teaching arithmetic (*Leg.* 747b–c). Eudemus' words clearly reflect a rationalist construction based on a well-known εὕρεσις–μίμησις kind of logic: practical arithmetic serves, first and foremost, the needs of merchants, of whom the Phoenicians were the most prominent. It does not really matter whether the author of the construction was Eudemus, Plato, or Herodotus. What matters is that *all* versions relating the origin of Greek science to Egypt, Babylon, or Phoenicia, whatever source they may come from, belong to a similar type of construction.

---

62 See above, 112f. A similar view on the origin of number from the circulation of heavenly bodies is found in the *Timaeus* (47a 1–6), but in the *Epinomis* the divine character of the heavenly bodies following their circular paths stands in the foreground.

63 974d 3–977b 8. For a detailed analysis, see Tarán. *Academica*, 69ff.

64 See above, 40.

Chapter 7

The history of astronomy

1. Eudemus' *History of Astronomy* and its readers

The *History of Astronomy*, Eudemus' last treatise on the history of science, can be appropriately analyzed by comparing it with the astronomical division of Theophrastus' *Physikōn doxai*. Astronomy, the only exact science Theophrastus covers, held an important place in his compendium. In Aëtius, the whole of book II and part of book III are related to cosmology. It is natural that the names figuring in Eudemus and Theophrastus partly coincide (Thales, Anaximander, Anaxagoras, the Pythagoreans), and so do many discoveries attributed to them. Interesting for us, however, are not only these coincidences, but also the differences found in Eudemus' and Theophrastus' material, as well as the criteria of selection. A comparative analysis of the *History of Astronomy* and the corresponding part of the *Physikōn doxai* allows us to state more precisely the specificity of their genres, which largely reflects the distinction between astronomy and physics as conceived by the Peripatetics and astronomers of that time.

Let us first attempt to bring together the little evidence on the *History of Astronomy* available to us and form a better idea of that treatise. The seven extant fragments of this work have come to us through five late authors: Theon of Smyrna (fr. 145), Clement of Alexandria (fr. 143), Diogenes Laertius (fr. 144), Proclus (fr. 147), and Simplicius, who cites it thrice (fr. 146, 148–149). The title of Eudemus' work is mentioned by four of these authors: Theon, Clement, Diogenes, and Simplicius, the latter again proving the most accurate.[1] The number of books in the *History of Astronomy* (Ἀστρολογικῆς ἱστορίας α′–ς′) as given in Theophrastus' catalogue,[2] is most likely in error. According to Simplicius, Eudemus discusses Eudoxus' theory in the second and probably final book of his work (fr. 148). The historian did, in fact, set forth the theory of Callippus and did mention Eudoxus' disciples Polemarchus and probably Menaechmus, but this could hardly have needed an additional book: Simplicius (fr. 149) stresses the brevity of Eudemus' rendering of Callippus' theory.

Hence, Simplicius' evidence appears to be the fullest and most detailed: he cites the title of Eudemus' work more correctly than the others, refers to a particular book of the treatise, and notes its clear and concise style. It is also important that Simplicius' three quotations come from different books: Anaxi-

---

1 Theon: ἐν ταῖς Ἀστρολογίαις, Clement: ἐν ταῖς Ἀστρολογικαῖς ἱστορίαις, Diogenes: ἐν τῇ περὶ τῶν ἀστρολογουμένων ἱστορίᾳ, Simplicius: ἐν τῷ δευτέρῳ τῆς Ἀστρολογικῆς ἱστορίας.

2 Fr. 137 No. 43 FHSG. See above, 166 n. 2.

mander and the Pythagoreans were obviously treated in the first book (fr. 146),[3] Eudoxus and his disciples in the second (fr. 148–149). Further, of all the excerptors of the *History of Astronomy*, Simplicius preserved the largest number of names: Anaximander, the Pythagoreans (fr. 146), Eudoxus (fr. 148), Meton, Euctemon, Callippus (fr. 149), and Polemarchus, while Theon reports about Thales, Anaximander, Anaximenes, and Oenopides (fr. 145), Clement and Diogenes about Thales (fr. 143–144), and Proclus about Anaxagoras (fr. 147).

All this leads us to suppose that Simplicius had the text of the *History of Astronomy* at his disposal, while the other aforementioned authors cited it secondhand. With Diogenes and Clement this is evident; Theon himself points to Dercyllides, a Platonist of the early first century AD, as his intermediate source.[4] Proclus obviously cited from memory; there is no evidence that he read Eudemus' work, though the possibility cannot be ruled out. As for Simplicius, one can hardly imagine that he praised the clear and laconic style of the *History of Astronomy* twice without being immediately familiar with it. The reference to the second book of the treatise could, of course, have been found in Simplicius' predecessor, but Simplicius was unlikely to have repeated it if he had known that the *History of Astronomy* had long ago been lost, in which case a reference to a particular book would make little sense. Let us recall that Eudemus' *Physics* is known to us almost exclusively from Simplicius,[5] who never fails to indicate pedantically the particular book he is citing.[6] It is also Simplicius to whom we owe the longest quotation from the *History of Geometry* (fr. 140, p. 57–66 Wehrli). Here he also refers to a particular book of this work (the second) and points out the brevity of Eudemus' exposition. If the commentator had at least two of Eudemus' works at his disposal, we cannot simply assume that the *History of Astronomy* was unavailable by that time.[7]

Generally, Simplicius explained the origin of his quotations, even if this was rather complicated.[8] Thus, while commenting on Aristotle's *Physics*, he notes

---

3 This evidence also suggests that Simplicius was familiar with Eudemus' work (see below, 248f.).

4 On Dercyllides' dates, see Tarrant, H. *Thrasyllan Platonism*, Ithaca 1993, 72ff.; Mansfeld, J. *Prolegomena: Questions to be settled before the study of an author, or a text*, Leiden 1994, 64f.

5 Of its more than hundred fragments, all but fr. 49 derive from Simplicius. Fr. 89 relates to the *History of Theology*, not to *Physics*.

6 "Beginning his *Physics*" (fr. 32, 34), "in the first book" (fr. 43–44, 50), "in the second book" (fr. 59, 62), "in the third book" (fr. 75, 81, 85–88), "in the fourth book" (fr. 101, 104–105).

7 For further arguments, see Schramm, M. *Ibn al-Haythams Weg zur Physik*, Wiesbaden 1961, 36ff. Cf. Knorr. Plato and Eudoxus, 319f.

8 On Simplicius' exactness and generosity in quoting, see Wildberg, C. Simplicius und das Zitat. Zur Überlieferung des Anführungszeichens, *Symbolae Berolinenses. Für D. Harlfinger*, Amsterdam 1993, 187–199; Baltussen, H. Philology or philosophy? Simplicius on the use of quotations, *Orality and literacy in ancient Greece*, Vol. 4, ed. by I. Worthington, J. M. Foley, Leiden 2002, 173–189.

that Alexander copied verbatim a quotation from Geminus' summary of Posidonius' *Meteorologica,* which takes its starting points from Aristotle, and then proceeds to cite this long passage (291.21–292.31) as if he were referring to Aristotle fourth-hand! In the case of Eudemus, the commentator's invaluable pedantry also provides some important details. In his account of Callippus' theory (fr. 149), he remarks that the latter's work is not available (οὔτε δὲ Καλλίππου φέρεται σύγγραμμα), referring subsequently to the summary of his theory in Eudemus (Εὔδημος δὲ συντόμως ἱστόρησε). This assertion would not make sense unless the *History of Astronomy*, unlike Callippus' book, was at Simplicius' disposal. Further, while citing Sosigenes, who in his turn excerpted from Eudemus, Simplicius makes clear that the evidence on Eudoxus comes from Eudemus, whereas that on Plato comes from Sosigenes (fr. 148).[9] Though we cannot rule out that Sosigenes quoted Eudemus and then 'amplified' him, prompting Simplicius to note the resulting discrepancy, a different explanation seems more likely: Simplicius found no mention of Plato in Eudemus. Another possibility would be that here Simplicius quotes an indirect source as if it were direct, unintentionally leaving us with no clue to figure out what this source was. But even so, his two other references to the *History of Astronomy* cannot come from Sosigenes. Fr. 146 on Anaximander and the Pythagoreans has nothing to do with the subject of Sosigenes' work, and fr. 149 is related to the Eudemian exposition of Callippus' system, which Sosigenes deliberately *omitted.*[10] Hence, even if, in the case of fr. 148, Simplicius purposely beguiled the reader into believing that he knew the *History of Astronomy* at first hand, in two other cases we have the means to check his assertions.

As a matter of fact, it is hardly surprising that this work was available to him. Simplicius read and quoted not only Parmenides, Anaxagoras, and Empedocles, but even such a rare text as Eudemus' biography (fr. 1), written by a certain Damas, who must have been Eudemus' disciple.[11] Eudemus' works constituted an important part of the Lyceum's heritage and were often used to comment on other works. Thus, Alexander and Philoponus referred to Eudemus' *Analytics* when commenting on Aristotle's logical treatises; Proclus cited the *History of Geometry* in his commentary on Euclid; and Damascius used the *History of Theology* in his commentaries on Plato's *Parmenides*. Simplicius, while commenting on Aristotle's *Physics,* relies on Eudemus' *Physics* and uses the *History of Astronomy* in his commentary on *De caelo*. Though the large number of his quotations from Eudemus' *Physics* cannot be compared with the three fragments from the *History of Astronomy*, it is quite comparable with one long quotation from the *History of Geometry*. In his *Physics*, Eudemus strictly followed Aristotle, generalizing and elaborating his ideas, whereas the *History*

---

9 See above, 87f. Discussing Hippocrates' quadrature of lunes, Simplicius also distinguishes between Eudemus' text and its exposition in Alexander (fr. 140).

10 See below, 233.

11 See above, 167 n. 4.

of Astronomy differs from Aristotle's *De caelo* both thematically and in genre. The first of these treatises was related to astronomical discoveries and their authors, while the second was a theoretical work in which *mathematical* astronomy occupied a modest place. Where Aristotle mentions the theories of *mathēmatikoi*, for example in *De caelo* 291a 29, the commentator adds historical information on Anaximander and the Pythagoreans that he borrowed from Eudemus. Particularly detailed is Simplicius' account of the theory of homocentric spheres developed by Eudoxus and his school, which Aristotle refers to in 293a 4f. and elaborates in more detail in *Met.* Λ 8. Most of Simplicius' information on the theories of Eudoxus and his disciples goes back to Eudemus, either directly or through Sosigenes.

The Peripatetic Sosigenes, the teacher of Alexander of Aphrodisias, was the author of the treatise Περὶ τῶν ἀνελιττουσῶν (sc. σφαιρῶν).[12] This book, repeatedly quoted by Simplicius in his long commentary on *De caelo* II, 12 (492.25–510.35), dealt not only with the retrograde spheres introduced by Aristotle, but also with various theories of 'saving the phenomena' in general. Sosigenes started with Plato's setting of the problem and proceeded to the solutions offered by Eudoxus and his pupils and, after them, by Aristotle. Subjecting Eudoxus' theory and its subsequent modifications to his quite professional criticism, Sosigenes then examines the theory of eccentrics and epicycles, criticizing it for its incongruity with Aristotle's philosophical postulates.[13] All the evidence suggests that, in his work, the *History of Astronomy* played a role analogous to that of the *History of Geometry* in Eratosthenes' *Platonicus*. The similarity with the *Platonicus* accounts perfectly for the fact that, in Sosigenes, Plato plays the same role as in Eratosthenes: he sets the problems to which the scientists, in turn, offer their solutions (3.1). Yet in contrast to Eratosthenes, who presented the consecutive solutions of the problem by Archytas, Eudoxus, and Menaechmus, Sosigenes concentrated mainly on the astronomical systems of Eudoxus and Aristotle, mentioning Eudoxus' students Callippus and Polemarchus as well as their younger contemporary Autolycus of Pitane only in passing.

The analysis of the ample quotations from Sosigenes found in Simplicius allows us to supplement the fragments of Eudemus' *History of Astronomy* with additional evidence and to shed some light on the fate of this work.[14] What is no less important, this analysis confirms that in several cases Simplicius derives his material directly from Eudemus. In addition to the already mentioned fr. 146 on Anaximander and the Pythagoreans and fr. 149 on Callippus' system, there

---

12 See Procl. *Hypotyp.* IV, 130.17f. On Sosigenes and his work, see Rehm, A. Sosigenes (7), *RE* 3 A (1927) 1157–1159; Schramm, *op. cit.*, 21f., 32ff.; Moraux. *Aristotelismus*, 344ff.

13 Schramm, *op. cit.*, 32ff.; Moraux. *Aristotelismus*, 355f.

14 On this, see also Mendell, H. The trouble with Eudoxus, *Ancient and medieval traditions in the exact sciences. Essays in memory of Willbur Knorr*, ed. by P. Suppes et al., Stanford 2000, 59–138.

is also the historical note inserted by Simplicius when he returns once again to the problem of 'saving the phenomena' and its solution offered by Eudoxus:

> Callippus of Cyzicus, who studied with Polemarchus, Eudoxus' pupil, arrived after him (i.e., Eudoxus) in Athens and stayed there with Aristotle, together with him correcting and augmenting Eudoxus' discoveries.[15]

The historical information on Callippus and Polemarchus, their origin in Cyzicus (cf. ibid., 505.21), their study with Eudoxus, and the subsequent arrival of Callippus in Athens undoubtedly goes back to Eudemus, who spent the late 330s and the 320s in the Lyceum and must have known Callippus personally.[16] Revealingly, we find here a number of features characteristic of Eudemus' works: he indicates the astronomers' origin, names their teachers, and alludes to discoveries they made in the wake of their teachers' theories.[17] In this passage, we owe to Simplicius himself only the mention of Aristotle's active part in correcting and augmenting Eudoxus' achievements in exact sciences: Callippus hardly needed such aid, nor was Eudemus himself inclined to exaggerate his teacher's achievements in the exact sciences.[18]

Earlier, this passage was understood to mean that Callippus was the disciple of Eudoxus' pupil Polemarchus, rather than of Eudoxus himself.[19] This interpretation was based on the wrong dating of Eudoxus' death in 355;[20] since Callippus' *floruit* falls, presumably, in 330 and his birth in 370, this made it impossible that he studied with Eudoxus. Eudemus' testimony, however, should rather be understood to mean that Callippus studied under Eudoxus together with Polemarchus; the most probable dating of Eudoxus (390–337) is quite consistent with this version. To all appearances, Callippus and Polemarchus belonged to the school founded by Eudoxus in Cyzicus; if they visited Athens together around 350–349, Callippus had the opportunity to make Aristotle's acquaintance; their relations in Athens in the late 330s must have been particularly intensive. We know nothing about Callippus' treatise on heavenly spheres;

---

15 Κάλλιππος δὲ ὁ Κυζικηνὸς Πολεμάρχῳ συσχολάσας τῷ Εὐδόξου γνωρίμῳ μετ' ἐκεῖνον εἰς Ἀθήνας ἐλθὼν τῷ Ἀριστοτέλει συγκατεβίω τὰ ὑπὸ τοῦ Εὐδόξου εὑρεθέντα σὺν τῷ Ἀριστότελει διορθούμενός τε καὶ προσαναπληρῶν (493.5–8).

16 Schramm, *op. cit.*, 37f.; Moraux. *Aristotelismus*, 348f.; Mendell. The trouble with Eudoxus, 89.

17 Cf. Ἑρμότιμος δὲ ὁ Κολοφώνιος τὰ ὑπ' Εὐδόξου προηυπορημένα καὶ Θεαιτήτου προήγαγεν ἐπὶ πλέον (Procl. *In Eucl.*, 67.20f. = Eud. fr. 133).

18 In fact, Eudemus, like Theophrastus, never mentioned Aristotle by name in his writings. The only exception known to me is the end of fr. 31 of his *Physics*, but here we cannot be sure that it comes from Eudemus (see above, 152). If it does, one has to note that, unlike Eudemian historico-scientific works, his *Physics* was written on Rhodes, after Aristotle's death (5.1).

19 Heath, T. L. *Aristarchus of Samos*, Oxford 1913, 212; cf. Rehm, A. Kallippos, *RE Suppl.* 4 (1924) 1431f.

20 See above, 95.

his theory has reached us only through the mediation of Aristotle and Eudemus, whose information might have been based on personal contacts with Callippus, rather than on a written source.[21] Even if this is not the case, it remains obvious that the work was already inaccessible to Sosigenes, so that the latter's knowledge of Callippus' system derives from Eudemus and Aristotle.

In fact, Sosigenes was not interested in Callippus, since his main targets were, first, Eudoxus, and second, Aristotle, whose system incorporated Callippus' modifications. Only after he treats these two systems in detail,[22] does Sosigenes proceed to discuss the phenomena, which Eudoxus' students knew but did not take into account and which Autolycus failed to save, namely the varying distances of the planets from the earth. It is in this verbatim quotation from Sosigenes (504.17–506.7) that Callippus, Polemarchus, and Autolycus appear for the first time, referred to by their full names.[23] The very tone of Sosigenes' dismissive remark on Callippus implies that the latter does not deserve any special treatment: "And what is there to say about the other phenomena, some of which Callippus of Cyzicus also tried to preserve after Eudoxus had failed to do so, even if Callippus did preserve them?" (504.20–23). This means that the earlier historical note on Callippus and Polemarchus (493.5–8), where Callippus is also called by his full name, comes directly from Eudemus, and not via Sosigenes.[24] Indeed, when Simplicius comes to Callippus' system (497.6–498.1), he refers not to Sosigenes but to Aristotle and Eudemus (fr. 149).

Sosigenes' further note on Polemarchus also seems to derive from Eudemus rather than from his direct acquaintance with Polemarchus' work. While criticizing Eudoxus' students, Sosigenes remarks that they were aware of the varying distances of the planets:

> For Polemarchus of Cyzicus appears to be aware of it, but to minimize it as being imperceptible, because he preferred the theory which placed the spheres themselves about the very centre in the universe.[25]

We do not know how detailed Eudemus' account of Polemarchus' theory was and whether he dwelled on the discrepancy between the observations and Eudoxian theory. Since for the historian the basic theory was always more important than its further technical elaborations, we can reasonably assume that in this

---

21 Rehm. Kallippos, 1434. According to Jaeger. *Aristotle*, 343 n. 1, the imperfect used by Aristotle in his story of Eudoxus and Callippus (*Met.* 1073b 17, 33) corresponds to the situation of a personal talk. See also Düring, I. *Aristoteles*, Heidelberg 1966, 148f.

22 Eudoxus: 493.11–494.20 (solar theory), 494.23–495.16 (lunar theory), 495.17–497.5 (planetary theory); Aristotle: 498.2–503.8, 503.28–32, 503.35–504.15.

23 Callippus of Cyzicus (504.20), Autolycus of Pitane (504.23), Polemarchus of Cyzicus (505.21).

24 For further arguments, see Schramm, *op. cit.*, 37f.

25 Πολέμαρχος γὰρ ὁ Κυζικηνὸς γνωρίζων μὲν αὐτὴν φαίνεται, ὀλιγωρῶν δὲ ὡς οὐκ αἰσθητῆς οὔσης διὰ τὸ ἀγαπᾶν μᾶλλον τὴν περὶ αὐτὸ τὸ μέσον ἐν τῷ παντὶ τῶν σφαιρῶν αὐτῶν θέσιν (505.21–23). – Transl. by T. Heath.

case he was as brief as in the case of Callippus. Meanwhile, the detailed description of Eudoxus' theory in Simplicius takes several large pages (493.11–497.5), while the digressions and explanatory notes of the commentator himself are quite insignificant here.[26] This description, which goes back through Sosigenes to the *History of Astronomy*, relies in turn on Eudoxus' work *On Velocities*, which is duly named in the very same passage (494.12). Since no other source refers to this title, we ought to accept Lasserre's conclusion: that the accounts of Eudoxus' theory of homocentric spheres, which are found in Sosigenes, Alexander, Simplicius, and other later commentators are based on Eudemus' detailed report and on Aristotle's short summary (*Met.* 1073b 17f.).[27]

Apart from Sosigenes, several references to the theories of Eudoxus and his pupils are found in Dercyllides, whose book *On the Spindle and Whorls in Plato's Republic* is quoted by Theon of Smyrna (*Exp.*, 198.9–202.7). The latter's excerpts from Dercyllides open with a valuable though distorted quotation from the *History of Astronomy* (198.14–199.8 = Eud. fr. 145), which will be discussed later. Dercyllides' own discourse that follows it contains two passages that can be interpreted as indirectly borrowed from Eudemus. The first of them says that Dercyllides "does not consider it necessary to see the causes for planetary movement in spiral lines, or lines similar to ἱππική" (200.23–25). In this ἱππική it is easy to recognize Eudoxus' hippopede mentioned by Simplicius in the material derived from Eudemus.[28] Further on, Theon says that Dercyllides

> reproaches those philosophers who, attaching the stars as inanimate objects to the spheres and to their circles, introduce the systems of many spheres, as Aristotle does, and among the mathematicians Menaechmus and Callippus, who have introduced some spheres as 'carrying' and others as 'rolling back'.[29]

Dercyllides' criticism clearly refers to the passage in the *Metaphysics* where Aristotle develops the theories of Eudoxus and Callippus.[30] Menaechmus, however, is not mentioned in Aristotle – moreover, the whole of Greek literature

---

26 For detailed analysis, see Schramm, *op. cit.*, 36ff.; Mendell. The trouble with Eudoxus, 87ff.

27 Lasserre. *Eudoxos*, 199. Alexander must have considered Eudoxus' theory in his commentary, now lost, on *De caelo* (see Ps.-Alex. Aphr. *In Met.*, 703 = Eudox. fr. 123). But in the section on homocentric spheres, Simplicius, who otherwise regularly refers to this commentary, makes no mention of Alexander.

28 *In Cael. comm.*, 497.2f. = fr. 124. See Lasserre. *Eudoxos*, 199.

29 αἰτιᾶται δὲ τῶν φιλοσόφων ὅσοι ταῖς σφαίραις οἷον ἀψύχους ἑνώσαντες τοὺς ἀστέρας καὶ τοῖς τούτων κύκλοις πολυσφαιρίας εἰσηγοῦνται, ὥσπερ Ἀριστοτέλης ἀξιοῖ καὶ τῶν μαθηματικῶν Μέναιχμος καὶ Κάλλιππος, οἳ τὰς μὲν φερούσας, τὰς δὲ ἀνελιττούσας εἰσηγήσαντο (*Exp.*, 201.22–202.2 = 12F2 Lasserre).

30 *Met.* 1074a 10: ὁ δὴ ἁπασῶν ἀριθμὸς τῶν τε φερουσῶν καὶ τῶν ἀνελιττουσῶν (sc. σφαιρῶν).

contains no further evidence of his astronomical theories. Provided that Dercyllides (or his source) did not make a mistake, mechanically adding to Callippus the name of his famous schoolmate, the mention of Menaechmus must go back to Eudemus' account of his theory.[31]

Let us now turn to the excerpt from the *History of Astronomy* that Theon borrowed from Dercyllides. This is a short catalogue of the main discoveries made by ancient astronomers:

> Eudemus relates in his *Astronomies* that Oenopides first discovered the obliquity of the zodiac[32] and the duration of the Great Year; Thales the eclipse of the sun and the fact that the sun's period with respect to the solstices is not always the same; Anaximander that the earth is suspended in space and moves about the middle of the cosmos; Anaximenes that the moon receives its light from the sun and how it is eclipsed. And others discovered in addition to this that the fixed stars move round the immobile axis that passes through the poles, whereas the planets move round the axis perpendicular to the zodiac; and that the axis of the fixed stars and that of the planets are separated from another by the side of a (regular) pentadecagon, i.e., 24° (fr. 145).

Though closely resembling the *Catalogue of geometers*, this excerpt is much more selective and less exact than the compilation by Porphyry and Proclus. The catalogue of astronomers includes only four names and, accordingly, covers only part of Eudemus' book I. Besides, some of the astronomers mentioned in Eudemus (the Pythagoreans, Anaxagoras) are left out here and the chronological order is broken: Oenopides is named before Thales, while one of his discoveries, the measurement of the obliquity of ecliptic, is assigned subsequently to some anonymous οἱ λοιποί.[33] Particularly disappointing is that nearly each of the catalogue's sentences contains mistakes – thus, the earth in Anaximander's system is said to move about the middle of the cosmos. These errors, even if we put some of them down to Theon, are too numerous to relate Dercyllides' evidence directly to the *History of Astronomy*. He is more likely to have used someone's excerpt from this work and contributed to the corruption of the original text himself. To judge from Dercyllides' allusion to Eudoxus' hippopede and his mention of Callippus and Menaechmus, the context of his source must have been much wider.

How often such excerpts migrated from one book to another may be seen from Ps.-Hero's *Definitions*, which contains a section coinciding, except for several trivial variations, with the passage in Dercyllides–Theon (166.23–

---

31 To be sure, in Eudoxus and Callippus there were neither retrograde spheres (in the Aristotelian sense), nor the term ἀνελίττουσαι σφαῖραι (*pace* Lasserre. *Léodamas*, 549), so that Dercyllides anachronistically associates this Aristotelian innovation with Eudoxus' students. The Peripatetic Adrastus (ca. 100 AD), on whose treatise Theon heavily relies, was not sure whether retrograde spheres come from Aristotle or from Eudoxus and Callippus (*Exp.*, 180.5–12).

32 See above, 171 f.

33 See above, 171 f., and below, 264 f.

168.12). Entitled τίς τί εὗρεν ἐν μαθηματικοῖς;, this section is part of a larger quotation from a rather superficial 'introduction to arithmetic' by the Peripatetic Anatolius.[34] Anatolius must have borrowed this passage from Theon (or Dercyllides), to be quoted in turn by Ps.-Hero. It should be noted that the heading "Who discovered what in mathematics?", adequately reflecting the subject-matter of Eudemus' works on the history of science, hardly fits the contents of the passage in question: the latter deals only with discoveries in astronomy, whereas Anatolius' introduction was related, on the whole, to geometry and arithmetic.

Hence, to the five authors who used evidence from the *History of Astronomy* we can add four more: Dercyllides, Sosigenes, Anatolius, and Ps.-Hero. Of these four, Sosigenes was the only one to have relied on Eudemus directly. There are grounds to believe that Alexander also made use of Eudemus in his lost commentary on *De caelo*, either directly or through his teacher Sosigenes.[35] All these authors belong to the same late period as the other sources that contain references to Eudemus. But whereas the *History of Geometry* was used by Eratosthenes and probably by Archimedes, the fate of the *History of Astronomy* in the Hellenistic period remains obscure; besides, there are no traces of its use by the *mathēmatikoi*. The first of these circumstances can be explained by the fact that the little of Hellenistic astronomical literature that has survived does not show any particular interest in early astronomy.[36] But why was the *History of Astronomy* chiefly cited in popular historico-philosophical compendia (Clement, Diogenes Laertius), commentaries on Plato's and Aristotle's philosophical works (Alexander, Proclus, Simplicius), mathematical handbooks for the readers of Plato (Dercyllides, Theon), and introductions to mathematics for beginners (Anatolius)? Even Sosigenes' treatise, apparently the most technical among these works, has, on the whole, a historico-critical character. Quotations from the *History of Geometry* in professional or almost professional works are found more often: Pappus' mathematical encyclopaedia and his commentary on Euclid's *Elements* X, Eutocius' commentary on Archimedes, and, finally, Porphyry's and Proclus' commentaries on Euclid's *Elements* I.[37] In the case of Eratosthenes and probably of Archimedes, Eudemus'

---

34 Ἐκ τῶν Ἀνατολίου (ibid., 160.8–168.12). Anatolius' book (cf. above, 63 n. 82) contained the following divisions: Who gave mathematics its name? What is mathematics? How many parts does mathematics consist of? What parts of mathematics are related to each other? Who discovered what in mathematics? The last heading is found also in Theon's manuscripts, but it is generally believed that these headings are interpolations.

35 See above, 234 n. 27.

36 See above, 185 n. 79. In any case, the 'catalogue of astronomers' in Dercyllides, as well as the mentions of Thales in Clement and Diogenes (Eud. fr. 143–144), must go back to the Hellenistic literature.

37 Porphyry quoted the *History of Arithmetic* in his commentary on Ptolemy's *Harmonics* (6.1).

material is used for properly mathematical ends, not only as commentary on ancient mathematicians. In contrast, none of the creators of astronomy showed any particular interest in the *History of Astronomy*, as far as we know.

In part, this difference can be explained by the agonistic spirit that surrounded the famous problems of doubling the cube and squaring the circle, almost until the very end of Greek geometry. This lent topicality to the solutions of these problems reported by Eudemus, particularly if the original sources were already inaccessible.[38] A further reason was the early appearance in geometry of such an authoritative summarizing work as Euclid's *Elements*, whose commentators did their best to find evidence of the origin of its separate theorems and books. The discoveries described in the *History of Astronomy* – such as the causes of solar and lunar eclipses, the order and number of planets, the central position of the earth, the obliquity of the ecliptic, etc. – were, in contrast, by the second part of the fourth century universally recognized and have belonged ever since to the history of astronomy, which was of little interest – and of little help – to astronomers themselves.[39] None of the astronomical works of the fourth century could, on the other hand, claim the status of Euclid's *Elements*. The theory of homocentric spheres was soon superseded by the theory of epicycles, so that Eudoxus' *On Velocities* was not widely read any longer. Euclid's and Autolycus' treatises on the movement of the heavenly sphere had indeed survived, but, popular as they were in late Antiquity, they did not attract any commentator's particular attention.[40] Even granting a role to sheer chance, we can suppose these factors to have been operative in gradually making the *History of Geometry* somewhat more topical for professionals than the *History of Astronomy*.

The differences in the reception of Eudemus' various works on the history of science should not, however, be overstated. In all three cases, we are dealing mostly with the same milieu of the late philosophical schools – Peripatetics, Platonists, and Neopythagoreans – in which knowledge of mathematics was part of professional education, though any original contribution to this science remained exceptionally rare. The level of mathematical knowledge in this philosophical community and the scientific interests of its members could vary considerably: the 'theurgist' Iamblichus with his arithmology was unlike the

---

38 Interestingly, Archytas' solution to the problem of doubling the cube is found in the mediaeval Arabic encyclopedia compiled by the Banu Musa brothers. It is, however, attributed to Menelaus; Eudemus and Archytas are not named (Knorr. *TS*, 101 ff.).

39 Even if the *History of Astronomy* contained some observational and computational data drawn from the early astronomers (Meton, Euctemon, Eudoxus, Callippus, etc.), they were already incorporated in the astronomical tradition and, accordingly, were not particularly important.

40 Neugebauer. *HAMA* II, 748 ff., 767 f. Euclid's *Phaenomena* and Autolycus' *On the Rotating Sphere* are considered in book VI of Pappus' *Collectio*, but unlike Eudemus he shows little interest in purely historical matters (cf. above, 173 n. 28, 190).

sober Eutocius, the commentator of difficult works by Apollonius, Archimedes, and Ptolemy. Yet in spite of their individual differences, all of these people regarded themselves as the heirs of a tradition going back to Pythagoras, Plato, and Aristotle, a tradition in which mathematics was an integral part. This community's interest in Eudemus' historical works, even if at times limited and superficial, contrasts with the almost total indifference that Greek mathematicians and astronomers showed toward them. This fact apparently confirms our preliminary conclusion: the historical view of science turned out to be unclaimed by ancient scientists themselves.

2. Thales and Anaximander

Having outlined the available material from the *History of Astronomy* and the fate of the book, let us now have a closer look at its contents and structure. As follows from the fragments and testimonies, Eudemus' book mentioned Thales, Anaximander, Anaximenes (?),[41] Anaxagoras, the Pythagoreans, Oenopides, Meton, Euctemon, Eudoxus, Callippus, Polemarchus, and probably Menaechmus. This list is nearly half as long as the one known from the *History of Geometry* and, naturally, contains only some of the names that must have figured in the original text. In comparison with the *History of Astronomy*, the astronomical division of *Physikōn doxai* omits such astronomers as Meton, Euctemon, Eudoxus etc.,[42] but refers to about twenty other physicists, apart from those found in Eudemus.[43] Thus, while Eudemus dealt with those who made discoveries in mathematical astronomy, Theophrastus registered the opinions of physicists as related to astronomy as a whole.

One cannot help noting that Eudemus' list falls into two roughly equal parts: from Thales through Anaxagoras, it comprises the same physicists as in Theophrastus; starting with Oenopides, they are followed by the mathematical astronomers. Incomplete as our data inevitably remain, this correlation is not purely by chance: it reflects the actual course of the development of Greek astronomy. Oenopides was not, of course, the first to study astronomy 'for its own sake'[44] rather than within the framework of a general cosmological theory, as did Anaximander and Anaxagoras before him and Democritus and Philolaus after him. It is very likely, however, that Oenopides was the author of the first special treatise on mathematical astronomy that gave impetus to the develop-

---

41 See below, 255f.

42 Eudoxus' students were excluded for chronological reasons. The reference to Eudoxus and Aratus in Aëtius (*Dox.*, 347.21f.) is a later addition; cf. below, 295f.

43 Xenophanes, Pythagoras, Alcmaeon, Hecataeus, Parmenides, Heraclitus, Empedocles, Melissus, Archelaus, Ion of Chios, Leucippus, Democritus, Antiphon, Philolaus, Diogenes, Ecphantus, Hicetas, Metrodorus, and Plato.

44 Cf. already Cleostratus of Tenedos (*DK* 6).

ment of this branch of science and contributed to the further divergence of physics and mathematics (7.5).

In the *History of Astronomy*, as in the *History of Geometry*, discoveries and theories were arranged chronologically, in accordance with the principle of *prōtos heuretēs*. Almost all our fragments and testimonies bear traces of this formula, so typical of Eudemus.[45] The absence of this terminology in Clement's testimony on Thales (fr. 143) may be the result of mere chance or, as in the account of Callippus' theory (fr. 149), may be explained by the context. At any rate, it is known that Eudemus counted Thales and Callippus among the *prōtoi heuretai*. Thus, in Diogenes Laertius (I, 23 = fr. 144) we read:

> Thales seems by some accounts to have been the first to study astronomy, the first to predict eclipses of the sun and to fix the solstices, so Eudemus in his *History of Astronomy*. It was this which gained for him admiration of Xenophanes and Herodotus and the notice of Heraclitus and Democritus. (transl. by R.D. Hicks)

Not surprisingly, Eudemus considered Thales the founder not only of Greek geometry, but of astronomy as well (πρῶτος ἀστρολογῆσαι). Given that the *Catalogue* assigned the invention of geometry to the Egyptians and that of arithmetic to the Phoenicians, the *History of Astronomy* was also likely to mention the oriental discoverers of this science. Assuming that Eudemus' opinion did not sharply diverge from the earlier (and later) tradition, these precursors of Greek astronomy must have been the Babylonians.[46]

In confirmation of Thales' achievements in astronomy, Eudemus cites as many as four witnesses, two of whom, Xenophanes and Herodotus, are even said to 'have admired' his discoveries. References to these authors may already have been contained in Eudemus' sources, e.g. in Hippias. Herodotus, the only witness available to us, does not, however, show any particular admiration, confining himself to a short note on Thales' prediction of a solar eclipse (I, 74). Whether the other three authors wrote only about the eclipse or mentioned that Thales studied solstices as well remains unknown; Democritus could well have touched upon the latter topic.[47] As for the eclipse prediction, Clement's testimony seems to indicate that Herodotus was one of Eudemus' main sources:

> Eudemus observes in his *Astronomical Histories* that Thales predicted the eclipse of the sun which took place at the time when the Medes and the Lydians were joined in battle against each other. The king of the Medes was Cyaxares, the

---

45 πρῶτος (fr. 144), εὗρε πρῶτος, ἐπεξεῦρον (fr. 145), πρώτου εὑρηκότος, πρώτους (fr. 146), πρῶτος (fr. 147), πρῶτος (fr. 148).

46 See Hdt. II, 109; [Pl.] *Epin*. 986e 9f.; Arist. *Cael*. 292a 7f. It should be noted, however, that in all three cases the Babylonians are mentioned along with the ubiquitous Egyptians; cf. below, 8.3.

47 Democritus mentioned Thales' parents and his Phoenician origin (68B 115a). Lebedev, A. Aristarchus of Samos on Thales' theory of eclipses, *Apeiron* 23 (1990) 80f., attributes to Democritus the *explanation* of solar eclipses that Aristarchus ascribed to Thales; cf. below, 240 n. 49.

> father of Astyages, and Alyattes, the son of Croesus, was the king of the Lydians. Herodotus in the first book (of his *History*) agrees with Eudemus (fr. 143).

Provided that Herodotus' text was not assigned to Eudemus as a result of contamination, the latter appears to have copied the historian's text almost entirely. Such unusual attention to historical detail apparently results from the absence of more detailed evidence of Thales' prediction (only briefly mentioned in Xenophanes and Heraclitus) and probably from chronological considerations. It is clear, at any rate, that Eudemus, facing serious problems while gathering evidence on Thales' astronomy, turned to sources he normally had no need of. Yet the work *On Solstices and Equinoxes* ascribed to Thales and mentioned by Diogenes Laertius (I, 23) before he cites Eudemus could hardly be among them.[48] Had Eudemus been familiar with this work and had he considered Thales its author, our information on the 'father of astronomy' would have been much fuller.[49] Meanwhile, the only discoveries Eudemus associates with Thales are the prediction of a solar eclipse and the determination (a more precise one) of solstices. The first discovery is mentioned in all three versions of Eudemus' report on Thales, the second in two of them. It is impossible to say for certain that the *History of Astronomy* credited Thales with no other discoveries. But these discoveries constitute the most reliable part of the tradition, unlike, for example, the advanced astronomical views attributed to Thales in Aëtius' doxography.[50]

According to Clement and Diogenes Laertius, Thales predicted (προειπεῖν) the eclipse; in Dercyllides–Theon (fr. 145) we find the following:

> Θαλῆς (sc. εὗρε πρῶτος) ἡλίου ἔκλειψιν καὶ τὴν κατὰ τὰς τροπὰς αὐτοῦ περίοδον, ὡς οὐκ ἴση ἀεὶ συμβαίνει.
>
> Thales (was the first to discover) the eclipse of the sun and the fact that the sun's period with respect to the solstices is not always the same.

---

48 As was supposed, e.g., by Tannery, P. *Recherches sur l'histoire de l'astronomie ancienne*, Paris 1893, 21; Heath. *Aristarchus*, 20; Burkert. *L & S*, 416. The close proximity of these two reports does not imply their common origin: Diogenes is notorious for his 'mosaic' technique of compilation. The work ascribed to Thales appeared, most probably, in the third century BC: Classen, C. J. Thales, *RE Suppl.* 10 (1965) 937.

49 Aristarchus of Samos, who expressly 'quoted' Thales' words on when and why solar eclipses occur (cf. above, n. 47), seemed to use this pseudo-epigraph. His passage is preserved on the papyrus: *The Oxyrhynchus papyri*, ed. by M.W. Haslam, Vol. 53, London 1986. See Mouraviev, S. POxy 3710, col. II 33–55, III 1–19, *Corpus dei papiri filosofici greci e latini*, ed. by F. Adorno, Florence 1992, 229–242; Lebedev. Aristarchus; Bowen, A. C., Goldstein, B. R. Aristarchus, Thales, and Heraclitus on solar eclipses: An astronomical commentary on P. Oxy. 53.3710 cols. 2.33–3.19, *Physis* 31 (1994) 689–729; Sider, D. Heraclitus on old and new months: *P. Oxy.* 3710, *ICS* 19 (1994) 11–18.

50 See e.g. *Dox.*, 340.7 (division of the celestial sphere into five zones), 353.20 (explanation of solar eclipses), 358.15 (explanation of lunar eclipses), 360b 14 (phases of the moon), 376.22 (spherical shape of the earth).

Inasmuch as the 'discovery' of the solar eclipse can be understood not only as a prediction but also as an *explanation* thereof, a number of scholars are ready to assign to Eudemus a report of Thales' knowledge of the true cause of eclipses.[51] Though an assertion that Thales explained the eclipse of the sun occurs in many post-classical sources, starting with Aristarchus of Samos,[52] this seems to me quite improbable.[53] First, Thales turned from 'predictor' of eclipses in Herodotus and Eudemus into its 'first discoverer' in Dercyllides–Theon only because the entire list of astronomers' achievements depends here grammatically on εὗρε πρῶτος. Due to the abridgement and redaction of the original, προειπεῖν, attested in the other two versions, was left out, so that a 'prediction' turned into a 'discovery'. The same corruption, though with a result diametrically opposite, befell Eudemus' evidence in Diogenes Laertius: πρῶτος ἀστρολογῆσαι καὶ ἡλιακὰς ἐκλείψεις καὶ τροπὰς προειπεῖν – "Thales was the first to study astronomy and to predict the solar eclipses and solstices" (I, 23). It is obvious that the meaningless 'prediction of solstices' appeared due to the loss of an appropriate verb (e.g., εὗρε κτλ., as in Dercyllides–Theon) at τροπάς, which made it depend on προειπεῖν.[54]

Second, everything we know about Thales and Greek astronomy of the sixth century suggests that he simply could not have had a theory offering a *correct* explanation of solar eclipses. Third, there is no reason to believe that Eudemus attributed such a theory to Thales at all, unlike the prediction, attested by two of the three sources. The 'explanation' implies knowledge that the moon, which eclipses the disc of the sun, reflects, rather than emitting light. Now, according to Dercyllides–Theon, Eudemus assigned the discovery of this fact (and the theory of lunar eclipses as well) not to Thales, but to Anaximenes (fr. 145). Most of the specialists, following an earlier and more reliable tradition, assume that it is in fact Anaxagoras who is meant here. We will return to this question later (7.4), concluding for now that we lack any evidence that would link the *History of Astronomy* with Thales' explanation of solar eclipses.

The debate concerning Thales' prediction has gone on for centuries.[55] Until the mid-20th century, the predominant opinion was that Thales' prediction could have relied on some Babylonian computational scheme.[56] A more de-

---

51 So Panchenko, D. Thales's prediction of a solar eclipse, *JHA* 25 (1994) 275–288; Bowen. Eudemus' history. In contrast to Bowen, Panchenko believes Thales to have *known* the true cause of eclipses. See also Waerden, B. L. van der. *Die Astronomie der Griechen*, Darmstadt 1988, 11.

52 11 A 2 (*Suda*), A 3 (Hesychius), A 17a (Aëtius), A 19 (Apuleius). On Aristarchus, see above, 239 n. 47, 240 n. 49.

53 So Tannery. *Recherches*, 33 n. 3; Boll, F. Finsternisse, *RE* 6 (1909) 2341 f.; Heath. *Aristarchus*, 18 f.; Dicks. Thales, 295 f.; Guthrie, W. K. C. *A history of Greek philosophy*, Vol. 1, Cambridge 1971, 49.

54 Cf. Heath. *Aristarchus*, 13 f.

55 For old literature on the question, see Demandt, A. Verformungstendenzen in der Überlieferung antiker Sonnen- und Mondfinsternisse, *AAWM* no. 7 (1970) 26 n. 1.

56 Usually the saros was mentioned here: Tannery, P. *Pour l'histoire de la science hel-*

tailed acquaintance with Babylonian astronomy has shown, however, that neither in the sixth century nor later was it able to make a *reliable* prognosis of a solar eclipse for a given latitude. This fact was particularly stressed by Neugebauer,[57] who questioned the authenticity of the whole tradition concerning Thales' and Pythagoras' scientific discoveries.[58] Because of Neugebauer's undeniable authority, the hypercritical attitude toward the tradition of Thales' prediction found a good number of adherents.[59] Meanwhile, this tradition goes back to Thales' younger contemporaries (Xenophanes); even if invented, it would still date to the sixth century. Yet to invent a story of a sage who predicted an eclipse, one has to know about the very *possibility* of such a prediction, based on the periodic character of this phenomenon (unlike, say, a divine revelation or a sign given to a prophet). Since Greek tradition before Thales does not know of any predictions of eclipses, the very idea could have been only of Babylonian origin.[60]

Neugebauer certainly knew but refused to take into account that Babylonian predictions concerned all *potential* lunar and solar eclipses for a given year. Among them, the Babylonians singled out by observations those that were actually visible, ignoring a great number of others that were either insignificant, or happened on a stormy night, or simply could not be seen at Babylon's latitude.[61] From at least the seventh century, Babylonian astronomers were predicting lunar eclipses, relying on various schemes of varying complexity, including saros, the period of 223 synodic months ($\approx$ 18 years).[62] The accuracy

---

lène, Paris 1887, 62f.; Ginzel, F. K. *Spezieller Kanon der Sonnen- und Mondfinsternisse*, Berlin 1899, 167f., 171f. (with references to earlier literature).

57 Neugebauer. *ES*, 142f.; idem. *HAMA*, 604.

58 Neugebauer. *ES*, 148.

59 See e.g. Dicks. Thales; Classen. Thales, 944f.; Samuel, A. *Greek and Roman chronology*, Munich 1972, 22 n. 4 ("The story is probably still not dead, but see Neugebauer, *Exact Sciences*, p. 142–43 if you want to bury it."); Longrigg, J. Thales, *DSB* 13 (1976) 295f.; Mosshammer, A. Thales' eclipse, *TAPA* 111 (1981) 145–155; Bowen, A., Goldstein, B. Meton of Athens and astronomy in the late 5th century B.C., *A scientific humanist: Studies in memory of A. Sachs*, ed. by E. Leichty et al., Philadelphia 1988, 40. Even those who admit that Eudemus followed an ancient tradition are not inclined to credit it (Demandt, *op. cit.*, 25f.).

60 The reconstructions of Thales' prediction founded on the hypothesis that in the seventh century the Greeks observed solar eclipses and recorded their dates (Hartner, W. Eclipse periods and Thales' prediction of a solar eclipse, *Centaurus* 14 [1969] 60–71; Panchenko. Thales's prediction) are not convincing: Stephenson, F. R., Fatoohi, L. J. Thales's prediction of a solar eclipse, *JHA* 28 (1997) 279–282.

61 Of 61 dated solar eclipses, only 21 were visible at the latitude of Babylon, though each of the 61 predicted dates corresponds to a real eclipse visible from the earth's surface: Steele, J. M. Solar eclipse times predicted by Babylonians, *JHA* 28 (1997) 133–139.

62 Aaboe, A. et al., Saros cycle dates and related Babylonian astronomical texts, *TAPS* 81.6 (1991) 21f.; Britton, J. P. Scientific astronomy in Pre-Seleucid Babylon, *Die*

of predictions of lunar eclipses by means of saros was high enough[63] and the method was further applied to solar eclipses as well,[64] though here it could not be equally successful. Hence, there is every reason to suppose that Thales, having known about the 223-month period between two solar eclipses, applied this scheme to the eclipse of May 18, 603, observed in Babylon and Middle Egypt,[65] and thus by lucky coincidence 'predicted' the eclipse of May 25, 585, which was practically full at Miletus' latitude.[66] Though the details of this story will forever remain unknown,[67] this explanation seems to better reconcile the evidence of the ancient tradition with contemporary knowledge of Greek and Babylonian astronomy of the early sixth century without assigning to Thales any special knowledge of the causes of eclipses, which at this time nobody possessed.[68]

Interestingly, after Thales, predictions of various natural phenomena were ascribed to many sages: Anaximander (12 A 5), Pherecydes, and Pythagoras (7 A 1, 6) predict earthquakes, and Anaxagoras even the fall of a meteorite (59 A 1, 11). But the ancient tradition is practically silent on further predictions

---

Rolle der Astronomie in den Kulturen Mesopotamiens, ed. by H. Galter, Graz 1993, 61–76; Hunger, H., Pingree, D. *Astral sciences in Mesopotamia*, Leiden 1999, 181 ff.; Steele, J. M. Eclipse predictions in Mesopotamia, *AHES* 54 (2000) 421–454; idem. *Observations and predictions of eclipse times by early astronomers*, Dordrecht 2000, 75 ff.

63 Steele, J. M., Stephenson, F. R. Lunar eclipses predicted by the Babylonians, *JHA* 28 (1997) 119–131. Of 35 dated predictions of lunar eclipses that took place between 731 and 77 BC, 19 were accurate, 12 nearly accurate, and 4 proved erroneous.

64 See the text: Aaboe et al., *op. cit.*, 25 f. As Steele. Eclipse predictions, 442 f., points out, the Babylonian astronomers approached solar eclipses in the same way as they did lunar ones, in spite of considerable differences in their frequency. He dates the first predictions of solar eclipses to the eighth-seventh centuries (ibid., 451).

65 Ginzel, *op. cit.*, 171 f.; Boll. Finsternisse, 2341.

66 Stephenson, Fatoohi, *op. cit.* It should be noted, however, that even the latest computer methods do not allow us to calculate the dates and visibility characteristics of solar eclipses in Antiquity with the desired accuracy: Thomann, J. Zur Nachrechnung antiker Sonnenfinsternisse, *Antike Naturwissenschaft und ihre Rezeption*, Vol. 9 (1999) 103–110.

67 Such as 1) how exactly did Thales learn about saros? 2) Why does Herodotus assert that Thales predicted the year of the eclipse, and not the month and day? 3) How do the traditions of Thales' prediction and of the battle between the Medes and Lydians relate to each other?

68 See Gigon. *Ursprung*, 52 f.; Guthrie, *op. cit.*, 47 f.; *KRS*, 81 f.; Zaicev. *Griechisches Wunder*, 181 f. Even such a sceptic as Dicks (Thales, 285; idem. *Early Greek astronomy*, London 1970, 43 f.) was inclined to endorse this conclusion. See also van der Waerden. *Astronomie*, 8 f. (cf. 11); Görgemanns, H. Sonnenfinsternisse in der antiken Astronomie, "*Stürmend auf finsterem Pfad ...*": *Ein Symposion zur Sonnenfinsternis in der Antike*, ed. by H. Köhler et al., Heidelberg 2000, 73 f. Cf. von Fritz. *Grundprobleme*, 134 n. 243.

of eclipses.[69] It seems that the Greeks did not use saros cycles to make possibility predictions, possibly because their application to subsequent solar eclipses observable in Greece proved unsuccessful. Besides, Greek astronomy after Thales was concerned with explanations, not predictions.[70]

In the case of solstices, Eudemus obviously could not credit Thales with the *discovery* of the 'turnings of the sun' (τροπαὶ ἡλίου) mentioned by Hesiod (*Op.* 479, 564, 663) and Alcman (fr. 17.5 Page). His report of Thales' discovery of the *inequality* of periods between the sun's passage through solstices (fr. 145) seems more trustworthy.[71] The time from the summer to the winter solstice is, indeed, four days shorter than that from the winter to the summer solstice.[72] It seems that Thales tried to estimate the solstices' dates[73] and, hence, the length of the solar year[74] more accurately than it had been known before. According to Meton's and Euctemon's calculations, which Eudemus could not ignore (cf. fr. 149), the length of seasons, starting with summer solstices, equaled 90, 90, 92, and 93 days; according to Callippus, 92, 89, 90, and

---

69 The only exception is Plutarch's note that Helicon of Cyzicus foretold a solar eclipse while accompanying Plato on his third visit to Sicily (*Dion.* 19, 4 = Lasserre 16T3; see Demandt, *op. cit.*, 24f., 29). According to Boll (Finsternisse, 2356f.; idem. Helikon (3), *RE* 8 [1912] 78f.), it was the eclipse of May 12, 361, almost full in Syracuse. But Helicon was Eudoxus' pupil, so he was born no earlier than 375/70, which excludes both his prediction of this eclipse and his trip to Syracuse together with Plato. Helicon could have made Plato's acquaintance only around 350, when Eudoxus and his pupils came from Cyzicus to Athens (see above, 98f.). Cf. Lasserre. *Léodamas*, 575f.

70 Attempts to explain eclipses were made by practically all the early Presocratics: Anaximander (12A11, 19, 21–22), Xenophanes (21A41), Alcmaeon (24A4), and Heraclitus (22A1, 12).

71 *A source book in Greek science*, ed. by M. R. Cohen, I. E. Drabkin, Cambridge, Mass. 1958, 92 n. 2; Szabó, Maula, *op. cit.*, 119f. It remains uncertain whether the evidence that Thales "was the first to determine the sun's course from solstice to solstice" goes back to Eudemus (D. L. I, 24: πρῶτος δὲ καὶ τὴν ἀπὸ τροπῆς ἐπὶ τροπὴν πάροδον εὗρε). Diogenes does not name his source.

72 The length of the seasons, starting with the summer solstice, is 92½, 88⅛, 90⅛ and 94½ days.

73 In Hesiod, τροπαί refers not to a particular day, but to the season when the winter or the summer solstice occurs. See Kahn, C. H. On early Greek astronomy, *JHS* 90 (1970) 113.

74 The connection between Thales' observation of solstices with the estimation of the solar year's length is indirectly confirmed by the fact he is credited with the division of the year into 365 days (D. L. I, 27). The intercalation period of 8 years introduced in the late sixth century by Cleostratus (6B4) presumed the year to be 365¼ days long (Samuel, *op. cit.*, 35f., 40; van der Waerden. *Astronomie*, 26f.). Neugebauer (*HAMA* II, 620f.), though discussing the octaëteris, leaves the question of its author open, for the existence of Greek astronomy in the sixth century does not fit his conception.

94 days.[75] The difference between semi-annual periods amounts, respectively, to 5 or 3 days. Thales' own estimate remains unknown.

Less plausible is the suggestion that Dercyllides–Theon (fr. 145) points to the inequality of the *four* seasons marked off by the solstices and equinoxes.[76] The problem is that Eudemus' evidence on Thales is silent on equinoxes.[77] Meanwhile, reconstructing Eudemus' reports on Thales, one should not ascribe to the latter anything that goes beyond the evidence safely attributable to the *History of Astronomy*, particularly if it is not confirmed by independent and reliable sources. In this case, late sources unanimously assert that the first to have determined solstices and equinoxes by means of the gnomon was Anaximander.[78] We do not know the extent to which this view was shared by Eudemus. It is obvious, in any case, that there is no compelling reason to link the equinoxes with Thales.

Eudemus has only two direct testimonies of Anaximander's astronomical discoveries. The first is found in Dercyllides–Theon's list of discoveries:

> Ἀναξίμανδρος δὲ ὅτι ἐστὶν ἡ γῆ μετέωρος καὶ κινεῖται περὶ τὸ τοῦ κόσμου μέσον.
>
> Anaximander (was the first to discover) that the earth is freely suspended and moves about the center of the cosmos (fr. 145).

The second is cited by Simplicius when he comments on Aristotle's statement that the relative position of heavenly bodies and the distances between them may best be studied in astronomical writings:[79]

> Ἀναξιμάνδρου πρώτου τὸν περὶ μεγεθῶν καὶ ἀποστημάτων (*sc.* τῶν πλανωμένων) λόγον εὑρηκότος, ὡς Εὔδημος ἱστορεῖ τὴν τῆς θέσεως τάξιν εἰς τοὺς Πυθαγορείους πρώτους ἀναφέρων.

---

75 These data are found in the second century BC papyrus known as *Ars Eudoxi* (col. XXIII). See Rehm, A. Das Parapegma des Euktemon, *SHAW* Nr. 3 (1913) 8ff.; Neugebauer. *HAMA* II, 627.

76 Tannery. *Science hellène*, 68; Heath. *Aristarchus*, 20; *KRS*, 83. Van der Waerden's explanation (*Astronomie*, 11f.) is manifestly erroneous.

77 Unlike solstices, which are more or less easy to observe, equinoxes are determined by calculations; besides, the word ἰσημερία is first attested in a Hippocratic treatise of the late fifth century (*De aer.* 11). Dicks, D.R. Solstices, equinoxes, and the Presocratics, *JHS* 86 (1966) 30f., alleged that the notion of equinoxes presupposes an elaborate astronomical theory that could not be available to the Greeks in the sixth century. This is, of course, incorrect; cf. Woodbury, L. Equinox at Acragas: Pind. Ol. 2. 61–62, *TAPA* 97 (1966) 608 n. 26; Kahn, *op. cit.*, 112ff.; Samuel, *op. cit.*, 23 n. 1. On comparatively simple ways to determine the equinoxes, see Nilsson, M.P. *Primitive time-reckoning*, Lund 1920, 313, 316.

78 12 A 1 (D.L.), A 2 (*Suda*), A 4 (Eusebius); see below, 249. Diogenes Laertius, on the other hand, assigns to Thales the 'discovery of the seasons' (I, 27). See Szabó, Maula, *op. cit.*, 33ff., 118f.

79 *Cael.* 291 a 29f.; cf. ὥσπερ καὶ δεικνύουσιν οἱ μαθηματικοί (291 b 9).

> Anaximander was the first to find an account of the sizes and distances (of the planets), as Eudemus says, adding that the Pythagoreans were the first who found the order of their position (fr. 146).

As has already been indicated, fr. 145 contains a distortion of Eudemus' text. The only Presocratic to have postulated the earth's movement around the center of the cosmos was Philolaus, not Anaximander. Since Montucla, who suggested the reading κεῖται for κινεῖται, practically no serious attempts have been made to prove that Anaximander represented the earth as moving.[80] Comparing fr. 145 with a parallel doxographical tradition that goes back to Theophrastus, we can restore the original meaning of Eudemus' evidence. According to Hippolytus (12 A 11.3), the earth in Anaximander's system "is freely suspended, being supported by nothing" (τὴν δὲ γῆν εἶναι *μετέωρον* ὑπὸ μηδενὸς κρατουμένην); as Diogenes Laertius says (II, 1 = 12 A 1), "the earth is in the middle, holding the central position" (*μέσην* τε τὴν γῆν *κεῖσθαι, κέντρου* τάξιν ἐπέχουσαν).

To all appearances, Eudemus and Theophrastus described the earth's position in Anaximander's system in terms close to Aristotle's note in *De caelo* 295b 11f., though somewhat more elaborately. μετέωρος, when applied to earth, does not mean that it literally 'is hanging in the air', i.e., suspended in it, as Simplicius erroneously believed (*In Cael. comm.*, 532.14). This term confers upon earth, for the first time, the status of a heavenly body like that of the sun, the moon, and the stars and constituting one system with them. The reports of the two Peripatetics differ only in that Theophrastus described the whole of Anaximander's teaching, along with the latter's erroneous *doxai*, while Eudemus focused his attention on those of Anaximander's discoveries that lead to contemporary notions of the cosmos.[81] The central position of the earth, with no other body to support it, was one of Anaximander's most brilliant astronomical insights. It goes far beyond even his most extravagant other notions, like 'heavenly wheels', which, though at odds with everyday experience, still stem from it, while the idea of earth afloat in the center of the cosmos clearly contradicts it. Since this idea is one of the cornerstones of Greek astronomy, Eudemus could not fail to name its discoverer.

Proceeding from the assumption that the *prōtos heuretēs* principle was among the essential criteria for selecting material for the *History of Astronomy*, it is unlikely that Eudemus would have mentioned those of Anaximander's statements he considered erroneous, such as that the earth has the shape of a

---

80 Cf. κεῖται/κινεῖται in the manuscripts of Arist. *Cael.* 291a 30. A similar confusion between κεῖσθαι and κινεῖσθαι (*Dox.*, 344a 11, b9) is noted by Conche, M. *Anaximandre. Fragments et témoignages*, Paris 1991, 203 n. 23. Cf. Arist. *Cael.* 295b 11–16: in Anaximander, the earth is at rest.

81 "Eudemus is trying to discover, in a good Aristotelian fashion, the nature of progress in science that led to the situation as he knew it; results are what he wants to record." (Burkert. *L & S*, 308).

column's drum. To be sure, the few extant fragments of the *History of Astronomy*, even when supplemented by parallel material from the *History of Geometry*, do not enable us to say how consistently Eudemus discarded erroneous ideas. It would be rash to allege that *all* the ideas that contradicted the views of the fourth-century specialists in astronomy remained outside the *History of Astronomy*. By its very nature, astronomy could not develop as victoriously as mathematics, and since even in the *History of Geometry* there is criticism of Antiphon's failed attempt to square a circle, we can expect the *History of Astronomy* to have contained similar material as well. In Antiphon's case, however, 'failed' means 'an attempt made by non-mathematical methods', whereas astronomy's criteria of truth were not as strict as those used in geometry. Even Eudoxian theory, the most scientific of the day, failed to 'save the phenomena' adequately and required further modifications. Hence, one can readily surmise that Eudemus, when sorting out the discoveries of the ancient astronomers, was compelled to apply less strict standards than those used in the *History of Geometry*.

These considerations are supported by the testimony already cited: Anaximander was the first to find an account of the sizes and distances of heavenly bodies (fr. 146).[82] What is instructive here are the words πρώτου ... τὸν λόγον εὑρηκότος.[83] Eudemus could not assert that Anaximander was the first to have found the *true* sizes of heavenly bodies and the *actual* distances between them: the figures accepted at the end of the fourth century differed considerably from those of Anaximander,[84] not to mention that his system placed

---

82 Further on, Simplicius notes that the size of the sun and the moon and their distances from the earth are estimated by observing eclipses (*In Cael. comm.*, 471.6–8). His suggestion that this method was also discovered by Anaximander is, of course, erroneous: such computations appeared in the third century BC.

83 Mansfeld, J. Cosmic distances: Aëtius 2. 31 Diels and some related texts, *Le style de la pensée. Recueil de textes en hommage à J. Brunschwig*, ed. by M. Canto-Sperber, P. Pellegrin, Paris 2002, 429–463, translates this phrase first as "Anaximander was first to discover the ratio (λόγος) of the sizes and of the distances", but then as "Anaximander was first to speak of the sizes and distances" (454, 459). Now, *λόγος περὶ* μεγεθῶν καὶ ἀποστημάτων cannot possibly mean 'ratio', either in Eudemus, or in Simplicius; in the latter λόγος περί normally means 'a theory/explanation of'. Hence, Mansfeld's assertion, built solely on his first translation, that the doxographical information on the sizes and distances in Anaximander derives from Eudemus, and not from Theophrastus, remains unsubstantiated.

84 In Anaximander, the sun is of the same size as the earth (12 A 21); Philip of Opus (*Epin.* 983a) and Aristotle (*Mete.* 345a 36) believed it to be greater than the earth. Eudoxus considered the sun to be 9 times greater than the moon (D 13 Lasserre) and (possibly) 3.3 times greater than the earth (Heath. *Aristarchus*, 331f.; Lasserre. *Eudoxos*, 211). In Anaximander, distances from the earth to the stars, the moon, and the sun must have been equal to 9, 18 and 27 radii of the earth, respectively (see below, 250f.). Mathematical astronomy of Aristotle's time (Eudoxus?) claimed that the

stars and planets under the moon (12 A 18). Still, Anaximander remains the first to put forth a theory, λόγος, on this subject, which proved a giant step forward and brought him, most deservedly, the glory of a discoverer. Following the general principle of *prōtos heuretēs*, Eudemus modifies it to include in the history of astronomy the pioneering theories whose further development contributed to the creation of the 'correct' picture of the world. A similar principle is predominant in the historico-scientific literature of the modern period as well, which concentrates chiefly on the precursors of successful scientific theories contemporary to it. Though from this perspective scientific progress does indeed often look like a teleological growth of a 'scientific fruit' from its 'seed' (admittedly, this was Eudemus' own view), attempts to reject this principle *altogether* generally result in relativism, which is no less harmful for the history of science than teleologism.

As for the second part of Simplicius' evidence, which relates to the Pythagorean discovery of the 'correct' order of the heavenly bodies, we will return later to it. It is sufficient to say here that Simplicius' words suggest an immediate familiarity with Eudemus' work, rather than dependence on a secondary source. Commenting on Aristotle's passage on the order and the sizes of the heavenly bodies, Simplicius picks from Eudemus' history the authors of these discoveries – Anaximander and the Pythagoreans. In the chronologically organized *History of Astronomy*, these names could hardly have stood side by side; in addition, Simplicius considers here two different discoveries, and not one and the same problem taken up by scientists of different generations (Hippocrates, Archytas, Eudoxus, etc.). Hence, it is not to Eudemus himself, but to his reader – here most naturally identified as Simplicius – that the comparison between Anaximander and the Pythagoreans apparently belongs. The other excerpts from the *History of Astronomy* either treat one figure only – Thales, Anaxagoras – or present a chronological list of discoveries, and that full of mistakes (Dercyllides–Theon). Unlike these, Simplicius' evidence is not only accurate in its account of facts, but also provides an important detail: Anaximander is credited not with the discovery of the true sizes and distances of the heavenly bodies, but with the first account of this subject. In lists that hand the information down through two or three intermediaries, such details normally tend to be lost.[85]

Is it possible to find additional evidence on Anaximander's astronomy that might derive from Eudemus? In searching for additional material from the *History of Geometry* (5.2), we relied among other things on the fact that Eudemus was one of the very few authors from whom the information about concrete dis-

---

sun's distance from the earth is many times greater than that of the moon (*Mete*. 345 a 36f.).

85 Though the possibility that Simplicius borrowed Eudemus' evidence from an earlier work, e.g. from Alexander's lost commentary to *De Caelo*, cannot be completely ruled out, this seems to me less likely; cf. above, 234 n. 27.

coveries by early Greek mathematicians could actually derive. Yet even in this field, the existence of 'rivals' whose works have perished has to be taken into account (3.1). In astronomy, which partly overlapped natural philosophy, the number of such rivals was considerably greater, which makes it difficult to distinguish Eudemus' evidence from what could go back to, say, Theophrastus. In principle, it is clear that we should proceed by ruling out, first, data from authors who made immediate use of doxographical sources (Aëtius, Achilles, Hippolytus) and, second, the doxographical data contained in the rest of the tradition. Of all the non-doxographical evidence, sources that use the *prōtos heuretēs* formula prove of particular importance to us.

Thus, Diogenes Laertius (II, 1) cites Favorinus as saying that Anaximander "was the first who invented the gnomon and set it up on a sundial (?) in Lacedaemon ..., in order to indicate the solstices and the equinoxes".[86] Though the *prōtos heuretēs* formula was, indeed, used in the *Physikōn doxai* as well, particularly often in its astronomical division,[87] the invention of the gnomon and its installation in Sparta is not a subject that Theophrastus was likely to treat. Favorinus' *Miscellaneous History* included a book on the first discoveries; it mentioned, among others, Pythagoras and Archytas; he was familiar with the biographical and scholarly tradition of Hellenistic authors that *could* have preserved Eudemus' evidence.[88] The tradition of the gnomon being installed in Sparta does not seem to have been invented,[89] but we cannot be certain that Favorinus' evidence (or, at least, a part of it) goes back to the *History of Astronomy*. Herodotus, as we know, says that the gnomon and polos were borrowed from Babylon (II, 109), which does not necessarily contradict Favorinus' source. If Anaximander was, indeed, the first whom the Greek tradition associated with the gnomon, he still could have been named its *prōtos heuretēs*. On the other hand, the story of the invention of the gnomon could hardly have figured in Anaximander's book, so that even if Eudemus mentioned it, its authenticity remains uncertain.

---

86 Εὗρεν δὲ καὶ γνώμονα πρῶτος καὶ ἔστησεν ἐπὶ τῶν σκιοθήρων ἐν Λακεδαίμονι, καθά φησι Φαβωρῖνος ἐν Παντοδαπῇ ἱστορίᾳ, τροπάς τε καὶ ἰσημερίας σημαίνοντα (12 A 1 = fr. 28 Mensching). For other variants, see: 12 A 2, 4. Mensching regarded Σκιόθηρα as a toponym; see also Classen, C. J. Anaximandros, *RE Suppl.* 12 (1970) 33; *KRS*, 103; Franciosi, F. *Le origini scientifiche dell' astronomia greca*, Rome 1990, 55. Cf. horologium quod appellant sciothericon Lacedaemone ostendit (Plin. *HN* II, 86 = 13 A 14a and below, n. 89).

87 See above, 161.

88 See above, 176 n. 39.

89 Guthrie, *op. cit.*, 75; Classen. Anaximandros, 33; Szabó, Maula, *op. cit.*, 33ff.; Franciosi, *op. cit.*, 54f. See also the parallel version on Anaximander' prediction of an earthquake in Sparta (12 A 5a). Pliny's story attributing the invention of the gnomon and its installation in Sparta to Anaximenes (13 A 14a) reflects the typical confusion between the two Milesians (Mensching, *op. cit.*, 114f.).

3. Physical and mathematical astronomies

The *Physikōn doxai* must, naturally, have contained much more information on Anaximander's cosmology than did Eudemus' selective history of astronomical discoveries. The evidence preserved in Aëtius confirms this relation, which was characteristic of other physicists as well: the cosmos is infinite (*Dox.*, 327b 10), perishable (331.12), and consists of a mixture of cold and hot (340a 4); heavenly bodies consist of fire enclosed by air (342b 7, 559. 25), the sun being the highest of all, followed by the moon and stars (345a 7); the stars are carried by circles and spheres (345a 22); the sun's wheel, full of fire, is 27 (or 28) times larger than the earth, and the sun itself is equal to the earth (348a 3, 351a 5, 560.4); solar eclipses occur when the opening in the rim of the wheel is stopped up (354a 3); the moon is also a circle full of fire, 19 times larger than the earth (355a 18); it gives off its own light (358a 6), and its eclipses have the same cause as solar ones (359a 13, 560.3); the earth is cylindrical in shape, and its depth is a third of its width; it stays in the center of the cosmos, held up by nothing (376.22, 559.22, 579.11). The detailed scheme used by Theophrastus to describe the Presocratics' cosmological doctrines included, along with purely physical aspects, what Aristotle related to μαθηματικὴ ἀστρολογία. Eudemus, meanwhile, applied his scheme of the *prōtoi heuretai* to the already selected astronomical material from which physics had been removed in advance.

Aristotle's notions, first set forth in his *Physics* (193b 22–194a 11), of boundaries between the two closely related sciences are certainly representative of the way mathematical astronomers of the fourth century understood their own subject.[90] In *De caelo* and the other works, Aristotle regularly refers to the expert knowledge of the *mathēmatikoi*;[91] in *Met.* 1073b 18f., he explicitly names Eudoxus and Callippus and develops their ideas. It seems very probable that the 'mathematicians' figuring in *De caelo* are none other than Eudoxus and his pupils, from whom Aristotle derived data of the sizes of heavenly bodies and distances between them, their velocities, the length of the earth's circumference, etc.[92] No less decisive was the influence of Eudoxus and

---

90 Elsewhere Aristotle says that ναυτικὴ ἀστρολογία, descriptive (and practical), establishes empirical facts, while μαθηματικὴ ἀστρολογία explains them (*APo* 78b 36f., see above, 73 n. 119). The line is drawn, accordingly, between ὅτι and διότι; *mathēmatikoi*, moreover, may not know the particular facts. The latter thesis seem to come from Aristotle himself, rather than from *mathēmatikoi*, who sought explanations consistent with observed data. Geminus' definition of astronomy (Procl. *In Eucl.*, 41.19f.) coincides practically with that of Aristotle: a science, closely related to physics, on heavenly bodies' motions, their sizes and shapes, their distances from earth and respective brightness, etc.

91 *Cael.* 291a 29f., b9, 297a 2f., 298a 15; see also *PA* 639b 7f., *Met.* 1073b 3f., *Mete.* 345a 36f.

92 Lasserre printed the references from *De caelo* as testimonies on Eudoxus (*Eudoxos* D7, 10–11).

his school on defining what is the subject of mathematical astronomy and precisely how it differs from physics. By this I mean not any formal definitions of astronomy, but the practices it had been following and the further course of its development. A pronounced tendency to develop astronomy by mathematical methods, while taking empirical data (φαινόμενα) fully into account, is a distinctive feature of Eudoxus' school. The celestial kinematics of Eudoxus and Callippus abstracted from the physical nature of heavenly bodies and explained their movements by means of mathematical models.[93] In the next generation, this tendency was reinforced by Autolycus and Euclid, who regarded astronomy as spherical geometry and applied their kinematic models to heavenly bodies' movements.

In his *History of Astronomy*, Eudemus considered the past from the standpoint of the mathematical astronomy of the late fourth century and selected material in accordance with its professional criteria. Conceived in this way, the history of astronomy included Thales' studies of solstices, Anaximander's theory of the sizes of heavenly bodies, the Pythagorean order of the planets, the estimate of the obliquity of the ecliptic by Oenopides, Eudoxus' program of 'saving the phenomena', and much else. It did not include the questions that contemporary astronomers left to the physicists (e.g., what does the sun consist of?), nor answers irrelevant to their science (e.g., Anaximander's explanation of eclipses). The second criterion of the selection of material – the search for discoveries that constituted an integral part of contemporary astronomy or could be treated as a stage in its progress – worked in the same direction as the first one, narrowing the factual scope of the *History of Astronomy* still further.

That Eudemus' astronomy was an idealized construction is hardly surprising. In a sense, so is any history of science, which is bound to rely on selected facts and to interpret them from contemporary positions. This does not necessarily imply that Eudemus modernized early Greek astronomy, attributing to it opinions that it *ought to have held* from the point of view of his own time.[94] Admittedly, this cannot be ruled out: Aristotle is known to have indulged in such interpretations,[95] nor is contemporary history of science immune from them. In regard to the *History of Astronomy*, however, any direct evidence for this is lacking. From the abundant data of explanations of celestial phenomena by earlier thinkers, Eudemus apparently selected the major discoveries that demon-

---

93 An Epicurean text of the late fourth century calls one of Eudoxus' students ἀστρολογογεωμέτρης (Philod. *De Epicuro* II (*PHerc.* 1289), 6 III). See Sedley, D. Epicurus and the mathematicians of Cyzicus, *CErc* 6 (1976) 27f.

94 It would be intriguing to speculate on the degree to which Eudemus' work reflected astronomers' views on the *history* of their science. What did the names of Anaximander, Pythagoras, Cleostratus, and Anaxagoras mean to them? Were they aware of their individual contributions to the development of astronomy? How original, in other words, was the historical approach to astronomy developed by Eudemus? The material to answer this question is, unfortunately, lacking.

95 Cherniss. *Aristotle's criticism*, 352ff.

strated the progress of mathematical astronomy from its modest origins to its then contemporary state of 'perfection'. The facts he reports are not only confirmed by other presently available data (or do not, at least, contradict them) – they form the actual basis of the modern history of early Greek astronomy. That allows us to characterize Eudemus' approach to the study of Greek science, albeit with reservations, as genuinely *historical.*

Current views on mathematical astronomy differ in many respects from those of Aristotle and Eudemus. An influential trend in the history of astronomy regards mathematical astronomy as a science that, proceeding from accurate, systematic, and preferably dated observations, builds quantitative models of heavenly bodies' movements and thus makes possible accurate predictions of their appearance in the firmament.[96] From this viewpoint, the history of Greek astronomy really starts not with Thales and Anaximander, but with, at best, Meton and Euctemon, if not Eudoxus or even, paradoxically, Hipparchus.[97] The 'speculations' of the Presocratics, who, unlike Babylonian astronomers, never conducted, let alone recorded, any systematic long-term observations, either do not enter this history at all or remain marginal at best.

This trend in the historiography of Greek astronomy is worth a special study.[98] Suffice it to say here that the history of astronomy as seen by Eudemus has much more appeal for me, if only because it better represents the views on the real tasks of science held by both Greek and the early modern astronomers.[99] Besides, it better corresponds to the facts.

Μαθηματικὴ ἀστρολογία as conceived by Aristotle and Eudemus should more properly be called *geometrical* astronomy, to distinguish it from the

---

96 Scientific astronomical theory is "a mathematical description of celestial phenomena capable of yielding numerical predictions that can be tested against observations" (Aaboe, A. Scientific astronomy in Antiquity, *The place of astronomy in the ancient world*, ed. by F. R. Hodson, London 1974, 23).

97 Dicks (Solstices) dates the beginnings of scientific astronomy to the last third of the fifth century (cf. Kahn, *op. cit.*). See also Aaboe. Scientific astronomy, 40f. (astronomy before Hipparchus is not scientific); Neugebauer. *HAMA* II, 571 (starts with Meton); idem. On some aspects of early Greek astronomy, Neugebauer, O. *Astronomy and history: Selected essays*, New York 1983, 361–369; Bowen, Goldstein. Meton, 54, 78f. (the astronomy of the fifth century is thoroughly practical; there was no theory); Toomer, G. J. Astronomie, *Le savoir grec*, ed. by J. Brunschwig et al., Paris 1996, 303f. (Meton is the first worthy of being called an astronomer, and even then owing to Babylonian influence). In Goldstein, B. Saving the phenomena: The background to Ptolemy's planetary theory, *JHA* 28 (1997) 1–12, the tendency to redate Greek mathematical astronomy as close to Ptolemy as possible has reached its culmination: this time it starts with Ptolemy himself!

98 For some notes on the subject, see von Fritz. *Grundprobleme*, 132f.

99 As G. Lloyd argues (Saving the appearances), instrumentalism was not characteristic of Greek astronomy, either in its early or in its later phase. On the 'realistic' orientation of astronomy from Copernicus to Newton, see von Fritz. *Grundprobleme*, 157, 192f.

Babylonian astronomy, which, mathematical as it was, relied on arithmetical schemes, not on geometrical models. The geometrical models of Eudoxus and Callippus (as well as of Autolycus and Euclid) were predominantly qualitative and offered little properly numerical data. The models of the Presocratics contained still less such data; those that are known to us are purely speculative.[100] This kind of astronomy was incapable of making accurate predictions and hardly sought to. It strove primarily to give *an account* of how exactly the cosmos is structured, what the real (and not apparent) motions of the heavenly bodies are, what their sizes and shapes are, why eclipses take place, etc.[101] An important role was also played by 'calendar' astronomy that tried to find the best scheme of a luni-solar calendar. Involved in this search were both practitioners of 'observational' astronomy (Cleostratus, Harpalus, Matricetas, Phaeinos, Meton, Euctemon), physicists (Philolaus, Democritus), and mathematicians (Oenopides).[102] Eudoxus' and Callippus' activity encompassed every major line of astronomical thought: 1) observations of stars, 2) problems related to the calendar, and 3) modeling the movements of the heavenly bodies.[103] Eudemus may well be supposed to have traced the development of each of these lines, with a focus on the second and particularly the third.

Having approached, by 450, the second stage of its evolution, Greek astronomy advanced still more quickly toward geometrization. This trend is mainly (though to a different degree) represented by Oenopides, Hippocrates, and Philolaus;[104] occasionally they employ physical arguments, too.[105] Eudemus

---

[100] The Pythagoreans and Empedocles believed, e.g., that the moon's distance from the sun is twice that from the earth (31 A 61). See also above, 247 n. 84.

[101] Von Fritz. *Grundprobleme*, 141 f.

[102] Eudemus assigns to Oenopides the discovery of the Great Year (fr. 145), the 59-year calendar cycle (41 A 8–9) that is also associated with Philolaus (44 A 22; Burkert. *L & S*, 314 n. 79). Theophrastus reports on observations by Cleostratus, Matricetas, and Phaeinos; it is from the latter that Meton allegedly learned about the 19-year cycle (*De sign.* 4 = 6 A 1). On Harpalus, see below, 270 n. 190. On Democritus, see 68 B 12, 14.1–3. Eudemus could have also mentioned the first calendar cycle, Cleostratus' octaëteris (cf. 6 B 4 *DK*, *Dox.*, 364 a 15 f. and above, 244 n. 74).

[103] The first line is represented by Eudoxus' Φαινόμενα and Ἔνοπτρον (fr. 1–120), the second by Eudoxus' and Callippus' calendar schemes and parapegmata (astronomical calendars with weather indications included) (fr. 129–269), the third by Eudoxus' *On Velocities* with Callippus' modifications (fr. 121–126).

[104] Oenopides (41 A 7, 13–14), Hippocrates (42 A 5; see Burkert. *L & S*, 305, 314, 332), Philolaus (44 A 16–17, 21). Archytas' astronomy is, unfortunately, almost completely unknown (cf. Zhmud. *Wissenschaft*, 219 f.), yet we are familiar with his attempts to apply mathematics to harmonics, mechanics, and, possibly, to optics (see above, 129 n. 45, 173 f., 216). As Eudoxus' teacher, Archytas may have been directly involved in the further geometrization of astronomy (see above, 97f., and below, 274 n. 202).

[105] Oenopides (41 A 10), Hippocrates (42 A 5), Philolaus (44 A 18–20). In the parapegmata, the first of which belongs to Meton, the connection between astronomical and

seemed to regard Oenopides as the first to develop the methodological principles of mathematical astronomy (7.5). By relating the *beginnings* of this science to Thales and Anaximander, he might have proceeded from the assumption that their discoveries laid the basis for further progress. The intrinsic logic underlying any science's development had already been formulated by Eudemus' teacher:

> For in the case of all discoveries the results of previous labours that have been handed down from others have been advanced bit by bit by those who have taken them on, whereas the original discoveries generally make an advance that is small at first, though much more useful than the development which later springs out them. For it may be that in everything, as the saying is, 'the first start is the main part': and for this reason also it is the most difficult; for in proportion as it is most potent in its influence, so it is smallest in its compass and therefore most difficult to see: whereas when this is once discovered, it is easier to add and develop the remainder in connection with it. This is in fact what has happened in regard to rhetorical speeches and to practically all the other arts: for those who discovered the beginnings of them advanced them in all only a little way, whereas the celebrities of to-day are the heirs (so to speak) of a long succession of men who have advanced them bit by bit, and so have developed them to their present form.[106]

Even denying Aristotle's teleological approach, one has to admit: without fundamental notions introduced in astronomy before the mid-fifth century, its further geometrization would have been impossible. In the first place, I mean the central position of the earth and its spherical shape, the notion of the heavenly sphere divided into zones by the equator, the tropics and the arctic and the antarctic circles, the independent movement and order of the planets, and explanations of solar and lunar eclipses. Interestingly, nearly all these ideas were in fact interpreted later as εὑρήματα, so that Eudemus could hardly have passed them by unnoticed.[107]

It is at this early period that deductive geometry, toward which mathematical astronomy was subsequently oriented, also took shape. The general form and methods of presenting the material in Autolycus' and Euclid's treatises was obviously modeled on mathematical *Elements*. Astronomy is presented here as a deductive theory that consistently demonstrates its theorems, proceeding from a number of definitions and axioms. The influence of geometry is also betrayed

---

meteorological phenomena has a regular character; see Rehm A. *Parapegmastudien*, *ABAW* 19 (1941).

106 Arist. *SE* 183b 17–32, transl. by W. Pickard-Cambridge.

107 Thus, he mentions the central position of the earth, the notion of the celestial sphere, the order of planets, and the explanation of lunar eclipses (fr. 145–146). The discovery of the earth's spherical shape was attributed to Pythagoras and Parmenides (D. L. VIII, 48 = Theophr. fr. 227e FHSG), the obliquity of the ecliptic to Pythagoras and Oenopides (*Dox.*, 340.21). The independent movement of the planets is associated with Alcmaeon (*Dox.*, 345.19 = 24 A 4). The doxographical tradition on Thales' discoveries in astronomy is unreliable (see above, 240 n. 50).

by the fact that these treatises "usually conceal any connection with astronomical applications and numerical data",[108] a feature most typical of deductive geometry. To be sure, the level of the axiomatization and demonstrativeness of the astronomical treatises remained inferior to geometrical ones: to coordinate the observed data and calculations with mathematical propositions and their deductive proof turned out to be an arduous task indeed. In addition, unlike geometry, the exposition of an astronomical theory *more geometrico* was not backed by an age-long tradition of *Elements*; one should recall that even Hippocrates, the author of the first *Elements*, relied heavily on Pythagorean mathematics. There is little doubt, however, that the spherical geometry of Autolycus and Euclid had precursors[109] – in form as well as content – and that the same method of exposition was characteristic of Eudoxus' *On Velocities*.

4. Anaxagoras. The Pythagoreans

Dercyllides–Theon states that Anaximenes was the first to discover the source of the moon's light and the causes of its eclipses (fr. 145). It has often been noted that this apparently contradicts the fact that Anaximenes attributed a fiery nature to all heavenly bodies including the moon (13 A 7.4, 14).[110] Already Tannery suggested that it is Anaxagoras who is meant here,[111] and this has never been seriously contested since.[112] Unlike Anaximander, Anaximenes introduced almost no new geometrical concepts, with one rather important exception: he 'moved' the stars to the outer place, beyond the moon, sun, and planets.[113] The related evidence is found in Theophrastus; as for Eudemus, we do not know whether he mentioned Anaximenes at all.

In regard to Anaxagoras, both his own words and reliable indirect tradition attest that he indeed thought the moon received its light from the sun and offered correct explanations for both lunar and solar eclipses.[114] Though the expression ἀλλότριον φῶς from Parmenides' poem (28 B 14–15) was often

---

108 Neugebauer. *HAMA* II, 748 ff.

109 Heiberg, J. L. *Litterargeschichtliche Studien über Euklid*, Leipzig 1882, 41 ff.; Hultsch, F. Autolykos und Euklid, *BSGW* 38 (1886) 128–155; idem. Astronomie, *RE* 2 (1896) 1842 f.; Tannery. *Géométrie*, 133 f.; idem. *Recherches*, 57 f.; Björnbo. *Studien*, 56 ff.; Heath. *History* 1, 348 f.; Mogenet, J. *Autolycus de Pitane*, Louvain 1950, 18 f. Even Neugebauer did not deny it (*HAMA* II, 750).

110 Bicknell, P. J. Anaximenes' astronomy, *AcCl* 12 (1969) 53–85; *KRS*, 156.

111 Tannery. *Science hellène*, 157 f.; idem. *Recherches*, 33 n. 4.

112 Heath. *Aristarchus*, 19; Bicknell, *op. cit.*, 59; *KRS*, 156; Wöhrle, G. Wer entdeckte die Quelle des Mondlichts?, *Hermes* 123 (1995) 244–247. Cf. Boll. Finsternisse, 2342.

113 13 A 7, 13. The order in which the sun, moon, and planets were arranged in his system is unknown.

114 Pl. *Crat.* 409 a–b = 59 A 76, see also 59 B 8, A 77 and below, 256 n. 117.

taken to mean that the moon shines by 'borrowed light',[115] Parmenides' cosmology, due to his "intentional use of ambiguity", hardly has a chance of being reliably reconstructed.[116] Even if Theophrastus credited him with the idea that the moon's light is received from the sun, he ascribed the explanation of eclipses to Anaxagoras.[117] Eudemus, to all appearances, shared this view. Interestingly, in Hippolytus, who in the case of Anaxagoras employs a reliable doxographical source, we find the formula *prōtos heuretēs*,[118] which is frequently attested in the astronomical division of the *Physikōn doxai.*[119] It is very likely that this time it also derives from Theophrastus. This makes it still more plausible that Eudemus, too, mentioned only Anaxagoras; it would have been strange if the views of the two Peripatetics had diverged on this point.

Proclus' commentary on *Timaeus* contains still another piece of Eudemus' evidence on the position of both luminaries in Anaxagoras' system. Commenting on the order of planets in Plato – the moon, the sun, Venus, Mercury (*Tim.* 38d) – Proclus dwells in detail on a later arrangement (shared by Ptolemy), which placed the sun after Mercury. He concludes with the following note:

> ὁ δ' οὖν Πλάτων εἰς τὴν πολλὴν κοινωνίαν καὶ τὴν ὁμοφυῆ πάροδον ἀπὸ τῆς αὐτῆς αἰτίας ἡλίου καὶ σελήνης <βλέπων> καὶ τὴν εἰς τὸν κόσμον πρόοδον αὐτῶν ὡς συνημμένην παραδέδωκε. καὶ οὐδὲ ταύτης ἦρξεν αὐτὸς τῆς ὑποθέσεως, ἀλλ' Ἀναξαγόρας τοῦτο πρῶτος ὑπέλαβεν, ὡς ἱστόρησεν Εὔδημος.
>
> At any rate, as Plato saw that there is a lot of common (between the sun and the moon) and that the passage of the sun and the moon is of a similar kind because of the same reason, he also handed down to us their progression into the cosmos as tied together. Meanwhile, it was not Plato who came up with this hypothesis – the first to conceive it was Anaxagoras, as Eudemus reports (fr. 147).

It seems to follow from this rather ambiguous passage that Eudemus did not write on the order of planets in Plato, but rather on the close relation between the sun and the moon in Anaxagoras, which he regarded as a certain innovation.[120] According to Hippolytus, Anaxagoras believed the moon to be closer to the earth than the sun (59 A 42.7). It remains unclear, however, whether this was his discovery or rather whether Eudemus could consider it a discovery at

---

115 See also Empedocles (31 B 45). Cf. *DK* I, 243n.; Heath. *Aristarchus*, 75f.; Wöhrle, *op. cit.*, 245.

116 Kahn, *op. cit.*, 105 n. 22; Burkert. *L & S*, 307 n. 40. Doxographical reports on Parmenides' moon are contradictory: now it has a fiery nature, now it is lit by the sun (*Dox.*, 335b 20, 356b 3, 357a 6, 358b 20 = 28 A 37, 42). The fragment on heavenly bodies (B 10) also allows many conflicting readings.

117 *Dox.*, 360b 14f., 23f. (= fr. 227e FHSG), 562.19f.

118 *Dox.*, 562.26 = 59 A 42: οὗτος ἀφώρισε πρῶτος τὰ περὶ τὰς ἐκλείψεις καὶ φωτισμούς (sc. τῆς σελήνης).

119 See above, 161.

120 Heath. *Aristarchus*, 85; Taylor, A. E. *A commentary on Plato's Timaeus*, Oxford 1928, 123.

all. The same order is found in Empedocles (31 A 61), who hardly followed Anaxagoras here.[121] A further question arises: how does this evidence agree with the fact that Eudemus ascribes to the Pythagoreans the discovery of the 'correct' order of heavenly bodies, the one starting with the moon and sun (fr. 146)? Does he mean the Pythagoreans who were a generation younger than Anaxagoras – Philolaus, for instance?[122] If this is so, and if Eudemus indeed ignored Empedocles, his words may be understood to mean that, while Anaxagoras correctly placed the two luminaries, the Pythagoreans discovered the true arrangement of all the heavenly bodies: the moon, the sun, the five planets, and the fixed stars. There is nothing uncommon in dividing a discovery into two parts: according to the *History of Geometry*, Hippocrates is the first to correctly approach the problem of doubling the cube, while Archytas is the first to solve it. And still, this version implies too many reservations and does not agree very well with what we know of the development of astronomy in the fifth century.

There is another possibility. Eudemus' words on the close relation between the sun and the moon may refer not to the fact that they are closer to the earth than the other planets,[123] but to Anaxagoras' explanation of eclipses. Proclus must have quoted Eudemus at second hand or from memory; he seems not to know about Eudemus' testimony that it was the Pythagoreans who introduced the right order of planets, and so takes Anaxagoras' explanation of eclipses as the introduction of this right order. Such an interpretation allows us to consider Eudemus' testimonies on Anaxagoras and on the Pythagoreans separately and drops the question of Empedocles' priority.

The astronomy of the fourth century accepted the following order of the heavenly bodies: the moon, the sun, Venus, Mercury, Mars, Jupiter, Saturn.[124] First, this order is attested in Plato,[125] who could have borrowed it from his Pythagorean friends. But Eudemus obviously had in mind not the Pythagoreans of Plato's time (e.g., Archytas), but either Philolaus or his precursors. That Philolaus did accept this order is beyond doubt,[126] but was it him Eudemus meant

---

121 Aristotle (*Met.* 984a 11f.) and Theophrastus (fr. 227a FHSG) considered Anaxagoras to have been older than Empedocles, yet Aristotle added that he was "later in his philosophical activity" (see above, 155 n. 153).

122 As Burkert suggested (*L & S*, 313).

123 No evidence on the position of planets in Anaxagoras has survived; it is said, however, that comets appear due to planets' collisions with each other (59 A 1.9, 81). See Burkert. *L & S*, 311. It cannot be excluded that he placed planets closer to the earth than the moon. Boll, F. Hebdomas, *RE* 7 (1912) 2566, pointed out that Proclus speaks of the close relation between the sun and the moon, not about the order of the heavenly bodies on the whole.

124 Boll. Hebdomas; Burkert. *L & S*, 300 n. 7. It relies on the periods of their revolution around the earth: the moon completes its revolution in 29½ days, the sun and the inner planets in a year, etc.

125 *Res.* 616d–617b; *Tim.* 38c–d, see also [Pl.] *Epin.* 987b–d.

126 Aët. II,7.7 = 44 A 16: stars, the five planets, the sun, the moon. Though the do-

to be the *prōtos heuretēs*? If so, why did he refer to anonymous Pythagoreans rather than to Philolaus? An answer to this could be prompted by Aristotle's account of Philolaus' astronomy: he never calls the latter by name, referring instead to the 'Pythagoreans' in general.[127] But Theophrastus and Meno (44 A 16–23, 27) had given up this manner of reference, and nothing suggests that Eudemus followed it. Aristotle, while criticizing the 'Pythagorean' system, put particular stress on what distinguished it from the 'normal' geocentric theory: in the center of Philolaus' cosmos there is the central fire (Hestia), around which all the heavenly bodies revolve, including the earth and the counter-earth. If Philolaus stands behind Eudemus' Pythagoreans, we have to infer that Eudemus separated the 'correct' part of Philolaus' system (the moon, the sun, and the five planets) from the 'erroneous' one (Hestia, the moving earth, and the counter-earth) and (contrary to Aristotle's criticism) decided to ignore the second and to regard the first as a discovery, which he attributed to Philolaus without mentioning him by name. This implication is obviously too heavy for the hypothesis to sustain.

Not a single ancient source states that Philolaus discovered the 'correct' order – and with good reason, because he did not. Nor could Eudemus believe that he did. Philolaus' innovations – Hestia, the counter-earth, the earth's rotation around Hestia – can only be understood as a modification of an earlier system in which the moon, the sun, the five planets, and the stars revolved around the earth.[128] This is what Eudemus understood to be the 'correct' order. The Pythagoreans who discovered it must have lived before Philolaus.[129] Most likely, Eudemus referred to the Pythagoreans 'in general' when he could not adduce any particular name of the *prōtos heuretēs*. In the *History of Geometry*, such references are related to the Pythagoreans of the first half of the fifth century in connection with, for example, the theorem of the sum of the angles of a

xography does not give the order of the five planets in Philolaus, its very silence testifies to this order having been the 'normal' one (Boll. Hebdomas, 2566; Burkert. *L & S*, 313).

127 *Cael.* 293a 18-b 30, *Met.* 986a 10f. See Zhmud. *Wissenschaft*, 268 ff.

128 Zhmud. Philolaus, 249f. Democritus' system seems to present a similar modification: the moon, Venus, the sun, other planets, the stars (68 A 86). Democritus studied with the Pythagoreans (A 1, 38) and must have known their mathematics (see above, 202 n. 162).

129 Interestingly, Burkert, *L & S*, 313f., who identified Eudemus' 'Pythagoreans' with Philolaus, was far from regarding the latter as the author of the 'correct order'. Instead, he postulated the existence of a common source for Philolaus and Democritus and connected it with the borrowing of the Babylonian data on planets that took place in the time between Anaxagoras and Philolaus. Meanwhile, the Greek order of planets has nothing in common with the Babylonian one, which was never based on the periods of the planets' revolution around the earth. The order accepted in Babylonian astronomy of the fifth century was the following: Jupiter, Venus, Saturn, Mercury, Mars; later it was replaced with a slightly different one: Jupiter, Venus, Mercury, Saturn, Mars (Neugebauer. *HAMA* II, 690).

triangle or the theory of the application of areas (fr. 136–137).[130] These parallels also point to Anaxagoras' contemporaries, rather then to the generation of Philolaus.

It has to be pointed out that the order of the heavenly bodies as well as their distances from the earth depend on the velocity (i.e., the period) of their revolution around the earth. In the last third of the fifth century, these problems became part of the standard course in astronomy. According to Xenophon, Socrates, while encouraging his students to familiarize themselves with practical astronomy, "strongly deprecated studying astronomy so far as to include the knowledge of heavenly bodies revolving on different orbits, and of planets and comets, and wearing oneself out with the calculations of their distances from the earth, their periods of revolution and the causes of these".[131] Though he had attended lectures on these subjects himself, Socrates believed that one could waste one's whole life on them, denying to oneself many important things. A similar notion of astronomy is found in Plato's *Phaedo* (98a): it is concerned with the sun, the moon, the rest of the stars (i.e. planets), their relative speed (τάχους τε πέρι πρὸς ἄλληλα), their revolutions, and their other changes. Astronomy is defined in *Gorgias* (451c 8–9) in the same way: the revolutions of stars, the sun, and the moon and their relative speeds. Archytas, finally, attributing to his predecessors (οἱ περὶ μαθήματα) a clear knowledge of τᾶς τῶν ἄστρων ταχυτᾶτος (47 B 1), suggests that lectures on the periods of heavenly bodies' revolutions and their distances from the earth must have been readily available by Socrates life-time. It is known, at any rate, that astronomy belonged to the four *mathēmata* taught by Socrates' contemporaries Hippias and Theodorus.[132] The question then arises whether the course of astronomy taught in Athens might possibly be connected with Philolaus, who lived at that time in Thebes.

A detailed examination of this question would lead us far from Eudemus' *History of Astronomy* to the history of early Greek astronomy as such. Our present task is rather to determine the time in which Eudemus places the Pythagoreans who discovered the 'correct' order of the heavenly bodies. We have repeatedly pointed to the major role he assigned to chronology, trying to place mathematicians and their discoveries in the most exact temporal succession. Since chronology was built into the very structure of his historico-scientific works, the question we have just raised can be reformulated as follows: are

130 In fr. 137, Eudemus emphasizes the antiquity of the discovery (see above, 199). To the same period belongs the contents of the (future) book IV of the *Elements*, which he attributed to the Pythagoreans (see above, 171). Fr. 142 from the *History of Arithmetic* also refers to the early phase of Pythagorean harmonics (6.1); its more accurate dating appears hardly possible.

131 *Mem.* IV,7.4–5, transl. by O. J. Todd. See Burkert. *L & S*, 315 n. 88.

132 See above, 63f. Meton and Euctemon were active in Athens around 430. In Aristophanes, the study of astronomy goes hand in hand with geometry (*Nub.* 194f.).

these Pythagoreans mentioned before or after Oenopides? To answer this, we first have to consider Eudemus' evidence on Oenopides.

5. Oenopides of Chios

Oenopides holds an intermediary position in the *History of Astronomy* between the physicists (Thales, Anaximander, and Anaxagoras), who figure in Theophrastus as well, and the mathematical astronomers that follow him (Meton, Euctemon, Eudoxus, etc.).[133] Does Oenopides himself belong to one of these two categories, or is he merely a link between them? At first sight, Oenopides ought to be numbered among the physicists: he figures, indeed, in the later doxography, which attributes to him physical principles. Yet to conclude that he was mentioned in the *Physikōn doxai* as well would be premature. In Aëtius, his name is mentioned three times; in none of these cases does the information derive with certainty from Theophrastus.

1) The passage of a Stoic origin, from the section "What is god?" (I,7.17): "Diogenes, Cleanthes, and Oenopides consider god to be the world's soul", obviously is a later insertion.[134]

2) The assertion that Pythagoras discovered the obliquity of the ecliptic, whereas Oenopides claimed the discovery as his own, is no less doubtful.[135]

---

133 See above, 238f. In the *History of Geometry*, his position is similar: preceded by Thales, Mamercus, Pythagoras, and Anaxagoras, he is followed by Theodorus, Hippocrates, Leodamas, Archytas, etc. According to the *Catalogue*, he was 'a little younger' than Anaxagoras, i.e., born around 490/85. We may presume that in the *History of Astronomy* he also followed Anaxagoras. This does not exclude, though, that the Pythagoreans who discovered the order of planets could have been mentioned before Oenopides and possibly before Anaxagoras as well. But unlike Anaxagoras, they could not serve Eudemus as a reliable chronological reference point. Burkert. *L & S*, 333, calls Oenopides Anaxagoras' pupil, though evidence on this is lacking.

134 To all appearances, this Oenopides is a different figure; his provenance from Chios is not indicated (the provenance of Oenopides the astronomer is given in most of the cases). Geminus mentions Zenodotus, "who belonged to Oenopides' school but was a pupil of Andron" (Procl. *In Eucl.*, 80.15f.). Since Zenodotus and Andron belong to the Hellenistic period (see above, 178f.), this Oenopides may also have been a Stoic (so Zeller, *op. cit.* III, 48 n. 1; von Fritz. Oinopides, 2271f. discarded this version for no good reason), particularly since he is mentioned along with the Stoics Cleanthes and Diogenes of Babylon. The idea that Diogenes of Apollonia is meant here and that the *Placita* does not refer to Diogenes the Stoic (*DK* 64 A 8), is based on a misunderstanding; originally Diels (*Dox.*, 676a) related this note to Diogenes the Stoic; see also *SVF* III, 216 fr. 31. God as the world's soul is a typically Stoic notion (*SVF* II, 217.24, 306.21, 307.8. 17, 310.18 etc.).

135 Aët. II,11.2, Ps.-Galen. *Hist. phil.*, 55 = *Dox.*, 340.25f., 624.2f. See Burkert. *L & S*, 306 n. 38. Cf. Bodnár, I. Oinopidès de Chios, *DPhA* 4 (2005) 761–767.

Theophrastus was hardly inclined to credit Pythagoras with astronomical discoveries,[136] and Eudemus connected the estimation of the obliquity of the ecliptic with Oenopides (fr. 145). The idea of Oenopides 'plagiarizing' Pythagoras is foreign to Peripatetic doxography and is more likely to go back to the Hellenistic tradition.[137]

3) Still more strange is the section on the Great Year (II, 32). It confuses two different notions, as is often the case: the first part of the section (II,32.1) is related to Plato's Great Year (*Tim.* 39b), i.e., to the period of the revolution of all the planets, while the second (II,32.2) speaks of different intercalation periods for a luni-solar calendar. Four different calendar schemes are mentioned here: the authors of the first three remain anonymous, while the fourth is attributed to Oenopides and Pythagoras.[138] The first scheme (8 years) stems, in fact, from Cleostratus, the second (19 years) from Meton, the third (76 years) from Callippus. None of them belonged to the physicists or was ever mentioned in Theophrastus. In addition, Callippus was two generations younger than Plato, who comes last in the *Physikōn doxai.* All this precludes attributing the latter evidence on Oenopides to Theophrastus.

Oenopides is unlikely to have set forth any physical doctrine or developed a cosmology of his own. His name is absent, in any case, from the corresponding sections in Aëtius. Though Sextus Empiricus (ca. 200 AD), followed by Ps.-Galen (fourth–fifth centuries AD?), does indeed attribute to him the idea of air and fire as first principles,[139] even a cursory look at Sextus' doxographical source proves its late origin.[140] There is absolutely no reason to relate evidence on Oenopides' principles back to the *Physikōn doxai,* let alone to Oenopides' own work.[141] Thus, the reliable doxographical tradition remains silent on his

---

136 Where Pythagoras and Parmenides have competing claims for discoveries in astronomy, Theophrastus sides with the latter (cf. D. L. VIII, 14 and 48; IX, 23 and *Dox.*, 345b 14). In Aëtius the division of the heavenly sphere into zones is connected with Thales, Pythagoras, and οἱ ἀπ' αὐτοῦ (*Dox.*, 345.7f.).

137 The first to have mentioned Oenopides along with Pythagoras was a historian Hecataeus of Abdera (ca. 300 BC) in his work on Egypt (*FGrHist* 264 F25, 96f. = Diod. I,96.2, cf. I,98.3). Revealingly, they are not yet connected with each other: Pythagoras borrows from Egypt geometry and arithmetic, and Oenopides the idea of the obliquity of the zodiac. See also the pseudo-Platonic *Rivals* (*Erast.* 132a = 41 A 2).

138 In fact, it belonged to Philolaus (44 A 22; Burkert. *L & S*, 314 n. 79) and not to Pythagoras.

139 Sext. *Pyrrh. hyp.* III, 30 = 41 A 5, *Adv. Math.* IX, 361; Ps.-Galen. *Hist. phil.* 18 = *Dox.*, 610.15. Section 18 of Ps.-Galen follows a source common with Sextus (*Dox.*, 246f., 249) and does not belong to the part of the compilation borrowed from Ps.-Plutarch (sections 25–133).

140 Named here along with the early physicists are the theologian Pherecydes, the Orphic Onomacritus, Strato of Lampsacus, and the physician Asclepiades of Bithynia (ca. 100 BC).

141 Fire and air make a highly unusual pair of principles, not being contrary to each

natural philosophy; even the cases cited above concern astronomy, not physics. As follows from Eudemus, Oenopides took up mathematics and astronomy; his writing was related to these subjects and has little in common with a doctrine of principles.[142] The only reliable fact testifying to Oenopides' interest in physical problems is his attempt to explain the floods of the Nile.[143] Yet it is this famous discussion, opened by Thales, that involved – along with physicists proper – the historians Herodotus and Ephorus, the author of the Periplus Euthymenes of Massalia, the typical mathematician Eudoxus, and even Euripides.[144]

Hence, the sum total of available evidence on Oenopides suggests that Eudemus and Theophrastus classified him as a *mathēmatikos*, whose doctrines did not need to be treated in the *Physikōn doxai*. This does not imply, of course, that Oenopides and other mathematicians could not take an interest in physical problems.[145] Rather, it is in the person of Oenopides, or more precisely, in his generation, that specialization in science (which naturally accompanies its rapid progress) was first becoming manifest. In the following generations, the results of this growing specialization are evident in the activities of such mathematicians as Hippocrates, Theodorus, Meton, Euctemon, Archytas, Leodamas, Theaetetus, and Eudoxus and his numerous pupils: philosophy lay either on the periphery of their interests (as with Archytas and even more so with Eudoxus) or beyond its horizon.

It is worth noting that Oenopides was among the few mathematicians mentioned in both the *History of Astronomy* and the *History of Geometry*. Eudemus, referring to Oenopides' own words, points out the connection between his geometrical and astronomical studies.[146] To all appearances, Eudemus was quite familiar with Oenopides' treatise on mathematical astronomy, whose traces disappear soon after the fourth century BC. Apart from the tradition concerning

---

other. None of the Presocratics ever suggested such a combination. The principles of the Stoics were the four elements, active fire and air and passive water and earth.

142 To believe Achilles, whose information goes back to Posidonius (*Dox.*, 230), the explanation of the Milky Way as the former path of the sun was one of such subjects (41 A 10).

143 Oenopides' theory is mentioned (without reference to his name) in Aristotle's *De inundatione Nili* (fr. 248, p. 196.19f. Rose). Though absent in Aëtius, it figures in many other authors (Diod. I,41.1 = 41 A 11, Sen. *QN* IV,2.26, etc.). See *Dox.*, 226f.; Gemelli-Marciano, L. Ein neues Zeugnis zu Oinopides von Chios bei Iohannes Tzetzes. Das Problem der Nilschwelle, *MH* 50 (1993) 79–93, and above, 143 n. 112.

144 *Dox.*, 226f.; Eudox. fr. 287–288 (from Γῆς περίοδος). It is hardly possible to say whether the subject of the Nile's floods was treated in Oenopides' astronomical treatise or in a special work.

145 See above, 132 n. 63. Interestingly, his explanation of the Nile's floods is rather natural-philosophical than geographical: the underground springs, being cold in summer and warm in winter, are dried in winter by underground heat, while in summer the water wells up, causing the floods. Cf. Hippon (38 A 10–11; cf. B 1) and [Hipp.] *De nat. puer.* 24–25.

146 See above, 171f.

the obliquity of the ecliptic and the Great Year, which figures in sources independent of Eudemus,[147] all evidence on Oenopides' mathematical astronomy goes back directly or indirectly to the *History of Astronomy*.[148] The reference to Oenopides found in a brief biographical note on Ptolemy is very likely to come from Eudemus as well:

> Πρῶτος δὲ παρ' Ἕλλησιν ὁ Χῖος Οἰνοπίδης τὰς ἀστρολογικὰς μεθόδους ἐξήνεγκεν εἰς γραφήν.
>
> Oenopides of Chios was the first among the Greeks who wrote down the methods of (mathematical) astronomy.[149]

This evidence was long neglected by specialists in early Greek astronomy and, for some reason, was omitted by Diels. He printed, in fact, only the continuation of this note that concerns Oenopides' chronology.[150] Only recently did Burkert mention this evidence and Franciosi give it its due recognition: "These words do not mean that 'Oenopides was the first to write on astronomical methods', but only that he treated astronomy systematically by applying to it the methods of geometry, including demonstrations and drawings."[151] That this information goes back (indirectly, of course) to the *History of Astronomy* is confirmed, first of all, by the typical Eudemian formula *prōtos heuretēs*, repeated nearly verbatim in the fragment on Eudoxus.[152] It is also important that this note does not simply point to a particular discovery (eclipses, ecliptic, etc.), but formulates a conclusion that would have been impossible without a comparison of Oenopides' work with that of his predecessors and, furthermore, that is in perfect agreement with all the rest of Eudemus' evidence. In addition, the

---

147 Aelianus says that Oenopides installed in Olympia a copper tablet with his astronomical calendar (*VH* X, 7 = 41 A 9). On the obliquity of the zodiac, see above, 261 n. 137.

148 Of the Hellenistic authors, Posidonius is the only one whose evidence may (though need not) derive from Oenopides' own book (see above, 262 n. 142).

149 This note is found as a scholium in several manuscripts of Ptolemy (Boll, F. *Studien über Claudius Ptolemäus*, Leipzig 1894, 53f.), as well as in an astrological manuscript of the 15th century (*Catalogus codicum astrologorum graecorum*. T. VIII, Pt. III, ed. by F. Cumont, Brussels 1912, 95).

150 41 A 1a (= Vit. Ptol. Neapol.): Oenopides' *floruit* falls at the end of the Peloponnesian war, he was the contemporary of Gorgias, Zeno of Elea, and Herodotus. This text appears to be based on a chorographical treatise that goes back to Apollodorus. The latter dated Gorgias' and Herodotus' *floruit* in 444/3 (foundation of Thurii) and Zeno's in 464/0. Herodotus' death was dated at the end of the Peloponnesian war (405/4); in our text it seems to be confused with Oenopides' and Herodotus' *floruit* (Jacoby, F. *Apollodors Chronik*, Berlin 1902, 231, 261f., 278f). The reconstructed date of Oenopides' birth ca. 484 agrees perfectly with Eudemus' evidence (see above, 260 n. 133).

151 Burkert. *L&S*, 314 n. 79; Franciosi, *op. cit.*, 96f.

152 καὶ πρῶτος τῶν Ἑλλήνων Εὔδοξος ὁ Κνίδιος, ὥς Εὔδημός τε ἐν τῷ δευτέρῳ τῆς ἀστρολογικῆς ἱστορίας ἀπεμνημόνευσε ... (fr. 148).

text on Ptolemy contains one more excerpt likely to go back to a chronologically arranged history of astronomy: "After Oenopides, Eudoxus won considerable fame in astronomy."[153]

A notion of Oenopides' methods in mathematical astronomy can be gained from constructions mentioned in the *History of Geometry*: how to draw a perpendicular to a given straight line from a point outside it (I, 12), how to construct a rectilinear angle equal to a given rectilinear angle (I, 23), and how to inscribe a regular pentadecagon in a circle (IV, 16).[154] The elementary character of the first two constructions is obviously at variance with the level of problems that occupied Hippocrates a generation later; usually this was explained by claiming that Oenopides was the first person who attempted to limit geometrical construction to the use of ruler and compass.[155] Meanwhile, since Oenopides himself considered problem I, 12 to be useful for astronomy and the same is said of problem IV, 16, it seems more natural to explain these constructions by the astronomical context of his work. The expression κατὰ γνώμονα, used by Oenopides to refer to the perpendicular (I, 12), "since the gnomon stands at a right angle to the horizon", suggests that his work treated astronomical instruments as well.[156] Oenopides, then, might have been the first to attempt to give an astronomical treatise the shape that, though familiar to us from Autolycus' and Euclid's works on spherical geometry, must have appeared much earlier.[157] We can surmise, accordingly, that his work, first, incorporated geometrical notions of the structure of the universe developed by the Greeks from Anaximander to Anaxagoras and, second, expounded them in conformity with the requirements of the deductive geometry of the mid-fifth century, removing them from the cosmological context to which they belonged in the works of physicists.[158]

Interestingly, Neugebauer in his *History of Ancient Mathematical Astronomy*, touching on Oenopides' calendar period,[159] remains silent on other problems that occupied Oenopides, though they proved to be of much greater importance for mathematical astronomy. The first of these problems is that of

---

153 μετὰ δὲ τὸν Οἰνοπίδην, Εὔδοξος ἐπὶ ἀστρολογίᾳ δόξαν ἤνεγκεν οὐ μικράν (further follows the synchronization of Eudoxus with Plato and Ctesias of Cnidus). Cf.: μετὰ δὲ τοῦτον Μάμερκος… ἐπὶ γεωμετρίᾳ δόξαν … λαβόντος (Procl. *In Eucl.*, 65.12f. = Eud. fr. 133).

154 Eud. fr. 138; Procl. *In Eucl.*, 283.7f., 269.8f.

155 Heath. *History* 1, 175; von Fritz. Oinopides, 2265f. As Knorr. *AT*, 15f., noted, this claim is groundless.

156 To obtain more or less reliable data, it was important to ensure that the gnomon was perpendicular to the horizontal surface (Szabó, Maula, *op. cit.*, 120).

157 See above, 255 n. 109.

158 The last task was carried out stage by stage (see above, 253 n. 105).

159 Neugebauer. *HAMA* II, 619. On Oenopides' calendar cycle, see also Heath. *Aristarchus*, 132f.; von Fritz. Oinopides, 2262f.; Bulmer-Thomas, I. Oinopides of Chios, *DSB* 10 (1980) 179f.

measuring the obliquity of the ecliptic. The ecliptical motion of the planets (including the sun and the moon) was known to the early Pythagoreans (24 A 4).[160] Once the concept of the ecliptic is introduced, the inclination of the ecliptic with respect to the celestial equator immediately follows from it. Oenopides' discovery consisted most probably in the measurement of the angle between them, and not in establishing the very fact of inclination.[161] But since Dercyllides-Theon removes Oenopides' name from its context and places it before Thales, this discovery appeared to be divided into two parts, the first of which, "discovered the obliquity of the zodiac", was attributed to Oenopides, while the second, "that the axis of the fixed stars and that of the planets are separated from another by the side of a (regular) pentadecagon", was attributed to the anonymous οἱ λοιποί (fr. 145).

The fact that the empirically found angle of the obliquity of the ecliptic was expressed via the side of the pentadecagon inscribed in a circle betrays the Pythagorean influence, as was rightly pointed out.[162] The regular fifteen-angled figure inscribed in a circle does, indeed, conclude book IV of the *Elements* (which belongs to the Pythagoreans);[163] it consists of a regular pentagon (from which the dodecahedron attributed to Hippasus was constructed) and an equilateral triangle, whose properties also attracted Pythagorean attention.[164] It is, in fact, only one of many traces that reveal that Oenopides' mathematical astronomy was largely indebted to the mathematics and astronomy of the early Pythagorean school. The Pythagoreans themselves may well have moved in the same direction at the same time as Oenopides: the unification of the four *mathēmata* into a group of 'related' sciences, accomplished by the mid-fifth century at the latest,[165] would not have taken place had not the Pythagorean astronomy already acquired features manifestly akin to geometry.

The sources attribute to Pythagoras himself the discovery of the earth's spherical shape, the identification of the Morning and the Evening star with Venus, and the division of the celestial sphere into zones.[166] His priority in the first two discoveries is contested by Parmenides, whose name is also associated with the division of the earthly sphere into zones.[167] Without entering into the

---

160 The other contenders for this discovery are 1) Anaximander (12 A 5, 22); see Guthrie, *op. cit.*, 96f.; cf. Couprie, D. The visualization of Anaximander's astronomy, *Apeiron* 28 (1995) 159–182; 2) Cleostratus (6 B 2); 3) Anaxagoras (59 A 1.9); see Dicks. *Early Greek astronomy*, 59.

161 Von Fritz. Oinopides, 2260f.; Burkert. *L & S*, 306 n. 38; Gundel. Zodiakos, 490; van der Waerden. *Pythagoreer*, 349; Szabó, Maula, *op. cit.*, 120f.

162 Heath. *Aristarchus*, 131 n. 4; von Fritz. Oinopides, 2261; Neugebauer. *HAMA* II, 629.

163 See above 171 n. 22.

164 See above, 170, 198.

165 See above, 63f.

166 Aët. II,12.1, III,14.1; Aristox. fr. 24; D. L. VIII, 48.

167 28 A 4, 44a; D. L. IX, 23, cf. 28 A 40a.

debate on priorities,[168] let us highlight the main point: the tradition relates all these discoveries to the threshold of the fifth century. By that time, the Pythagoreans already knew that the planets move in a direction opposite to that of the fixed stars.[169] The discovery of the earth's spherical shape led to the formation of the main astronomical model of antiquity, which consisted of two concentric spheres, the stellar and the terrestrial one.[170] The division of the celestial sphere into zones implies some notion of the celestial equator and two tropic circles crossed by the oblique circle of the zodiac. Even if the early Pythagoreans and Parmenides did not fully elaborate these geometrical notions, they are present as such in Hippocrates: the terrestrial sphere is inside of the celestial one, both of them are divided into zones; the planets move in circular orbits along the ecliptic, and the horizon divides these orbits into unequal segments.[171] This means that Oenopides, an older contemporary and likely a teacher of Hippocrates, was directly involved in the elaboration and dissemination of geometrical astronomy and of the 'double-sphere' model of the cosmos.

The passage in Dercyllides–Theon that starts (a, b) and ends (e) with Oenopides' discoveries helps us to define more precisely the scope of the problems he worked on:

> Oenopides was the first to discover (a) the obliquity of the zodiac and (b) the period of the Great Year ... (he is followed by Thales, Anaximander, Anaxagoras). And others discovered in addition to this that (c) the fixed stars move round the immobile axis that passes through the poles, (d) whereas the planets move round the axis perpendicular to the zodiac; and that (e) the axis of the fixed stars and that of the planets are separated from one another by the side of a (regular) pentadecagon (fr. 145).[172]

---

168 See Zhmud. *Wissenschaft*, 211f. Parmenides' teacher was the Pythagorean Aminias (D. L. IX, 21).

169 Aëtius associates this idea with Alcmaeon (24 A4, cf. A 12), who could hardly have been its author (Heath. *Aristarchus*, 49f.). In Aristotle's *Protrepticus*, Pythagoras stresses the importance of observing the sky (fr. 18, 20 Düring). Oenopides followed the Pythagorean idea: the sun moves in the direction opposite to the revolution of the heavenly sphere (41 A 7).

170 Goldstein, B. R., Bowen, A. C. A new view of early Greek astronomy, *Isis* 74 (1983) 330–340, esp. 333f. assign the introduction of this model to Eudoxus, admitting, however, that neither of its components was new. Their main argument in favor of such a late date is the cosmological context in which these components still remain in Plato. There are no grounds, however, to postulate such a context for the professional astronomical works on which Plato himself relied. In the same way, Aristotle places the homocentric spheres of Eudoxus and Callippus in a physical and even theological context, certainly alien to their original context.

171 42 A 5. See Burkert. *L & S*, 305, 314. A still higher level of knowledge is presupposed by the astronomy of Meton and Euctemon (7.6). Cf. Bowen, Goldstein. Meton, 54f.

172 In his commentary, Wehrli pointed out that Theon repeats verbatim the last part of fr. 145 (from "that the stars move" to the end) immediately following his long ex-

To whom did Eudemus attribute the discoveries (c) and (d) made by 'the others' in the period between Anaxagoras and Oenopides? If one compares this evidence with the accounts of Alcmaeon's (24 A 4) and Philolaus' (44 A 16) astronomy as well as with Eudemus' testimony that the Pythagoreans discovered the correct order of the planets (fr. 146), one has to credit this school also with the notion of the independent movement of the planets along the ecliptic. If Eudemus treated the last three discoveries (c, d, e) in chronological order, as is typical for him, this list suggests two important conclusions: 1) The Pythagoreans discovered the order of planets before Oenopides, i.e., in the first half of the fifth century. 2) Oenopides' last discovery (e) not only closes the list, it also relies on the two preceding ones, namely that the stars move around the axis of the celestial sphere (c) and the planets around the axis perpendicular to the zodiac (d). It is very probable then that Oenopides' work systematically treated these and the other propositions of geometrical astronomy. That is what Eudemus actually had in mind when he mentioned Oenopides' priority in expounding astronomy in a methodical way.

If our picture of Oenopides' astronomy does not seriously disagree with reality, his contribution to this science is comparable to the first geometrical *Elements* written by Hippocrates. Our information on both works is equally meager. Hippocrates' *Elements*, not a line of which has survived, is mentioned only once, in Proclus' *Catalogue*, whereas the level of geometry it represents is known to us from a single long quotation from the *History of Geometry*, preserved in Simplicius (fr. 140, on squaring the lunes). Against this background, Eudemus' evidence on Oenopides' book that opened a new period in the development of astronomy does not after all seem that insignificant.

6. From Meton to Eudoxus. 'Saving the phenomena'

It is still not known whether the *History of Astronomy* mentioned Hippocrates. Though in the *Meteorology* Aristotle criticizes one of Hippocrates' astronomical theories, the ancient sources are silent on his discoveries in this science. More prominent in the history of astronomy were his two contemporaries, Meton and Euctemon of Athens, whose *floruit* is traditionally dated by the solstice they observed on June 28, 432. Meton and Euctemon made systematic observations in different regions of Greece, created the first astronomical calendars, the so-called parapegmata, suggested a new 19-year calendar cycle, and determined the inequality of the four astronomical seasons. Quite famous in their time,[173] they were the earliest of the Greek astronomers whose dated ob-

---

cerpt from Dercyllides (cf. *Exp.*, 199.3–5 and 202.8–10). Hence, Wehrli concluded that the Eudemian origin of this passage is doubtful. Now, Theon is much more likely to have repeated the sentence quoted previously from Dercyllides than to have edited the quotation in his own manner and repeated it verbatim once again.

173 Meton figured in Phrynichus (*Schol. Ar. Av.* 997 a) and Aristophanes (*Av.* 992–1019);

servations are cited by Ptolemy.[174] Eudemus, it seems, had much to tell about their discoveries. Unfortunately, the fragments of the *History of Astronomy* contain only a brief reference to one of them – the inequality of the four seasons, mentioned in connection with Callippus' modification of Eudoxus' system:

> According to Eudemus, Callippus asserted that, assuming the periods between the solstices and equinoxes to differ to the extent that Euctemon and Meton held that they did, the three spheres in each case (i.e., for the sun and moon) are not sufficient to save the phenomena, in view of the irregularity which is observed in their motions (fr. 149).

According to Meton's and Euctemon's calculations, the seasons are 90, 90, 92 and 93 days long, starting with the summer solstice.[175] This shows, contrary to what Eudoxus' system has postulated,[176] that the sun's motion along the ecliptic is unequal. Did Meton and Euctemon themselves realize their calculations' significance for astronomy, not just for the calendar? Did Eudemus mention these calculations outside the quoted passage? There is a definite answer to the last of these questions. In Eudemus' report, Callippus mentions Meton's and Euctemon's calculations as something already familiar, without citing any numerical data; Eudemus had obviously already discussed the matter. In Eudemus' time, the problem of the irregularity of the heavenly bodies' movements played so important a role that he could hardly have passed over the *prōtoi heuretai* of this fact in silence, even if they themselves had failed to come to the conclusions that would be drawn from their discoveries a hundred years later. The logic of the development of science, as Aristotle understood it, allows us to discern in the discoveries of the past what their authors could not conceive of.[177] Still, one can attribute to Eudemus such an interpretation of Meton and Euctemon's discovery only by admitting that the level of their knowledge was insufficient to account for the solar anomaly by the varying speed of its motion, as it was later suggested by Callippus.

Though the fragments of Meton and Euctemon still have not been collected, which they certainly deserve, even the sundry evidence of their astronomical studies testifies, along with the reconstruction of Euctemon's parapegma,[178] to

---

the latter called him 'Thales' and associated him with the quadrature of the circle; cf. Plut. *Nic.* 13, 7–8, *Alcib.* 17, 5–6; Ael. *VH* 13, 12.

174 Ptol. *Alm.*, 203.7f. (notes the inaccuracy of observations), 205.15f., 207.9f. Still more frequent are the references in his *Phaseis*, where Ptolemy says that Meton and Euctemon conducted their observations in Athens, on the Cyclades, in Macedonia, and in Thrace (67.2f.).

175 See above, 251 and n. 95.

176 Eudoxus considered the number of days in the four seasons to be practically the same: 91, 92, 91 and 91 (*Ars Eudoxi*, col. XXIII; Neugebauer. *HAMA* II, 627f.).

177 See above, 251 and n. 95.

178 Rehm. Das Parapegma des Euktemon; idem. *Parapegmastudien*, 27f.; Pritchett,

the manifest progress made in mathematical astronomy since Oenopides' time. The sources often refer to Meton as the 'geometer',[179] which probably reflects the scientific character of his studies, rather than any particular contribution to geometry. With Euctemon's parapegma, Greek astronomy starts to divide the ecliptic into twelve zodiacal signs, with the sun staying for thirty or thirty-one days in each of them. After A. Böckh, it has been generally admitted that Euctemon already distinguished between the real and the visible rising and setting of stars,[180] which presupposes calculations made by means of a celestial globe.[181] The instrument used by Meton and Euctemon in their observations was the polos, i.e. a concave hemisphere with a gnomon in its center and the circle of the celestial meridian with solstices, equinoxes, etc., marked on its surface.[182] The geometrical character of Meton's and Euctemon's astronomy is manifest, so that they could hardly fail to draw conclusions from the anomaly

W. K., Waerden, B. L. van der. Thucydidean time-reckoning and Euctemon's seasonal calendar, *BCH* 85 (1961) 17–52.

179 ἄριστος ἀστρονόμος καὶ γεωμέτρης (*Schol. Ar. Av.* 997a). Kubitschek, W. Meton, *RE* 15 (1932) 1465, believed that these words might have been induced by Aristophanes' *Birds* (see above, 267 n. 173), yet Meton also figures as the geometer in the scholia that derive, through Achilles as intermediary, from Posidonius and are hardly dependent on Aristophanes: Pasquali, G. Doxographica aus Basiliusscholien, *NGWG* (1910) 197.2 (= fr. 3b Lasserre).

180 Böckh, A. *Über die vierjährigen Sonnenkreise der Alten*, Berlin 1863, 82f., 96f.; Rehm. *Parapegmastudien*, 10; idem. Parapegma, *RE* 18 (1949) 1335f.; Pritchett, van der Waerden, *op. cit.*, 37f.; van der Waerden. *Astronomie*, 80; Wenskus, O. *Astronomische Zeitangaben von Homer bis Theophrast*, Stuttgart 1990, 29. It is with this distinction that Autolycus begins his book *On Risings and Settings* (I, 1). It was obviously known much earlier. See also Gemin. *Eisag.* XIII, 6ff.

181 Cf. Bowen, Goldstein. Meton, 54f. The tradition ascribes the invention of the celestial globe to Atlas, Musaeus, Thales, Anaximander, Anaximenes, and Eudoxus (Schlachter, A. *Der Globus*, Berlin 1927, 9ff.; Eudox. fr. 2, cf. T 14). In Aristophanes, to Strepsiades' question: "Tell me, for the gods' sake, what is this?" the pupil answers: Ἀστρονομία μὲν αὑτηί (*Nub.* 200f.). The scholiast explains: σφαῖραν δείκνυσιν. See Schlachter, *op. cit.*, 14; Franciosi, *op. cit.*, 64, 114f.; Gisinger saw here an allusion to a terrestrial globe or a book entitled *Astronomy* (Schlachter, *op. cit.*, 107), but the map of the earth is mentioned later, while the book does not account for Strepsiades' puzzlement. Worthy of notice is Plato's remark (*Tim.* 40d) that without a 'visual model' (δίοψις … τῶν μιμημάτων; cf. *Leg.* 669e, 796b) it is impossible to grasp the complexity of the planets' movement. It is not clear whether he meant a celestial globe or some other three-dimensional model (Heath. *Aristarchus*, 155; Taylor A. E., *op. cit.*, 241f.), but his remark undoubtedly belongs to the pre-Eudoxian period.

182 πόλος is mentioned in Herodotus (II, 109) and twice in Aristophanes (fr. 169, 227 K.–A.), who directly refers to the polos installed by Meton in Athens. Later it was often called ἡλιοτρόπιον (*Sch. Ar. Av.* 997a). See Rehm, A. Horologium, *RE* 8 (1913) 2417ff.; idem. *Parapegmastudien*, 28f.; Franciosi, *op. cit.*, 112f. On preserved sundials, see Gibbs, S. L. *Greek and Roman sundials*, New Haven 1976.

that they discovered in the sun's movement. A kinematic model used in Plato's *Republic* (617c) to account for the non-uniform movement of the planets[183] implies that pre-Eudoxian astronomy did possess some means to explain the irregularity of the planetary motions.[184]

While Meton's and Euctemon's *floruit* falls in the 430s, Eudoxus was born about 390, and not a single specialist in mathematical astronomy is known to have been active in the period between them.[185] This lacuna, particularly strange considering the flourishing state of mathematics at that time,[186] poses a number of questions to the historian of science. This 'lost' generation includes Archytas and, a bit younger, Plato, whose astronomical knowledge is quite solid and corresponds to the level of the science contemporary with him.[187] Did Plato study astronomy with Meton's and Euctemon's contemporaries, such as Theodorus, or does he owe at least part of his knowledge to Archytas? Eudoxus is known to have studied geometry with Archytas, but who taught him astronomy? Could Archytas have been an intermediary between the scientists of the last third of the fifth century and Eudoxus' generation?[188] Was the astronomical literature of the late fifth century rich enough to ensure the transmission of knowledge without direct study with specialists in the field?[189] Or was astronomy taught by the mathematicians known to us, who, like Archytas, could not boast any independent achievements in astronomy?

All these questions are relevant to the *History of Astronomy* inasmuch as its fragments mention only names already familiar to us. The *Catalogue of geometers,* meanwhile, lists six names that are not attested elsewhere: Mamercus, Neoclides, Leon, Theudius, Athenaeus, and Hermotimus, five of whom lived in the late fifth and the fourth century. Two of the four astronomers named at the beginning of Theophrastus' *On Weather Signs* do not figure in other sources either.[190] Hence, the *History of Astronomy* could also have contained names of scientists unfamiliar to us or information about the astronomical studies of eminent mathematicians. The variety of Eudoxus' astronomical works and the ma-

---

183 See Knorr. Plato and Eudoxus, 316 f.

184 I owe this parallel to I. Bodnár.

185 Though Philolaus' followers Hicetas and Ecphantus suggested the idea of the earth's rotation around its axis (50 A 1; 51 A 5), they were physicists, not *mathēmatikoi.*

186 Mentioned in the *Catalogue* between Hippocrates and Eudoxus are Leodamas, Theaetetus, Archytas, Neoclides, and Leon.

187 See Heath. *Aristarchus*, 134ff.; Dicks. *Early Greek astronomy*, 92ff.

188 See above, 97, 253 n. 104.

189 Special treatises on astronomy were written by Oenopides, Meton, Euctemon, and Democritus. Philip continued the tradition of compiling the parapegmata started by Meton (see above, 102f.).

190 Cleostratus, Matricetas, Phaeinos, and Meton (*De sign.* 4 = *DK* 6 A 1). Censorinus says that, between the time of Cleostratus and Eudoxus, the octaëteris was improved by Harpalus, Nauteles, and Menestratus (*De die nat.*, 18.5 = 6B4 *DK*); of these, only Harpalus is known to us as an engineer and astronomer (ca. 480). See *DK* I, 42n.; Diels, H. *Antike Technik*, Berlin 1920, 4; Heath. *Aristarchus*, 292f.

ture quality of his theories are totally at variance with the idea that this science stagnated during the preceding four decades. In this period, observation yields more and more empirical data that contradict the Pythagorean idea of the regular circular movement of celestial bodies. The most important of these data are: the anomaly in the sun's movement discovered by Meton and Euctemon; the deviation of the moon and the planets from the plane of the ecliptic; the retrograde movements and stops of the planets on their way along the ecliptic; and differences in planetary brightness indicating differences in their distances from the earth. These anomalies led Eudoxus to develop a theory of homocentric spheres, based on his own vast observations, as well as on a brilliantly conceived kinematic model.[191]

Before going back to Eudemus' testimony concerning Eudoxus' program of 'saving the phenomena' (which Sosigenes ascribed to Plato), we have to look into the origin of two notions important to Plato, namely the uniform circular movement of the planets and their divine nature, to which they owe the perfect circularity of their orbits.[192] The first idea is attested in Philolaus, but goes back to the earlier Pythagorean astronomy.[193] Ascribing the circular movement, postulated by Anaximander for the sun and the moon, to the ecliptical motion of the planets as well, the Pythagoreans must have aspired to bring the movement of all heavenly bodies under the same principle, rather than to do justice to empirical observations. The circle being at the time the only figure that could account for the movements of the planets from a *geometrical* point of view, their numerous deviations from circular orbits were simply ignored. The same geometrical logic demanded that this circular movement had to be uniform, since even much later Greek astronomy could not deal mathematically adequately with an irregular movement without reducing it to a combination of several uniform movements.[194] Anaximander, in turn, postulated the daily cir-

---

[191] To be sure, Eudoxus' model could not account for the varying distances of the planets, though Polemarchus seemed to be aware of them (see above, 233).

[192] *Crat.* 397c–d; *Leg.* 885e, 886e, 887e. A still more detailed argument for the divine nature of heavenly bodies is to be found in the *Epinomis* (986e 9f.). Aristotle, discussing the circular movement of heavenly bodies, also used a combination of metaphysical and theological arguments, which were to be repeated until the end of Antiquity (*Met.* 1072a 6f., 1073a 36–39; *Cael.* 269a 3–270b 30; *Mete.* 339b 20–27; fr. 12 Rose).

[193] Zhmud. *Wissenschaft*, 214f., 218ff. As a matter of fact, the circular movement of the planets follows from their order as established by the Pythagoreans: otherwise the latter is meaningless. See also Geminus' evidence: "The hypothesis underlying the whole astronomy is that the sun, the moon and the five planets circulate at uniform speeds in the direction opposite to that of the heavenly sphere. The Pythagoreans were the first to approach such questions, and they assumed that the motions of the sun, the moon and the five planets are circular and uniform." (*Eisag.* I, 19). It is unlikely that Geminus should have projected onto the Pythagoreans what Eudemus ascribed to Eudoxus (Mittelstraß, *op. cit.*, 156f.).

[194] Mendell. The trouble with Eudoxus, 65.

cular movement of the sun and the moon not because he regarded it as the most appropriate for their divine nature. The decisive factors for him were, first, the uniform circular movement of the stars, which, unlike the circular movement of the planets, is an *empirically observed* fact, and second, the natural logic of the kinematic model, which does not allow a different conception of a continuous movement.

To the experts in astronomy, observations and mathematical arguments must have been of still greater importance. Euclid refers to them in his preface to the *Phaenomena:*

> Since the fixed stars are always seen to rise from the same place and to set at the same place, and those which rise at the same time are seen always to rise at the same time and those which set at the same time are seen always to set at the same time and these stars in their courses from rising to setting remain always at the same distances from one another, while this can only happen with objects moving with circular motion … we must assume that the (fixed) stars move circularly, and are fastened in one body, while the eye is equidistant from the circumferences of the poles … For all these reasons, the universe must be spherical in shape, and revolve uniformly about its axis … (transl. by T. Heath.)

While the notion of uniform circular movement emerged as a result of the observation of stars and was later applied to other heavenly bodies, the notion of the divine origin of the planets had a different source. Absent from traditional Greek religion[195] and not found in Ionian physics (cf. 59 A 79), it was attested, among the Pythagoreans, only in Alcmaeon.[196] Plato was the first to formulate it explicitly; he associated it with the spherical shape of the universe and the circular movement of heavenly bodies.[197] The Pythagorean saying calls the circle and the sphere 'the most beautiful';[198] with Plato they become 'the most perfect' and therefore inherent in heavenly bodies (*Tim.* 33b–34a). It is obvious that Plato projected this doctrine onto contemporary astronomy that brought to-

---

195 In the sixth and fifth centuries BC, the sun and moon did not belong to popular deities; they figured in very few myths and had no cults dedicated to them. They were, of course, regarded as deities, but not of a higher rank than, e.g., the god of wind or the goddess of dawn. Stars, let alone planets, were not regarded as deities; the very word 'planet', which means a vagrant star, was very far from suggesting a uniform circular movement. The planets did not have any divine names; these were borrowed from the Babylonians and first mentioned in Plato's *Timaeus* (38d) and later in the *Epinomis* (986e–987a); see Cumont, F. Les noms des planets et l'astrolatrie chez les Grecs, *ACl* 4 (1935) 5–43. Plato noted that many of the 'barbarians' believe in the divinity of heavenly bodies (*Crat.* 397c–d).

196 24 A 12.

197 Nilsson, M. P. *Geschichte der griechischen Religion*, 3rd ed., Vol. 1, Munich 1967, 839ff. Archytas, unlike Plato, considered circular movement to be characteristic of nature as a whole (φυσικὴ κίνησις, 47 A 23a), not of heavenly bodies alone (see above, 97).

198 D. L. VIII, 35. See Burkert. *L & S*, 168 n. 18, 169 n. 23, 171 n. 41.

gether heavenly bodies that were quite heterogeneous from the viewpoint of religion: the sun and the moon, on the one hand, the stars and the planets, on the other. Acquaintance with Pythagorean astronomy that ascribed uniform, circular movement to *all* the heavenly bodies could strengthen Plato's belief in the divine nature of the planets, but could hardly serve as its main source.

Motives that were significant for Plato could not have played any important role for Eudoxus, whose theory deliberately abstracted itself from the 'nature' of heavenly bodies, no matter whether physical or divine. Eudoxus did not discuss whether the true movement of heavenly bodies was circular: astronomy had found the answer to that long before. The problem was different: to find a mathematically correct model that would reduce the *apparent* irregularities in planetary motion to the *true* circular movement.

> Eudoxus of Cnidus, as Eudemus reports in the second book of his *History of Astronomy* and as Sosigenes repeats on the authority of Eudemus, is said to have been the first of the Greeks to deal with this type of hypotheses. For Plato, Sosigenes says, set this problem for students of astronomy: 'By the assumption of what uniform and ordered motions one can save the apparent motions of the planets'? (fr. 148).[199]

Plato's supposed role in posing this problem has been sufficiently discussed above (3.1). In Plato's dialogues the expression σῴζειν τὰ φαινόμενα, as well as the very idea that a mathematical model has to be verified by empirical observations, are totally lacking, though he was certainly not opposed to observations as such. In Aristotle, on the contrary, expressions similar but not identical in meaning (ἀποδιδόναι τὰ φαινόμενα, ὁμολογεῖν τοῖς φαινομένοις, ὁμολογούμενα λέγειν τοῖς φαινομένοις) occur quite often,[200] whereas the conviction underlying them belongs to the fundamentals of his philosophy. Eudemus, no doubt, also knew and shared the scientific principle expressed by the formula σῴζειν τὰ φαινόμενα. But does this really mean that this principle goes back to Aristotle and Eudemus, rather than to Eudoxus and Callippus?

Let us specify: what we are discussing here is not so much the general thesis, shared both by many Presocratics[201] and by Aristotle, that phenomena are the visible aspect of hidden things (ὄψις ἀδήλων τὰ φαινόμενα), but rather a

---

199 καὶ πρῶτος τῶν Ἑλλήνων Εὔδοξος ὁ Κνίδιος, ὡς Εὔδημός τε ἐν τῷ δευτέρῳ τῆς ἀστρολογικῆς ἱστορίας ἀπεμνημόνευσε καὶ Σωσιγένης παρὰ Εὐδήμου τοῦτο λαβών, ἅψασθαι λέγεται τῶν τοιούτων ὑποθέσεων, Πλάτωνος, ὥς φησι Σωσιγένης, πρόβλημα τοῦτο ποιησαμένου τοῖς περὶ ταῦτα ἐσπουδακόσι, τίνων ὑποτεθεισῶν ὁμαλῶν καὶ τεταγμένων κινήσεων διασωθῇ τὰ περὶ τὰς κινήσεις τῶν πλανωμένων φαινόμενα.

200 *APo* 89a 5; *Cael.* 306a 7, 309a 26; *GC* 325a 26; *GA* 760b 33; *Met.* 1073b 36; *EE* 1236a 26. Cf. 'inversed' formulas: βιάζεσθαι τὰ φαινόμενα, ἐναντία λέγειν πρὸς τὰ φαινόμενα (*Cael.* 315a 4; *EE* 1236b 22).

201 Anaxagoras (59 B 21 A), Democritus (68 A 111). See Regenbogen, O. Eine Forschungsmethode antiker Wissenschaft (1930), *Kleine Schriften*, Munich 1961, 141ff.; Diller, H. Ὄψις ἀδήλων τὰ φαινόμενα, *Hermes* 67 (1932) 14–42.

more specific theory that explained the celestial motions by reducing their apparent variety to a limited number of mathematical regularities. This astronomical theory was anticipated by the mechanics of Archytas, which reduced the action of various tools and devices to the principle of unequal concentric circles and provided a mathematical analysis of their movement, of linear and angular velocities in particular.[202] Another parallel to this theory is to be found in the mathematical harmonics of the same Archytas. It is worth noting that the Peripatetic Aristoxenus criticized the mathematical treatment of music by the Pythagoreans, reproaching them for their neglect of phenomena.[203] Theophrastus' criticism was similar (fr. 716 FHSG). Hence, the principle ὁμολογεῖν τοῖς φαινομένοις does not necessarily imply the *mathematical* treatment of phenomena; moreover, the very concept of φαινόμενα is much broader in Aristotle than in Eudoxus.

As a result, the formula σῴζειν τὰ φαινόμενα became associated, not with the explication of phenomena in general, but rather with the astronomical program, so it was very likely related to Eudoxus' theory from the very beginning.[204] The expression ἀποδιδόναι τὰ φαινόμενα appears twice in the very passage of the *Metaphysics* that discusses the modifications Callippus introduced into Eudoxus' theory.[205] Callippus himself must have explained the sense of these modifications to Aristotle and Eudemus.[206] Eudemus' *History of Astronomy* ascribes the origin of the principle of 'saving the phenomena' to Eudoxus (fr. 148) and cites this formula as pronounced by Callippus personally (fr. 149).[207] The author of the treatise *Phaenomena* must have been fully aware when he suggested the principle of 'saving the phenomena', which was soon to become the most important scientific principle of astronomy.[208] Admittedly, Eudoxus' *Phaenomena* dealt only with the fixed stars and was a descriptive rather than a mathematical treatise. But Euclid's *Phaenomena*, the next treatise bearing this title, is a mathematical work that demonstrates why the appearances produced by the motion of the celestial sphere are as they are: it thus remains in the tradition of Eudoxus' *Phaenomena* and *On Velocities*.

---

202 See above, 97, 253 n. 104.

203 *καὶ τούτων ἀποδείξεις πειρώμεθα λέγειν ὁμολογουμένας τοῖς φαινομένοις, οὐ καθάπερ οἱ ἔμπροσθεν, οἱ … ἀλλοτριολογοῦντες … νοητάς … κατασκευάζοντες αἰτίας …, πάντων ἀλλοτριωτάτους λόγους λέγοντες καὶ ἐναντιωτάτους τοῖς φαινομένοις* (*Elem. harm.* I, 41.17 f.). See above, 129 n. 46; Mittelstraß, *op. cit.*, 144 f.

204 Mittelstraß, *op. cit.*, 141 ff.

205 1073 b 36, 1074 a 1. It is generally believed that Aristotle inserted Λ 8 in the *Metaphysics* after 330.

206 See above, 233 n. 21.

207 Simplicius mentions several times the 'saving of phenomena' by Heraclides Ponticus, Eudoxus' contemporary: *In Cael. comm.*, 444.33 f. = Her. Pont. fr. 106 (Heraclides and Aristarchus), 519.10 = fr. 108; *In Phys.*, 292 = fr. 110 (from Geminus). Cf. above, 103 f. and Düring. *Aristoteles*, 150 f.

208 Düring. *Aristoteles*, 142 ff., 152 f.

It is revealing that the history of the formula σῴζειν τὰ φαινόμενα can be traced from Simplicius back nearly to the time of Eudoxus himself. It is often found in Sosigenes,[209] before him in Theon and Adrastus,[210] and still earlier in Posidonius and Geminus,[211] though the latter preferred similar expressions, like ἀποδιδόναι τὰ φαινόμενα and, especially, συμφωνεῖν τοῖς φαινομένοις.[212] Almost the same expressions are found in Hipparchus' commentary on the *Phaenomena* by Aratus and Eudoxus,[213] yet our formula occurs also in Hipparchus, as well as in his older contemporary Attalus, who also commented on Aratus.[214] Both of them seem to take the principle of 'saving the phenomena' for granted, which betrays the influence of the astronomy of Eudoxus, rather than of the *History of Astronomy* by Eudemus.[215] Plutarch's evidence on Aristarchus and Cleanthes (*De facie* 923 A) brings us still closer to the fourth century, in which Aristotle and later Eudemus testified to the birth of the Eudoxian astronomy. Indeed, Aristotle's remark in the *Prior Analytics* looks as if he were commenting not just on the development of Greek astronomy, but specifically on Eudoxus' astronomical career:

> It is the business of experience to give the principles which belong to each subject. I mean for example that astronomical experience supplies the principles of astronomical science: for once the phenomena were adequately apprehended, the demonstrations of astronomy were discovered (ληφθέντων γὰρ ἱκανῶς τῶν φαινομένων οὕτως εὑρέθησαν αἱ ἀστρολογικαὶ ἀποδείξεις).[216]

A detailed description of Eudoxus' system of homocentric circles based on his treatise *On Velocities* must have occupied the larger part of the second book of Eudemus' *History of Astronomy*. Simplicius cites it in a shortened, yet adequately extensive exposition by Sosigenes.[217] Only in two cases does Simplicius refer to Eudemus' text directly. Once, he points out a disagreement be-

---

209 See Simpl. *In Cael. comm.*, 488.23, 492.28. 30, 493.3f., 497.21, 499.15, 501.24 (ἀποδιδόναι τὰ φαινόμενα), 502.9, 504.18f., 505.18, 509.16, 510.31.

210 *Exp.*, 150.20, 166.5, 175.1, 175.15, 180.9, 198.14, etc.

211 Simpl. *In Phys.*, 292 = Her. Pont. fr. 110 = Posid. fr. 18 E–K.

212 *Eisag.*, 10.20, 118.26, 122.9–11. 23, 142.14. Cf. ἐπιμαρτυρεῖν τοῖς φαινομένοις (178.4).

213 συμφωνεῖν τοῖς φαινομένοις, διαφωνεῖν πρὸς τὰ φαινόμενα, συμφώνως ἀποδιδόναι τῷ φαινομένῳ, etc. (Hipparch. *In Arat.*, 4.9. 13, 24.10. 15, 34.9, 70.6, 106.6, 128.18–20, 138.21–22).

214 Hipparch. *In Arat.*, 176.10; Attalus, fr. 28 (*Comm. in Arat. reliquiae*, p. 23.37 Maass).

215 Cf. Mendell. The trouble with Eudoxus, 84ff.

216 Transl. by A. Jenkinson. On the relation of the observational data (τὰ φαινόμενα) to astronomical theory see also *APo* 78b 35–79a 6 and esp. *PA* 639b 6f., 640a 13f.

217 *In Cael. comm.*, 493.11–497.8; Schramm, *op. cit.*, 36ff. The preference Simplicius gave to Sosigenes' text is accounted for, in particular, by the fact that he used it as a source for a large section devoted to the discussion of different astronomical theories (492.31–510.23).

tween Sosigenes and Eudemus whether the program of 'saving the phenomena' belongs to Eudoxus or to Plato (fr. 148). In the second case, the explanation of reasons why Callippus introduced additional spheres, which interests Simplicius, happens to be absent from Sosigenes, who did not enter into a detailed description of these modifications because, in his view, they had not improved the ability of Eudoxus' theory to save the phenomena,[218]

> But Eudemus briefly stated what were the phenomena in explanation of which Callippus thought it necessary to assume the additional spheres. According to Eudemus, Callippus asserted that, assuming the periods between the solstices and equinoxes to differ to the extent that Euctemon and Meton held that they did, the three spheres in each case (i.e. for the sun and moon) are not sufficient to save the phenomena, in view of the irregularity which is observed in their motions.[219] But the reason why he added the one sphere which he added in the case of each of the three planets, Mars, Venus, and Mercury was shortly and clearly stated by Eudemus (fr. 149, transl. by T. Heath).

As for the rest of the material, Simplicius must have found Sosigenes' exposition perfectly adequate. It is worth noting, however, that Simplicius' understanding of Eudoxus' system seems more superficial than Eudemus'. That, at least, is the opinion shared by most of the scholars who, relying on Simplicius' text, have tried to reconstruct Eudoxus' system.

Attempts at such a reconstruction have been made since the beginning of the 19th century; an Italian astronomer G. Schiaparelli proved the most successful; his reconstruction (1875) was considered exemplary throughout the 20th century. Heath, however, was probably wrong to maintain that, in the absence of new evidence, the reconstruction of Schiaparelli "will no doubt be accepted by all future historians ... as the authoritative and final exposition" of the Eudoxian system.[220] Recently it has been criticized by two historians of Greek astronomy.[221] Leaving this discussion for the specialists, I still venture to hope that its results will not change the traditional opinion of Eudemus as a trustworthy and competent historian of early Greek astronomy and mathematics.

---

218 See above, 233; Schramm, *op. cit.*, 46. Cf. Knorr. Plato and Eudoxus, 320; Mendell. The trouble with Eudoxus, 114f.

219 In fact, Callippus' division of the year differs from that of Meton and Euctemon (see above, 244). Eudemus' report implies that Callippus' correction of the Eudoxian system is earlier than his division of the year (I owe this point to H. Mendell).

220 Heath. *Aristarchus*, 194. On Schiaparelli's predecessors see ibid., 194 n. 1–2.

221 Mendell, H. Reflections on Eudoxus, Callippus and their curves: hippopedes and callipopedes, *Centaurus* 40 (1998) 177–275; idem. The trouble with Eudoxus; Yavetz, I. On the homocentric spheres of Eudoxus, *AHES* 51 (1998) 221–278. No new evidence has appeared since the time of Schiaparelli. What has appeared is new computer programs, allowing us to model the astronomical phenomena easily. As Mendell points out, these programs became a major factor in the revision of Schiaparelli's reconstruction (The trouble with Eudoxus, 59).

Chapter 8

Historiography of science after Eudemus: a brief outline

1. The decline of the historiography of science

In Antiquity, Eudemus' history of science was an exception, yet there are no grounds to explain this by the unique character of its author. His works are part of one of the historiographical projects initiated by Aristotle. A retrospective view of the three Eudemian histories of exact sciences, his *History of Theology*, and Theophrastus' physical and Meno's medical doxography allows us to state more clearly the conclusion we have already formulated here: without Aristotle, the ancient historiography of science would hardly have been realized. This conclusion does not, of course, invalidate the other factors and causes considered in previous chapters. It only serves to emphasize the unique circumstances under which the ancient historiography of science was born.

First of all, in Antiquity, unlike the Arabic Middle Ages or the European Renaissance, the history of science arose not in the scientific milieu where it 'should' have arisen, but in the framework of a philosophical school, close as the latter stood to the science of the day. Second, it originated not in the course of restoring a disrupted scientific tradition,[1] but at the moment when Greek mathematics and astronomy, having laid their foundations, were soon to achieve their most glorious heights. The path Greek science had taken by the end of the fourth century was not long and complicated enough to suggest an 'objective' interest in the historiography of science on the part of scientists themselves. What they did need was the systematization and summarizing of the principal results achieved during the previous period. This is precisely the task accomplished by Euclid in his *Elements*, *Phaenomena*, *Sectio canonis*, and *Optics*, which eclipsed all the similar writings of his forerunners. A parallel process was at work in the natural sciences: Aristotle's and Theophrastus' summarizing works on physics, zoology, and botany remained generally unsurpassed in Antiquity.

Peripatetic historiography on the whole, of which Eudemus' treatises on the history of science were an integral part, can also be regarded as the *historical systematization* of the achievements of the Greek culture, which many of the fourth-century authors considered to be nearing its perfection. It is from this standpoint that Aristotle and his pupils wrote historical surveys of theoretical (physics, mathematics, theology) and practical (music, poetry, rhetoric) sciences in progress, listing glorious discoveries and names that represented the

---

[1] See above, 3f.

rise of these sciences from their first beginnings to the latest spectacular attainments. In the physical and medical doxography, Theophrastus and Meno went still further, presenting a *systematic overview* of what had been accomplished in the preceding epoch. Dicaearchus in the *Life of Hellas* considered the development of culture as a whole, from its primeval state to his own time, noting not only economic progress, but moral decay as well.

Thus, we can say that the history of science emerged not when it became necessary, but when it became possible. This circumstance obliges us to do justice to Aristotle and his pupils, whose interest in the development of knowledge allowed the history of science to appear long before it became needful for scientists. Hence the third particular feature of the Peripatetic history of science: the scientific community's lack of professional interest predetermined its subsequent fate in Antiquity. Unlike the other historiographical genres emerging in the Lyceum, such as biography, doxography, and the history of culture, the history of science received almost no continuation, and especially not in the professional community. Biography, which did not avoid scandalous details, addressed a wide educated audience; the histories of philosophy and medicine later found the new forms answering the intellectual needs of the followers of Hellenistic philosophical and medical schools. The history of culture also had its own successors.[2] The history of science, even if it was interesting for Greek mathematicians, did not engage them enough to take up Eudemus' work.

It is revealing that Eratosthenes, probably the only Hellenistic scientist who shared to a certain extent Eudemus' interests, applied the latter's historico-scientific approach to geography, not astronomy or mathematics. At the very beginning of his *Geography* Eratosthenes opposed the tendency, popular in Hellenism and especially among the Stoics, to derive all knowledge, including geography, from Homer,[3] which made the boundary Aristotle drew between science and myth a relative one. Though a poet himself, Eratosthenes approached geography as a scientist and started it, therefore, not with Homer, as Strabo would later insist (I,1.2), but with Anaximander and Hecataeus of Miletus. Anaximander was the first to draw the map of the earth, while Hecataeus was the first to write a prose work on geography and also drew a geographical map (fr. I B 5 Berger). Among Eratosthenes' predecessors, Strabo (I,1.1) also names Democritus, Eudoxus, Dicaearchus, and Ephorus, while Agathemerus (*Geogr.* 1, 1) mentions Hellanicus of Lesbos, Damastes of Sygeum, Democritus, Eudoxus, and Dicaearchus. Since all these names are listed in chronological order and mentioned in Eratosthenes' fragments, his list of persons who contributed to geography must have been as follows: Anaximander, Hecataeus,

2 On the influence of Dicaearchus' *Life of Greece*, see Ax, W. Dikaiarchs *Bios Hellados* und Varros *De vita populi Romani*, *Dicaearchus of Messana*, 279–310.

3 Berger, H. *Die geographischen Fragmente des Eratosthenes*, Leipzig 1880, 19ff.; Geus, *op. cit.*, 264ff. Cf. Strab. I,1.10; 2, 15.

Hellanicus, Damastes, Democritus, Eudoxus, Dicaearchus. A new epoch in geography starts with the campaigns of Alexander, which opened for the Greeks vast regions of Asia and Northern Europe (fr. I B 10–11 Berger). We deal here, obviously, with the fragments of a historical overview of the development of geography from its first discoverers until the third century.

Joining praise with criticism, Eratosthenes did his best to point out his forerunners' individual contributions to geography, their εὑρήματα. In geography, however, unlike mathematics, incontestable discoveries exist side by side with inaccurate, even erroneous notions:

> The ancients considered the oecumene round in shape; in its middle lay Greece and in the middle of Greece Delphi, because it holds the navel of the earth. Democritus, a very learned man, was the first to understand that the earth (*sc.* oecumene) is oblong and the ratio between its length and width is 3:2. In this he was followed by the Peripatetic Dicaearchus. Eudoxus, on the contrary, believed the ratio of length to width to be 2:1, and Eratosthenes more than 2:1 (Agathem. *Geogr.* 1, 2 = fr. II C 1 Berger).

This doxographical overview, which in Agathemerus immediately follows the list of geographers who were Eratosthenes' predecessors, goes back to Posidonius (fr. 200a E.-K.), and, through the latter, probably to Eratosthenes. It is not clear whether it was part of the historical introduction to *Geography*, since it could belong to book II as well.[4] It is obvious in any case that, in a scientific treatise on geography, Eratosthenes could not consistently select the material on the same principles as Eudemus did in the *History of Astronomy*. While Eudemus focused his attention on the major discoveries exemplifying the progress of science, Eratosthenes combined this method with the doxographical one, which allowed him to mention all the opinions relevant to a given problem and to criticize those he disagreed with.

Unlike Eratosthenes' historical introduction to *Geography*, his dialogue *Platonicus* derives directly from the material of Eudemus' *History of Geometry* (3.1). It does not seem, however, to develop Eudemus' work in either content or form. Selecting from the whole of the previous history of geometry the single problem of doubling the cube (interesting for him since he took part in solving it), Eratosthenes invents a fictitious story for it, based on the Academic legend of Plato as the architect of science. Eratosthenes is unlikely to have done any research in the sources: the solutions of Archytas, Eudoxus, and Menaechmus he quotes are already found in Eudemus. Neither in the *Platonicus*, where the subject imposed certain limitations on him, nor in his letter to the king Ptolemy III does Eratosthenes mention any of the other solutions, that of Philo of Byzantium for example, which, in principle, he could have known.[5] *Platonicus* presents the history of mathematics in the dramatic form of a popular philosophical dialogue with mathematical content, in which its main hero, Plato, is the

4 Berger, *op. cit.*, 142f.

5 Knorr. *TS*, 144f.

mouthpiece of its author's ideas. In spite of the popularity *Platonicus* enjoyed in Antiquity,[6] the genre itself does not seem to have had any direct followers, though it may have exercised a certain influence on the mathematical anthologies of late Antiquity, which contained the solutions of such famous problems as those of squaring the circle, doubling the cube, etc. In fact, *Platonicus*, but for its dialogical form, could be regarded as the earliest known example of such an anthology. Yet even here Eratosthenes must have followed the historiographical tradition of the Lyceum, represented, for example, by Aristotle's work on a still more famous scientific problem – that of the Nile's floods.[7]

Unlike the mathematics of the Imperial age, whose self-awareness increasingly depended on its awareness of its distant past, the exact sciences of the Hellenistic epoch had little reason to turn to the pre-Euclidean period. This lack of historical sensibility can partly be accounted for by the very style of mathematical and astronomical treatises formed and generally adopted by the fourth century: formal and utterly impersonal,[8] it left no place for historiographical references or outlines. The surviving works by Euclid and Autolycus include no names; the same is true of the Aristotelian *Mechanics* and Aristarchus of Samos' *On the Sizes of the Sun and the Moon and on Distances to Them.* The mathematical treatises of the second half of the third century often opened with a brief introduction written in the form of a letter to a fellow scientist. In the individual introductions to the books of his *Conics*, Apollonius cites several names of his contemporaries and predecessors,[9] failing, however, to mention Menaechmus and Aristeas the Elder, who laid the foundations of the theory of conic sections. Archimedes' introductions to his works, though more detailed, treat his predecessors (Democritus, Eudoxus, Aristarchus) in the context of analysis focused on individual problems – an approach typical of scientific literature up to the present day, not of historiography. The introduction to the work by Diocles, a contemporary of Apollonius, also deals with a particular problem: the author suggests using conic sections to solve the two tasks formulated by Pythion of Thasos in his letter to Conon of Samos and by Zenodorus during his visit to Diocles himself.[10]

In Hypsicles' *Anaphoricus* the introduction and, consequently, the names are absent, while the introduction to his work on geometry, which later constituted book XIV of Euclid's *Elements*, is very similar to Diocles' one. Accord-

---

6 Geuss, *op. cit.*, 175 n. 155.

7 See above, 143.

8 Asper, M. Mathematik, Milieu, Text. Die frühgriechische(n) Mathematik(en) und ihr Umfeld, *Sudhoffs Archiv* 87 (2003) 1–31.

9 Eudemus, Naucrates, Euclid (I); Philonides (II); Attalus, Thrasydeus, Conon, Nicoteles (IV). Reference is rarely made to particular theories and discoveries.

10 Diocles. *On burning mirrors*, 3–7, 15. Further on, the text makes two mentions of Archimedes (136; 149). Menaechmus, Aristeas the Elder, and Euclid, as well as Diocles' numerous forerunners in the solution of the problem of doubling the cube, are not named at all (186–207).

ing to Hypsicles, his father and Basilides of Tyre who visited the father in Alexandria, pointed to a mistake in Apollonius' work on regular polyhedra, after which Hypsicles himself found the correct proof in the second edition of Apollonius' work and decided to develop it. We learn hardly anything more from Hipparchus' commentary to Eudoxus' and Aratus' *Phaenomena*. In his introduction, Hipparchus sets himself the task of correcting the mistakes of Eudoxus and his follower Aratus that were left unnoticed even by Attalus, the best of the earlier commentators; apart from these three, only Philip of Opus and Pytheas are briefly mentioned in the commentary (28.3, 30.8). The three astronomical treatises by Theodosius of Bithynia (late second century), including his *Sphaerics*, a textbook compiled of material partly going back to the fourth century, contain practically no references.[11]

Taken as a whole, the surviving texts of Hellenistic mathematicians and astronomers demonstrate that they could not care less about presenting their predecessors' contribution in a historical perspective. What interested them as a rule was their precursors' errors and failures, rather than the problems the latter had succeeded in solving. Even such summarizing works as Euclid's *Elements* or Apollonius' *Conics* aimed primarily and, in fact, solely at the systematic arrangement of vast, often heterogeneous material and at giving it an impeccable mathematical form.[12] Although for scientists, particularly those who work in the productive period of their science's evolution, each of these aims seems natural, both are diametrically opposed to the perspective from which Eudemus' history of science was written. And there lies one of the reasons why the historical view of science appeared to be unclaimed by Greek scientists themselves.

Another factor was, probably, that of narrow specialization. Mathematicians and astronomers of the Hellenistic epoch did not, as a rule, venture out of their domain. One of the few exceptions is the historian and geographer Eratosthenes, who is particularly notable for his interest in the history of science. As a result, at least in the surviving texts, that second-order discourse is lacking that is characteristic, for instance, of Arabic culture, in which scientists were more versatile (even to the detriment of originality) and the history of science served as an introduction to the discipline and included reflections on science's methods, its present, past, and future, and its relationship with power, society, and the science of other peoples. Unlike the innumerable references to the mythical 'Chaldaeans', the properly *historical* information on Babylonian astronomy borrowed by the Greeks is insignificant and does not bear comparison

---

[11] Heath. *History* 2, 246ff.; Bulmer-Thomas, I. Theodosius of Bithynia, *DSB* 13 (1976) 319–321. Only his *On Days and Nights* mentions Meton and Euctemon, as well as Euclid's *Phaenomena* (II, 10, 18).

[12] Although Apollonius does not generally conceal his debt to his predecessors, Pappus (*Coll.* VII, 676.25ff.) reproaches him with boastfully attributing other people's achievements to himself, contrasting him, strangely enough, with the just and noble Euclid, who, as far as we know, never referred to anyone at all.

with the well-documented history of the Arabs' reception of Greek astronomy. The authors of the Imperial age, who are closer to philosophy, gave some attention to this subject, but the level, character, and even the volume of this material is negligible compared to that found in Arabic sources, even from the point of view of a person who, like myself, has only second-hand knowledge of it.

Unlike mathematicians and astronomers, the Greek engineers of the third century, Ctesibius, Philo of Byzantium, and Biton, the authors of treatises on siege and shooting engines, show a much greater interest in the history of their discipline.[13] Thus, Ctesibius' *Belopoeica*, which has survived in Hero's retelling,

> describes thoroughly and in detail the construction of the earliest non-torsion arrow-shooting engine, the *gastraphetes*, and then writes a unique and important constructional theory of torsion catapults from the first primitive design to advanced machines built in accordance with the formulae for calibration. Thus he covers technical developments with regard to torsion engines in the period ca. 350–270 B.C.[14]

Though the treatise does not mention any names (they might have been present in Ctesibius, but are left out in Hero), on the whole his preface and his text testify to the author's intention to present the research and experiments of his forerunners in progress, rather than as a sum of finished achievements. The same approach can be traced in the textbook on artillery written by Philo, Ctesibius' follower:

> In old days, some engineers were on the way to discovering that the fundamental basis and unit of measure for the construction of engines was the diameter of the hole. This had to be obtained not by chance or at random, but by a standard method which could produce proportion at all sizes ... The old engineers, of course, did not reach a conclusion, as I say, nor did they determine the size, since their experience was not based on a sound practical foundation. But they did decide what to look for. Later engineers drew conclusions from former mistakes, looked exclusively for a standard factor with subsequent experiments at a guide, and introduced the basic principle of construction, namely the diameter of the circle that holds the spring. Alexandrian craftsmen achieved this first, being heavily subsidized because they had ambitious kings who fostered craftsmanship (*Belop.*, 106f.).

Biton's book is more technical and descriptive, but even he does not fail to name the inventors of each of the six mechanisms he is writing about. He even reports where each of them was invented, so that we learn, for example, that

[13] This feature has already been noted (Cuomo. *Pappus*, 95f.; eadem. The machine and the city: Hero of Alexandria's Belopoeica, *Science and mathematics*, 165f.), without, however, being adequately explained. It might, to a certain extent, be related to the style of these treatises, which is rather discursive and free from the strict limits imposed on the mathematical works by the axiomatico-deductive method of presenting the results.

[14] Marsden, E.W. *Greek and Roman artillery. Technical treatises*, Oxford 1971, 1.

Zopyrus of Tarentum built the middle gastraphetes in Miletus and the mountain gastraphetes in Cumae.[15] That the names of engineer-inventors were widely known in the Hellenistic epoch is attested by *Laterculi Alexandrini*, the second-century papyrus from Egypt that was part of a school library. Along with the names of the highest mountains and the longest rivers, etc., it contains the names of famous lawgivers, sculptors, and architects. Under a special rubric come the names of the seven famous engineers (μηχανικοί) of the classical and Hellenistic epochs, accompanied by brief indications of their discoveries.[16] This was the information each schoolboy was supposed to learn.

Similar lists were also compiled on the basis of excerpts from Eudemus' works.[17] Yet we do not know of any other lists of the *prōtoi heuretai* in the exact sciences, while those that go back to Eudemus contain practically no new names.[18] It is also revealing that the late authors who pass these lists on (Dercyllides, Theon of Smyrna, Anatolius, Porphyry, Proclus) were not original scientists, but philosophers, compilers, and commentators – as were, in fact, most of those in whom we find Eudemus' quotations.[19] Does this mean that the mathematicians and astronomers of Antiquity failed to accumulate a 'critical mass' indispensable for the existence of an established genre, which for philosophical biography and doxography was provided by the numerous adepts of philosophical schools?

To answer this and related questions on the causes of the decline of the historiography of science it is important to appreciate the real scale of ancient science. R. Netz estimates the number of the Greek mathematicians known to us by name at 144 persons and believes that the whole number of mathematicians active in Antiquity did not exceed one thousand.[20] The contrast with the modern picture of mass science is striking: the number of specialists in exact sciences who are active in St. Petersburg now is almost as large as that of all the ancient mathematicians and astronomers. It seems that, in Antiquity, to maintain a discipline alive, it sufficed if every century a few persons practiced it seriously, i.e., achieved new results, the others serving only to pass the new knowledge on. The development of a discipline, not necessarily a mathematical one, often appeared to be suddenly interrupted, as was the case with Aristotle's zool-

---

15 Biton, 61f., 65 = p. 74, 76 Marsden.

16 Diels, H. *Laterculi Alexandrini: aus einem Papyrus ptolemäischer Zeit*, Berlin, 1904, 8–9.

17 Theon. *Exp.*, 198.14f.; Procl. *In Eucl.*, 64ff.; Ps.-Heron. *Def.*, 108.10–25, 166.23–168.12.

18 Not a single name was added to the list of astronomers, while that of geometers was augmented by Euclid alone. After Euclid, Proclus (*In Eucl.*, 68) mentions Archimedes and Eratosthenes, but only to establish Euclid's chronology. He makes no mention of their discoveries in mathematics.

19 See above, 236.

20 Netz, R. Greek mathematics: A group picture, *Science and mathematics*, 196–216.

ogy, Theophrastus' botany,[21] Archimedes' hydrostatics later, and Ptolemy's mathematical astronomy and geography and Diophantus' 'algebra' still later: the whole of Antiquity brought to these subjects hardly a single scientist worthy of their originators.[22] In this respect, the fate of the history of science does not seem to be unique; on the contrary, its origin, development, and decline fully conform with the general regularities that become manifest as we approach Greek science in its entirety.

Since the small number of mathematicians active before Eudemus (who, as we remember, cited the names of 20 geometers) did not preclude the emergence of the history of science, the small number of Greek mathematicians in general can hardly account for the lack of followers of Eudemus. We will adduce for comparison one more figure found in Netz: the number of pagan (non-Christian) philosophers of Antiquity known by name amounts to 316.[23] Even supposing the ratio between mathematicians and philosophers to have been one to three or four, rather than one to two, the sudden break in the development of the history of science, in contrast to the flourishing philosophical doxography and biography, remains unexplained. A more important factor seems to have been the degree to which philosophy and medicine, on the one hand, and mathematics, on the other, were institutionalized. Philosophy and medicine generally existed, from their very origin and to the end of Antiquity, within the framework of schools, while in mathematics schools were an exception.[24] Pythagorean mathematics developed within the framework of a philosophical school that stemmed, in turn, from a political society, *hetaireia*.[25] In the chain of generation from Pythagoras to Hippasus, Theodorus, and Archytas, only one link, that between Hippasus and Theodorus, is missing. The further progress in mathematics is associated with Theodorus' student Theaetetus and, in particular, with Archytas' student Eudoxus, the founder of the first Greek mathematical and astronomical school in Cyzicus.[26] At the very end of the fourth century, Eudoxus' followers were challenged by Epicurus, who succeeded in winning some of them, Polyaenus of Lampsacus in particular, to his side.[27] Hence, this school, starting with Eudoxus (born ca. 390), had lasted for at least three gen-

---

21 Lennox, J. G. The disappearance of Aristotle's biology: A Hellenistic mystery, *Apeiron* 27 (1994) 7–24.

22 Hydrostatics was pursued by Menelaus, whose work has not survived.

23 Netz. Greek mathematics, 205, with reference to Runia, D. Aristotle and Theophrastus conjoined in the writings of Cicero, *Cicero's knowledge of the Peripatos*, ed. by W.W. Fortenbaugh, P. Steinmetz, New Brunswick, 1989 (Rutgers University Studies in Classical Humanities, Vol. 4), 23–38.

24 On the poor institualization of Greek mathematics, see Asper, *op. cit.*, 17 ff.

25 Zhmud. *Wissenschaft*, 78 ff.

26 Among the members of the school were Menaechmus, Dinostratus, Callippus, Polemarchus, Athenaeus, Helicon, Amyclas, Theudius, and Hermotimus. See above, 98 f.

27 Sedley. Epicurus, 23 f. On Polyaenus, see below, 287.

erations. After the beginning of the third century, we do not hear of it anymore, nor of other mathematical schools that lasted for such a long time and were so large in number. "Mathematics during the Hellenistic period – as Toomer justly remarked – was pursued not in 'schools' established in 'cultural centres', but by individuals all over the Greek world, who were in lively contacts with each other both by correspondence and in their travels."[28]

For the Lyceum, Eudoxus and his school represented science as such; it is from here that Eudemus drew information about new scientific theories and, still more important, the criteria for what is scientific and what is not.[29] The absence of such schools in the later period had a negative effect on the relations between philosophy and mathematics, as well as on the history of science, particularly because activities of the philosophical and medical schools were recorded in a rich historiographical literature. In the Hellenistic period there appear two special historiographical genres related to philosophical schools: the successions of philosophers (Διαδοχαί) with the main focus on biography, and the literature on philosophical schools (Περὶ αἱρέσεων),[30] which lay closer to doxography. Both these genres are found in medical historiography as well. Dated to the period from 50 BC to 50 AD are Ischomachus' Περὶ τῆς Ἱπποκράτους αἱρέσεως (*FGrHist* 1058 F 1) and the three works with the same title of Περὶ τῆς Ἡροφίλου αἱρέσεως.[31] A renowned physician, Soranus of Ephesus (first half of the second century AD), wrote *Successions of Physicians* (Διαδοχαὶ ἰατρῶν), from which we have Hippocrates' biography (*FGrHist* 1062 F 1–4). On the whole, the ancient historiography of medicine is represented by quite a number of works, and the introduction to Celsus' *On Medicine*, based on Hellenistic models, demonstrates the interest of contemporary physicians in the general history of medicine as well.[32]

Since the Renaissance, the history of science, philosophy, and medicine has generally been pursued by specialists who taught at the universities or had, at least, studied in them. In Antiquity the functions of the universities were partly performed by philosophical and medical schools, which served as higher education and research centers. In the exact sciences such centers were, to all evi-

---

28 Toomer. *Diocles. On burning mirrors*, 2. Netz. Greek mathematics, 215f., also stresses the absence of stable mathematical schools, but overlooks evidence on Pythagoras' and Eudoxus' schools.

29 See above, 250f.

30 They stem, respectively, from Sotion and Hippobotus: Sotion. *Die Schule des Aristoteles*, Suppl. II, ed. by F. Wehrli, Basel 1978; Gigante, M. Frammenti di Ippoboto, *Omaggio a Piero Treves*, Padua 1984, 151–193; Giannattasio Andria, R. *I frammenti delle "Successioni dei filosofi"*, Naples 1989.

31 Staden, H. von. Rupture and continuity: Hellenistic reflections on the history of medicine, *AHM*, 143–187 (one of them, written by Apollonius Mys, included almost thirty books).

32 Mudry, P. *La préface du De Medicina de Celse*, Rome 1982; Staden, H. von. Celsus as historian?, *AHM*, 251–294.

dence, absent. We know little about the secondary school in the Hellenistic epoch, but the tradition of teaching courses in mathematics privately or, in some cases, in public gymnasia as well, does not seem to have been extinct,[33] while these sciences themselves formed part of the pedagogical ideal of the time, ἐγκύκλιος παιδεία.[34] But higher education and, accordingly, the attitude toward professional scientific research were usually imparted to young people in rhetorical and philosophical schools. These schools seldom regarded mathematics with Isocrates' indulgence, let alone with Plato's enthusiasm, and hardly encouraged their students to work further in science. The decline of interest in science becomes manifest in the Hellenistic philosophical milieu; I regard this as one of the reasons why the history of science as a genre had a fate different from that of doxography or biography.

The first histories of the exact sciences were written by a Peripatetic philosopher and addressed (primarily, at least) to his philosopher colleagues, rather than scientists. Those who read and used these histories in the Imperial age were, for the most part, philosophers as well. It follows that the reasons for the lack of immediate successors to Eudemus are to be sought in the changes of philosophical climate, which were indicative of still deeper processes in the culture as a whole. The spectacular achievements of the Greek scientists of the third and second centuries in mathematics, astronomy, mechanics, physics, geography, physiology, and anatomy do not seem to show that general interest in science was on the decline, particularly when we take into account the appearance of popular scientific literature addressed to the educated public. Nevertheless, in the leading philosophical schools of Hellenism, oriented toward values extrinsic or even opposed to scientific knowledge, the positive attitude toward mathematics characteristic of the first-generation Academics and Peripatetics was radically abandoned. Neither of these schools fully supported the cognitive ideals of the classical period; their attitude toward mathematics and astronomy was more or less indifferent, sceptical, and even hostile.[35] The

33 Nilsson, M. *Die hellenistische Schule*, Munich 1955, 16, 52; Morgan, E. *Literate education in the Hellenistic and Roman worlds*, Cambridge 1998, 6 n. 14, 33ff.; Cribiore, R. *Gymnastics of the mind: Greek education in Hellenistic and Roman Egypt*, Princeton 2001, 41f., 180f.

34 Marrou, H.-I. *A history of education in Antiquity*, Madison, Wis. 1982; Kühnert, F. *Allgemeinbildung und Fachbildung in der Antike*, Berlin 1961 (the best collection of literary evidence); Fuchs, *op. cit.*

35 Vlastos, G. Zeno of Sidon as a critic of Euclid, *The Classical tradition*, ed. by L. Wallach, Cornell 1966, 148–159; Sedley. Epicurus; Mueller, I. Geometry and scepticism, *Science and speculation. Studies in Hellenistic theory and practice*, ed. by J. Barnes et al., Cambridge 1982, 69–95; Angeli, A., Colaizzo, M. I frammenti di Zenone Sidonio, *CErc* 9 (1979) 47–133; Romeo, C. Demetrio Lacone sulla grandezza del sole, *CErc* 9 (1979) 11–35; Angeli, A., Dorandi, T. Il pensiero matematico di Demetrio Lacone, *CErc* 17 (1987) 89–103; Barnes, J. The size of the sun in Antiquity, *ACD* 25 (1989) 29–41; Erler, M. Epikur, *Die Philosophie der Antike*, Vol. 4,

decline in the interest in mathematics in the Academy after Xenocrates is no less manifest[36] than the wane of scientific interest in the Lyceum after Strato, who never pursued the exact sciences personally. The Cynics', Cyrenaics', and Sceptics' negative attitude toward science is generally known.[37] Epicurus denied mathematics, mathematical astronomy, and the entire ἐγκύκλιος παιδεία; the mathematician Polyaenus of Lampsacus, whom Epicurus converted to Epicureanism, wrote a special treatise on the fallacy of geometry as a whole.[38] The Epicureans continued their polemic against geometry, though two of them, Philonides, a friend of Apollonius of Perga, and Basilides of Tyre seem to have succeeded in combining their Epicureanism with mathematical studies.[39]

The attitude toward theoretical sciences among the Stoics was more complicated.[40] Zeno in his early *Republic* declared ἐγκύκλιος παιδεία to be useless (D. L. VII, 32), probably under Cynic influence. That he changed his attitude toward the exact sciences later is plausible, though not evident.[41] Cleanthes accused Aristarchus of Samos of impiety because of the latter's heliocentric hypothesis.[42] Chrysippus expressed a more favorable opinion of ἐγκύκλιος παιδεία; he compared some theorems of geometry with Platonic Forms; his view of geometrical solids and figures contradicts mathematics;[43] all the elementary astronomical data assigned to him had already been included in physics in Plato's and Aristotle's time.[44] Judging by this scanty evidence, during the first two centuries, *mathēmata* were irrelevant for the Stoics; what they wrote on this subject was often at variance with the views of professionals. Unlike Plato

169f.; Cambiano, G. Philosophy, science and medicine, *The Cambridge history of Hellenistic philosophy*, ed. by K. Algra et al., Cambridge 1999, 585–613.

36 Still, Arcesilaus, the scholarch of the Academy in the mid-third century, studied with the mathematicians Autolycus and Hipponicus (D. L. IV, 29–32).

37 Kühnert, *op. cit.*, 99f.

38 Cic. *Acad. pr.* II, 106; *De fin.* I, 20, 71–72; cf. D. L. X, 6; Epic. fr. 117, 163, 227, 229 Usener.

39 Erler, M. Philonides; Basilides und Thespis, *Die Philosophie der Antike*, Vol. 4, 251f., 280. Apollonius mentions Philonides (*Conic.* II, praef.) and Hypsicles mentions Basilides in the introduction to book XIV of the *Elements*. To be sure, as Cambiano remarks (Philosophy, 589), "it is not possible to determine whether, at the time Apollonius and Hypsicles mention them, they had already become attached to Epicureanism".

40 See e.g. Jones, A. The Stoics and the astronomical sciences, *The Cambridge companion to the Stoics*, ed. by B. Inwood, Cambridge 2003, 328–344.

41 Kühnert, *op. cit.*, 76f. Zeno's circle included Aratus, a popularizer of Eudoxus, on whose poem the Stoic Boethus of Sidon wrote a commentary (Gemin. *Eisag.* XVII, 48).

42 Plut. *De facie* 923 A; *Quaest. conv.* 1006 C.

43 D. L. VII, 129 = *SVF* III, 739; Procl. *In Eucl.*, 395.13f., Plut. *Mor.* 1079 D, E – 1080 A = fr. 458, 460 Hülser. Cf. also *SVF* II, 365, 482, 489.

44 *SVF* II, 527, 580, 625, 648–692. The same concerns Zeno (*SVF* I, 119–120).

and Aristotle, the Stoics neither strove to put *mathēmata* at the service of their on the whole practically-oriented philosophy, nor regarded them as a model for imitation.[45] They rather saw such a model in τέχνη, which, according to Zeno's definition, included only knowledge useful for life (fr. 392–397 Hülser).

The adoption of new cognitive ideals transformed the cognitive space as well. The Aristotelian unified field of theoretical sciences that included μαθηματική along with φυσική and θεολογική (4.2) was now out of the question. Mathematics loses the autonomy it had in the eyes of Aristotle and Eudemus and turns into philosophy's maidservant:

> But certainly everybody knows that philosophy gave to all individual sciences the principles and the seeds from which then apparently their theorems arose. For although equilateral and non-equilateral triangles, circles and polygons as well as the other figures were additionally discovered by geometry, geometry did not discover the nature of the point, the line, the surface and the body, which are namely the roots and cornerstones of the mentioned figures ... These definitions are left for philosophy, as the whole topic of the definitions is incumbent upon a philosopher.[46]

If the point of view exposed here by Philo of Alexandria is indeed that of the Stoics, the latter must have been completely at variance both with the position of the Lyceum and the views of the mathematicians themselves.[47] Following his teacher, Eudemus admitted that mathematics had principles of its own on which its entire edifice rested; leaving to metaphysics the task of *investigating* some of them was but a natural division of labor.[48] The idea of somebody looking for an alternative definition of line seemed to him ridiculous.[49] The Stoics did not find it ridiculous in the least.[50]

The first Stoic to pursue science seriously was Posidonius; later he was also considered to have been the best expert in *mathēmata* among the Stoics.[51] This

---

45 According to Stoic dogma, ἐγκύκλια μαθήματα belong not to ἀγαθά, but to ἀδιάφορα (*SVF* III, 136). Cf. Mansfeld, J. Intuitionism and formalism: Zeno's definition of geometry in a fragment of L. Calventius Taurus, *Phronesis* 28 (1983) 59–74.

46 Phil. Alex. *De congr.* 146–147 = fr. 416 Hülser; Cambiano. Philosophy, 592f. – Philosophy is the mistress (δέσποινα) of ἐγκύκλια μαθήματα (Clem. Alex. *Strom.* I, 5, 30 = fr. 6 Hülser). This metaphor is found already in Aristippus of Cyrene (D. L. II, 79) and the early Stoic Aristo of Chios (*SVF* I, 350).

47 See above, 118, 168.

48 Arist. *Met.* 1005a 19–29, 1025b 4f., 1061b 19–21; Eud. fr. 32 and esp. 34.

49 "For mathematicians display their own principles and give its definition to every thing they talk about, so that a person who does not know all this would look ridiculous if he tried to investigate what a line is and every other mathematical object." (fr. 34).

50 See above, 287 n. 43 and, in particular, Posid. fr. 195–199 E.-K. with commentary; for Stoic definitions of line and figure, see also Simpl. *In Arist. Cat.*, 264.33ff. = fr. 459 Hülser; D. L. VII, 135.

51 T 83–84 E.-K; Bréhier, E. Posidonius d'Apamée, théoricien de la géométrie, *Etudes de philosophie antique*, Paris 1955, 117–130; Kouremenos, T. Posidonius and Gemi-

fully fits with the obvious influence exercised on him by Plato and Aristotle. In his works, Posidonius treated the problems of astronomy and mathematical geography at great length and wrote a special treatise to defend Euclid's geometry against the attacks of the Sceptical Academy and the Epicureans, Zeno of Sidon in particular (fr. 46–47 E.-K.). Despite Posidonius' genuine interest in science, his view of its relations with philosophy remains typically Stoic: *mathēmata* are but auxiliary means at philosophy's service.[52] Proceeding from Aristotle's passage on the difference between physics and mathematical astronomy,[53] Posidonius shifts the accents radically. Physics establishes the basic principles and deduces from them its propositions on the size, shape, and order of heavenly bodies; astronomy borrows from physics these principles and investigates the same problems proceeding from observations. Physics explains the causes, while astronomy is a descriptive discipline; its various hypotheses attempt to 'save the phenomena' without providing the true explanation of their causes. Among the principles that the astronomer is obliged to borrow from philosophy is the proposition that the movement of heavenly bodies is simple, regular, and ordered. Proceeding from this, the scientist proves this movement to be the circular one.[54]

While Aristotle considered *mathēmata* the most exact of sciences and advised listening to mathematicians,[55] Posidonius believed that the proper func-

---

nus on the foundations of mathematics, *Hermes* 122 (1994) 437–450. Of all the Stoics only two Posidonians – Geminus of Rhodes and Diodorus of Alexandria – wrote on mathematics and astronomy: Steinmetz, P. Die Stoa, *Die Philosophie der Antike*, Vol. 4, 710f. (On Posidonius' later follower Cleomedes, see below, 292 n. 73). To be sure, whether and to what extent they were Stoic philosophers remains unclear. Diodorus, e.g., is regularly called *mathēmatikos* (Achil. *Isag.*, 30.20, 41.17 Maass; Pasquali, *op. cit.*, 196.26 and below, 290 n. 56). Rather, they appear as mathematicians influenced by Posidonius, in particular by his meteorology. Both wrote on this subject, admittedly, not a usual one for a mathematician, but their purely philosophical works are unknown.

52 Fr. 90 E.-K.; Kidd, I. G. Philosophy and science in Posidonius, *A & A* 24 (1978) 7–15. Posidonius' theory of the origin of culture (fr. 284 E.-K.) maintains that many if not all the practical arts were discovered by 'wise men', a kind of primitive philosophers. Cf. Seneca's objections to making philosophy responsible for technical progress, too (*Ep.* 90, 5f.).

53 *Phys.* 193b 22f.; see above, 132 n. 59.

54 ληπτέον δὲ αὐτῷ ἀρχὰς παρὰ τοῦ φυσικοῦ, ἁπλᾶς εἶναι καὶ ὁμαλὰς καὶ τεταγμένας κινήσεις τῶν ἄστρων, δι' ὧν ἀποδείξει ἐγκύκλιον οὖσαν τὴν χορείαν ἁπάντων … (fr. 18 E.-K.).

55 *Cael.* 306a 27 and above, 250. According to Ptolemy, only mathematics provides firm and reliable knowledge (see above, 118). This view was shared by Aristotle's commentators. Explaining why of the three parts of philosophy only the middle one is called *mathēmata*, Elias says: because only *mathēmata* can provide reliable demonstrations; ταῦτα γὰρ μανθάνομεν ἀκριβῶς, τὰ δὲ ἄλλα εἰκάζομεν μᾶλλον ἢ μανθάνομεν (*In Porph. Isag.*, 28.24f.). David closes the same considerations with an

tion of the astronomer is to subordinate his research to the results of the physicist. Only by following the latter's lead will he produce a description of the heavenly motions that is an accurate representation of reality.[56] This demonstration of philosophy's superiority over science is manifestly Platonic in inspiration, and χορεία, related by Posidonius to the movement of heavenly bodies, refers us directly to Plato's *Timaeus* (40c). Projecting Posidonius' statement into the domain of the history of science, we easily reproduce the situation described later by Sosigenes: Plato, proceeding from the fundamental principle of planets' regular and ordered movement, sets a problem for astronomers, and Eudoxus is the first to advance the hypothesis of 'saving the appearances'.[57] Sosigenes' words could well suggest a direct influence of Posidonius, had not this notion of science's direct dependence on philosophy been a commonplace in his time. We find the same attitude both in the Platonist Dercyllides[58] and the Peripatetic Adrastus.[59] It is hard to say whether Posidonius himself was familiar with the legend of Plato as the architect of *mathēmata*. Geminus, in whom this fragment from Posidonius is found,[60] seems to have shared his views. In his *Introduction to Phaenomena*, however, it is not Plato or Eudoxus but the Pythagoreans whom he mentions in this connection, though the arguments in favor of the regular circular movement that he assigns to them bear an indelibly Platonic character.[61] Hence, by the early second century AD, the general thesis that astronomy is directly dependent on the basic principles established by philosophy has taken root in the Stoic, Peripatetic, and Platonic

---

example: even Aristotle, who teaches us the same in logic, derived his logical arguments from mathematics (*Proleg. Phil.*, 59.23ff.). See Gutas. Paul the Persian, 274f.

56 On the other hand, the mathematician Diodorus of Alexandria, following Posidonius in his definition of differences between mathematics and physics, believed that the two sciences were closely linked and could not do without each other in scientific research: διαφερούσας γοῦν ταύτας ἐν ταῖς ζητήσεσιν ἐπιπεπλέχθαι τὴν ἑτέραν δεομένην τῆς ἑτέρας (Achil. *Isag.*, 30.20–29 Maass).

57 Eud. fr. 148. See above, 86f., 273f.

58 He insists that the astronomers must build their theories on the fundamental ἀρχαί taken from Plato and reproaches all those who deviate from these principles (Theon. *Exp.*, 199.9–202.7). Kidd, *op. cit.*, 11, suggests that Dercyllides' position, especially in *Exp.*, 200.4–12, may reflect Posidonius' ideas, but the Platonist's program seems to be of a more general character. See the next footnote.

59 His entire astronomy is deduced from 'physical' principles (Theon. *Exp.*, 147.19ff.). All irregularities in the planetary motions are apparent and κατὰ συμβεβηκός, they have to be explained by different astronomical hypotheses: φυσικὸν μὲν καὶ ἀναγκαῖον, καθάπερ τὰ ἀπλανῆ, καὶ τῶν ἄλλων οὐρανίων ἕκαστον ἁπλῆν καὶ μίαν καθ' αὑτὸ φορὰν ὁμαλῶς φέρεσθαι καὶ εὐτάκτως (150.21f.).

60 See above, 229f.

61 Gemin. *Eisag.* I, 19–21 (see above, 271 n. 193): it is hard to imagine the divine and heavenly bodies moving quickly and slowly in alternation; their immortal nature implies circular regular movement alone. Theon (*Exp.*, 150.12f.) attributes to Pythagoras almost the same idea.

schools, so that its historical corollary – Eudoxus fulfilled Plato's methodological requirements – seems only natural.

The new interest in exact sciences, notable in Posidonius, is present in Geminus' works as well. It has to be pointed out again that his introductions to astronomy and mathematics are of a systematical character; they are neither directly related to the history of science, nor make any use of Eudemus' works.[62] Apart from the reference to the Pythagoreans as *prōtoi heuretai*, his introduction to astronomy does not include even the briefest historical overview. The problems treated in his introduction to mathematics correspond on the whole to the scientific interests of Posidonius, whose works were among Geminus' main sources. This textbook acquainted the reader with the foundations of mathematics, its methodology, and philosophical discussions around it, with particular emphasis on the classification and elucidation of mathematical notions. All Geminus' fragments that contain the mathematicians' names show that he referred to them to illustrate his theoretical propositions.[63] The Hellenistic mathematicians Amphinomus, Menaechmus, Zenodotus, and Theodorus figure in the context of methodological discussions of the notions of 'element', 'theorem' and 'problem'.[64] Apollonius, Nicomedes, Hippias, and Perseus are mentioned in connection with the classification of curves.[65] Archimedes is cited once as an illustration of the subject of mechanics and another time as an example of the mathematicians who call all the axioms postulates.[66] Geminus reproaches Apollonius for his attempts at demonstrating the axioms; his book about unordered irrationals is far too complicated to serve as an introduction to mathematics.[67]

Quite often Geminus refers to the authority of philosophers, considering their opinion no less than that of mathematicians.[68] Plato and Aristotle are

---

62 See above, 185f. – The idea of Geminus as a historian of mathematics seems to be ineradicable. Since Barocius, the translator of Proclus' commentary to Euclid (1560), listed among Proclus' sources *libros geometricarum enarrationum Gemini*, the historians of mathematics, beginning with Ramus, started to ascribe to Geminus a history of geometry that never actually existed. By the mid-19th century, this misunderstanding was finally cleared up (Nesselmann, G. *Algebra der Griechen*, Berlin 1842, 4f.; Schmidt. Philologische Beiträge, 79ff.), only to reappear in a slightly modified form in Tannery and those who followed him.

63 On Geminus' material in Proclus, see Tittel, *op. cit.*, 112f.

64 Procl. *In Eucl.*, 72.3ff., 77.2ff., 79.3ff., 200.22.

65 "This is the way in which other mathematicians also are accustomed to distinguish lines, giving the property of each species. Apollonius, for instance, shows for each of his conic lines what its property is, and Nicomedes likewise for the conchoids, Hippias for the quadratrices, and Perseus for the spiric lines." (Procl. *In Eucl*, 356.6f, transl. by G. Morrow; cf. 105.5, 15 on Apollonius).

66 Procl. *In Eucl.*, 41.6, 17; 181.18; cf. Papp. *Coll.* VIII, 1026.5f.; Eutoc. *In Archim. de plan. aequil.*, 266.1.

67 Procl. *In Eucl.*, 183.18; 74.22.

68 Plato (Procl. *In Eucl.* 41.8, 103.21, 117.17, 192.12), Aristotle (33.21, 104.22, 188.7,

called 'founders of geometry'; their views on the classification of lines have more weight than those of Apollonius.[69] Mathematical discoveries are touched upon only once, in connection with the classification of various conic sections.[70] Interesting for the historian is also the note on how the geometers before Apollonius defined the cone and the three kinds of conic sections.[71] On the whole, Θεωρία τῶν μαθημάτων fully corresponded to its name and was oriented not toward the history of mathematics, nor its particular problems, as later in Pappus, but toward its methodology and philosophy, which is confirmed by the ample use made of it in Proclus' philosophical commentary to Euclid. Pappus and Eutocius, who addressed their works to a professional audience, show comparatively little concern for Geminus.[72]

Posidonius' and his students' involvement with the exact sciences remained only a short episode in the history of Stoicism; later Stoics returned to their usual natural philosophy.[73] In the first century BC, mathematics gains a more solid ground in other philosophical schools – re-emerging Aristotelianism, Neopythagoreanism, Middle Platonism and later, particularly, Neoplatonism. Though the late philosophical schools never showed any systematic interest in the history of scientific knowledge, as distinct from an antiquarian interest, it is to them that we owe the larger part of the surviving evidence on the history of science in various kinds of introductions and commentaries, for example, in Dercyllides, Nicomachus, Adrastus, Theon of Smyrna, Cleomedes, Porphyry, Iamblichus, Proclus, and Simplicius. Still more valuable from this point of view are the works of commentators and systematizers of mathematical sciences, such as Hero, Menelaus, Sosigenes, Sporus, Pappus, and Eutocius. But the fate of historico-scientific tradition in the Imperial period is outside the scope of this book. The rich variety of sources of this time needs a detailed

---

192.10, 202.11), Speusippus (77.16, 179.15), Xenocrates (279.5), Chrysippus (395.14), Posidonius (80.21, 143.8, 176.6), Stoics (77.3), Epicureans (322.6).

69 Procl. *In Eucl.*, 192.5f., 103.21f.

70 "Some of these sections, in particular the conic, were discovered by Menaechmus… others by Perseus, who composed an epigram on his discovery." (Procl. *In Eucl*, 111. 20f.).

71 Eutoc. *In Apollon. Con.*, 168.17f. Another such note (ibid., 170.4f.), on the history of the theorem on the equality of the angles of the triangle to two right angles, is based on a misunderstanding (see above, 198).

72 Pappus mentions him once (*Coll.* VIII, 1026.8), Eutocius thrice (*In Archim. de plan. aequil.*, 266.1; *In Apollon. Con.*, 168.17, 170.25).

73 A partial exception is Posidonius' follower Cleomedes, who lived most probably between 50 BC – 200 AD (Bowen, A. C., Todd, R. B. *Cleomedes' lectures on astronomy. A translation of* The Heavens *with an introduction and commentary*, Berkeley 2004, 2ff.). His work *On the Orbits of the Heavenly Bodies* related to Stoic cosmology and physics and includes some glimpses of mathematical astronomy and geography. Thus, he describes in detail Posidonius' and Eratosthenes' methods of measuring the earth's circumference (I, 7).

study of its own;[74] a brief overview of them would hardly add anything substantial to what we have said above on different occasions. Let us return, then, to the beginnings of Hellenism and consider the extent to which the material of the history of science was represented in biography and doxography.

2. Biography and doxography

The first biographies, written by Aristoxenus and Dicaearchus, dealt with two groups of celebrities: philosophers and poets. Clearchus of Soli (fr. 30–31) and Phanias of Eresus (fr. 11–13) added to them the politicians. Biography emerged as a genre complementary to doxography (where biographical data were reduced to a minimum), addressed to a wider audience and, accordingly, more free. The biography of a philosopher was not expected to include a more or less detailed exposition of philosophical, let alone scientific, doctrines.[75] The fact that Pythagoras and Archytas, who turned out to be the heroes of Peripatetic biography, were also scientists, did not influence the development of the genre in the least.

In the third and the second centuries, the biography continued to flourish, including in the Lyceum, so that the very name 'Peripatetic' begins to denote a 'biographer'. Thematically, biography remained practically unchanged: Neanthes, Satyrus, Hermippus of Smyrna, Antigonus of Carystus, Sotion, Heraclides Lembos, and Antisthenes of Rhodes continued to write of philosophers, poets, and politicians, to whom Hermippus added lawgivers and orators. In Antiquity, a scientist (*mathēmatikos*) *qua* scientist had never become an object worthy of a biographical description, unless, like Archytas, Eudoxus, or Aratus, he was also a philosopher, a politician, or a poet. To all appearances, the biographies of Euclid, Eratosthenes, Apollonius, Hipparchus, Hero, and Ptolemy, not to mention dozens of the less eminent scientists, were never written.[76] Physicians and grammarians proved more lucky.[77]

---

74 For a valuable comparison of Pappus and Eutocius, see Knorr. *TS*, 225 ff. On Pappus, see also Cuomo. *Pappus*.

75 Mejer, J. *Diogenes Laertius and his Hellenistic background*, Wiesbaden 1978, 90 ff.

76 Biographical evidence of Hypatia (ca. 355–415) has been preserved in her pupil Synesius of Cyrene and in a short note by Socrates Scholasticus (*Hist. Eccl.* VII, 15), who does not mention her mathematical works (cf. *Suda*. s.v. Hypatia). Unlike other eminent representatives of the Neoplatonic school in Alexandria, Hypatia was not favored with a separate biography. In the biography of his teacher Isidore of Alexandria (ca. 490), Damascius notes: "Isidore was much more distinguished than Hypatia, not only in the way that a man is than a woman, but also as the true philosopher is than the geometer." (*Vit. Isid.* fr. 164).

77 Besides Soranus (see above, 285), biographies of physicians were published by his younger contemporary Hermippus of Berytus (*FGrHist* 1061 T 4), and those of grammarians by Asclepiades of Myrlaea in Bithynia (second–first centuries BC). See Pfeiffer, R. *History of classical scholarship*, Oxford 1968, 158, 162, 272 f. In the

The only exception I know of is Archimedes. Eutocius refers twice to his biography,[78] written by a certain Heraclides, who was most probably the disciple Archimedes mentioned in the epistle to Dositheus that serves as an introduction to the book *On Spirals*.[79] In the first of these fragments, Heraclides states that Archimedes was the first to discover (πρῶτον ἐπινοῆσαι) the theorems on conic sections, but did not publish them, while Apollonius of Perga, who lived in the time of Ptolemy Euergetes, claimed this theory as his own. The accusations of plagiarism, which Eutocius rightfully repudiates, were a current motif in intellectuals' biographies, so that Heraclides here follows the canons of the biographical genre. The second fragment, related to the book *On Measuring the Circle*, is more extensive:

> This book, as Heraclides in the *Life of Archimedes* says, is indispensable for the necessities of life (πρὸς τὰς τοῦ βίου χρείας ἀναγκαῖον),[80] since it shows that the circumference is three times longer than the diameter, and the resulting excess is less than 1⁄7 of the diameter but more than 10⁄71 of it. This, he says, is but an approximation, while Archimedes, making use of certain spirals, found a straight line which is exactly equal in length to the circumference of a given circle.

Summing up the main result of *On Measuring the Circle*, Heraclides compares it with a more rigorous mathematical demonstration found in Archimedes' *On Spirals*, the first version of which Heraclides himself handed over to Dositheus. If in the rest of the biography Heraclides discussed Archimedes' other works, both published and unpublished, or at least adduced their most important results, his life of Archimedes bore the traits of the scientific biography known to us in modern literature. Such a biography turns out to be closer to Eudemus' works on the history of science than to the Hellenistic philosophical biography. Unlike the stories of Archimedes' part in the defense of Syracuse, repeated with an increasing number of fabulous details by Polybius, Titus Livius, and Plutarch,[81] this type of scientific biography required from its reader a certain degree of mathematical competence. Does not this seem to account for the fact that other examples of this kind are unknown, while the biography by Heraclides is quoted by Eutocius alone?[82]

---

late 16th century, Bernardino Baldi (*Vite*, 36f.) was still complaining that people write biographies of the grammarians, orators, sophists, etc., but not of the mathematicians.

78 *In Archim. de dimens. circ.*, 228.19f.; *In Apollon. con.*, 168.5f. = *FGrHist* 1108 F 1–2.

79 Archim. *Spir.*, 2.1f., 4.27f. = *FGrHist* 1108 T 1a–b (with commentary).

80 On the motif of mathematics' practical utility, see above, 48 n. 16.

81 Polyb. VIII, 5, 3–5; Liv. XXIV, 34; Plut. *Marc.* XIV, 7–XVII, 7, XIX, 4–6.

82 We may conjecture, however, that the information on Archimedes' scientific and technical discoveries, in particular the invention of the so-called Archimedes' screw (Diod. Sic. I, 34, 2; V, 37, 3), goes back to this book. The well-known story of Archimedes, who discovered his famous law in the bathtub while examining King Hiero's crown (Vitr. IX, praef. 9–12), is considered by many to be apocryphal. Still, it explains the meaning of the discovery so vividly, exactly, and fully that it is only

Let us consider, for comparison, two biographies from Diogenes Laertius' compendium, which combines the characteristics of several genres (biographical successions, doxography, and the literature on philosophical schools) and, as a result, contains a great variety of material. The biography of Eudoxus, included by Diogenes Laertius for some unclear reasons in the chapter on the Pythagoreans (VIII, 86–91),[83] briefly mentions, among his other εὑρήματα ("was the first to arrange the couches at a banquet in a semicircle"), only one mathematical discovery, that of "curves" (90), without explaining its meaning. No mention is made of Eudoxus' two major works of astronomy, *Phaenomena* and *On Velocities*; from the *Circuit of the Earth* there is only one quotation, unrelated to geography. Though Eudoxus is called an astronomer, geometer, and physician, the nature of his achievements in mathematics, astronomy, and geography interested neither the biographer himself nor, apparently, his readers. The biography of Archytas is similar. The account of his scientific discoveries – the mathematization of mechanics and the solution to the problem of doubling the cube – is compressed into two short sentences (D. L. VIII, 83).[84]

In considering the biographies from Diogenes Laertius' collection, we should of course take into account the specific characteristics of this author. Yet there is no evidence that the preceding biographical tradition was substantially different: it could be interested in εὑρήματα, not in scientific theories. Later biography made no new contribution in this respect. Proclus' pupil and biographer Marinus does not say a word about his mathematical and astronomical works, but considers it relevant to cite his teacher's horoscope. Even Neopythagorean biography, such as Porphyry's *Life of Pythagoras*, which gave no less attention to *dogmata* than to *bios*, still passed over scientific theories in silence. Iamblichus' extensive *On the Pythagorean Life* also briefly mentions *mathēmata*, without much detail, Iamblichus having reserved this subject for the subsequent volumes of his Pythagorean series, which comprised ten books.

Unlike biography, doxography after Theophrastus underwent considerable changes and, as a result, included a wealth of new historico-scientific evidence. In the first part of the first century BC, Posidonius' school, notable for the revival of interest in physical problems, produced a short version of *Physikōn doxai* in six books,[85] which Diels called *Vetusta placita*. Included in it were the opinions of Aristotle, Peripatetics, Academics, Epicureans, and especially Stoics. Since Posidonius' concept of physics was much broader than

natural to think of its author as a person both close to Archimedes and versed in science.

83 Eudoxus studied mathematics with Archytas, but he never was a Pythagorean, nor did he write any philosophical works.

84 See above, 176.

85 Diels, H. Über das physikalische System des Straton, *Sitzungsb. der Preuss. Ak. d. Wiss.* (1893) 102.

that of Aristotle and Theophrastus,[86] doxography was enriched thematically as well. According to Posidonius, the founder of this much extended physics is not Thales any longer, but Homer (fr. 48–49, 137, 222 E.-K.); it includes mantic (fr. 7, 26–27 E.-K.); medicine draws still closer to physics;[87] *mathēmata* turns from an independent branch of theoretical sciences into physics' auxiliary instrument (fr. 18, 90 E.-K.). In accordance with the new criteria of selecting the material, *Vetusta placita* included opinions absent from Theophrastus: 1) those of ancient theologians and poets (Aët. I, 6); 2) of mathematicians, i.e., astronomers, and sometimes astrologers, whose teachings interested Posidonius (cf. fr. 111–112 E.-K.); 3) of the classical (Hippocrates, Polybus, Diocles) and especially of Hellenistic physicians (Herophilus, Erasistratus, Asclepiades).

Leaving out theologians, doctors, and Hellenistic physicists, let us note that in Aëtius (who in the first century AD revised *Vetusta placita*, condensing them to five books) we find the *doxai* of seven astronomers:[88] Oenopides, Eudoxus, Aratus, Aristarchus, Eratosthenes, Hipparchus, and Seleucus,[89] and of two astrologers: Berosus (ca. 300) and Epigenes of Byzantium (ca. 250). Fairly often (nine times) *mathēmatikoi* figure as a separate category of specialists.[90] In some cases it is possible to find out who specifically is referred to as *mathēmatikoi*: thus, cited in II,31.2 is the distance between the earth and the moon, which goes back (though with errors) to a treatise by Aristarchus.[91] In many cases, however, the statements assigned to mathematicians are so general that looking for a concrete author does not make any sense.[92] Apart from the astronomical *doxai*

86 See above, 289 f.

87 Athenaeus of Attaleia, the founder of the Pneumatist medical school, was Posidonius' pupil. See Kudlien, F. Posidonius und die Ärzte-Schule der Pneumatiker, *Hermes* 90 (1962) 419–429 (with doxographical reports on Posidonius' medical views). According to the post-Posidonian account of Stoic natural philosophy (D. L. VII, 133), aetiology, one of the latter's three parts, has two subdivisions, in one of which medical inquiries have a share, "in so far as it involves investigations of the ruling principle of the soul and the phenomena of the soul, seed and the like". On Stoic interest in medicine, see Hankinson, R. J. Stoicism and medicine, *The Cambridge companion to the Stoics*, 295–309.

88 Philip of Opus figures only as a source of evidence on the Pythagoreans (II,29.4).

89 In II,1.5 he is called an Erythraean and hence figures in Diels' index as Seleucus Erythraeus, but it follows from III,17.9, where the mathematician Seleucus is mentioned, that the person in question is Seleucus of Seleucia (ca. 150), the only astronomer who supported Aristarchus' heliocentric hypothesis.

90 In one case *mathēmatikoi* appear to be astrologers (Aët. V,18.5), in all the other cases, astronomers. See Diels' index for *mathematici* (*Dox.*, 686).

91 Mansfeld. Cosmic distances, 441.

92 In some of the mathematicians, the order of planets is the same as in Plato, others place the sun before the inner planets (II,15.5); Alcmaeon and the mathematicians believe that the planets move from the West to the East (II,16.2–3); Plato and the mathematicians believe that the inner planets move along the same path (ἰσοδρό-

that have an individual or collective author, we find in Aëtius those 1) assigned to anonymous οἱ μέν, οἱ δέ, or 2) that have no author at all. Related to the first category are the three calendar schemes mentioned in the section on the Great Year (II,32.2) and belonging to Cleostratus, Meton, and Callippus (7.5), as far as we can tell. In the sections on the Milky Way (III,1.2) and the comets (III,2.1) the *doxai* of Hippocrates of Chios appeared to be attributed to "some of the Pythagoreans".[93] Coming under the second category, for example, is section II,31.1, which adduces the data on the sidereal period of planets, which most probably go back to Eudoxus.[94]

All this material, which has not yet attracted the attention of specialists, needs a separate study. We may note, as a preliminary, that all the mathematicians who figure in Aëtius lived before 100 BC, so that it was the compiler of *Vetusta placita* who included them in the doxography. In *Vetusta placita*, the opinions of *mathēmatikoi* 'in general' were added, as a rule, to similar opinions already expressed by earlier physicists, while more individual *doxai* figured under the name of their author. Aëtius is unlikely to have added anything to this material; he must, on the contrary, have omitted a few names, thus making anonymous some of the *doxai* that initially had an authorship.[95]

3. From *inventio* to *translatio artium*: scheme and reality

Popular in the post-classical epoch, the theme of *origo artis* included the invention of various sciences, but was hardly related to the history of science in the form given to it by Eudemus. Most authors who touched upon it knew little of science and, as a rule, satisfied their curiosity with the help of the scheme, familiar to us (2.3), of εὕρεσις–μίμησις (*inventio–translatio*). With time, the second of these companion notions, which relates to the transmission of knowledge from one people (or author) to another, grows steadily in importance.[96]

μους) as the sun (II,16.7); the Morning Star and the Evening Star are the same planet Venus (ibid.); Plato, Aristotle, the Stoics, and the mathematicians explain in the same way the phases and eclipses of the moon (II,29.6); why the moon appears earth-like (II,30.7); the Pythagoreans and the mathematicians on the mirror reflection (VI,14.3).

93 The Milky Way: Arist. *Mete.* 345b 9f. = 42 A 6; Olymp. *In Mete.*, 68.30f.; the comets: *Dox.*, 231; cf. Arist. *Mete.* 342b 29ff. = 42 A 5; *Schol. in Arat.*, 546.21–22 Maass. The *doxa* of Hippocrates, preserved in the scholia, probably derives from Achilles, who relied on *Vetusta placita*.

94 The moon 30 days, the sun, Venus, and Mercury 12 months, Mars 2 years, Jupiter 12 years, Saturn 30 years. The same planetary periods figure in *Ars Eudoxi* (col. V) and in Sosigenes ap. Simpl. *In De caelo*, 495.26ff. = fr. 124 Lasserre.

95 See above, n. 93.

96 Worstbrock, *op. cit.*

The earliest version of the origin of astronomy and arithmetic is known from Aeschylus, one of whose tragedies ascribes their invention to Palamedes, another to Prometheus.[97] Under Herodotus' influence (II, 109), this mythical version yields to the historical (or, rather, pseudo-historical) one that regarded astronomy and geometry as coming from Egypt and Babylon, and arithmetic from, probably, Phoenicia.[98] There seems to have been no radical disagreement between 'serious' and 'not serious' genres and authors concerning the Oriental origin of mathematical sciences.[99] This idea remained predominant until the end of Antiquity and was inherited from it by Byzantine, Arabic, and, later, European historiography. Everybody seemed to agree that geometry was first invented in Egypt. Named as the person who brought it to Greece were either Thales or Pythagoras.[100] More complicated is the origin of astronomy, which, according to the three main versions, derived from Egypt (with Thoth-Hermes),[101] Babylon and Phoenicia,[102] or Greece. The last version added to the traditional inventors of astronomy, Palamedes and Prometheus, two more: Atlas and Endymion.[103] The invention of arithmetic and counting was attributed not only to their traditional Greek *prōtoi heuretai* Palamedes and Prometheus, but also to the Egyptians and the Phoenicians.[104] The orientalizing tendency showed even in the stories of the origin of medicine, whose first discoverers, Asclepius and Chiron, mentioned by Homer himself (*Il.* IV, 193–219), seemed incontestable.[105] Apart from the standard genealogy of

---

97 Aisch. *Prom.* 457–460, fr. 303a Mette; see above, 37.

98 See above, 40. Porphyry (*VP* 6) sets forth the most current of versions: the Greeks (i.e., Pythagoras) borrowed geometry from Egypt, astronomy from Babylon, arithmetic from Phoenicia. The same in Iulian. *Contra Galil.* I, 178a–b.

99 Isoc. *Bus.* 22–23; Pl. *Phaedr.* 274c 7–d 2, *Leg.* 747a–c; [Pl.] *Epin.* 986a 3f.; Arist. *Met.* 981b 23f.; Eud. fr. 133; Aristox. fr. 23. The most radical version is presented in *Busiris*: that Egyptian priests invented medicine and philosophy and also pursued astronomy, arithmetic, and geometry. It is doubtful, however, that Isocrates would have taken this version seriously.

100 Thales: Eud. fr. 133; D. L. I, 24; Ps.-Heron. *Def.*, 108.11. Pythagoras: Hecat. Abder. (*FGrHist* 264 F 25, 96f.); Anticlides (*FGrHist* 140 F 1); Callim. ap. Diod. X, 6, 4; Iambl. *De comm. math. sc.*, 66.21f.

101 Isoc. *Bus.* 22–23; Pl. *Phaedr.* 274c 7–d 2; Hecat. Abder. (*FGrHist* 264 F 15); D. L. I, 11; Clem. Alex. *Strom.* I, 16, 74; *Schol. in Dionys. perieget.*, 233.

102 [Pl.] *Epin.* 986a 3f.; Strab. 16, 2, 24; D. L. I, 11; Clem. Alex. *Strom.* I, 16, 74; Elias. *In Porph. isag.* 30.8–11.

103 Prometheus: Aesch. *Prom.* 457–460. Atlas: Ps.-Eupolemus ap. Euseb. *Praep. Ev.* IX,17.9; Diod. III, 60; IV, 27; Vitr. VI, 10, 6; Plin. *NH* II, 31; D. L. Prooem., 1. Endymion: Mnaseas (third century BC) ap. *Schol. in Apol. Rhod.*, 265.10, 275.22; *De incredibilibus*, 11; Alex. Aphr. *Probl.* 1, 134.

104 Isoc. *Bus.* 28, Pl. *Phaedr.* 274d 1f.; *Leg.* 747a–c; Eud. fr. 133; Aristox. fr. 23; Strab. 16,2.24; Iambl. *In Nicom.* 10.9; Procl. *In Eucl.*, 65.3f.

105 Cels. *De med.* I, 2; Plut. *Quaest. Conv.* 647 A; *Schol. in Il.* IV, 219; Eustath. *Comm. ad Hom. Il.* I, 733.3–11.

medicine featuring Chiron and/or Asclepius – Asclepiades – Hippocrates, we find in Hecataeus of Abdera the Egyptian version, which later figures, along with the Greek one, in Ps.-Galen.[106]

The Hellenized Jews very early advanced their own version of the origin of philosophy and sciences. The Peripatetic Aristobulus (second century BC) believed that Pythagoras, Plato, and Aristotle had borrowed many of their teachings from the Jews.[107] In his work *On the Jews*, Alexander Polyhistor, referring to the Jewish historians of the second and first centuries, stated that Chaldaean astronomy was invented by Abraham while he stayed in Babylonia; it was Abraham again who transferred this science to the Phoenicians and Egyptians.[108] A similar version is found in Josephus Flavius (*AJ* I, 167–168): astronomy and arithmetic were invented by the Chaldaeans, Abraham taught these sciences to the Egyptians, who in turn passed them over to the Greeks. This view was later vigorously advocated by Clement of Alexandria, Eusebius, and Theodoretus, who supported it with a wealth of material borrowed from the catalogues of discoveries: not philosophy alone, but nearly all the arts and sciences were invented by the 'barbarians', the most ancient of whom were the Jews.[109] Tatian, having named the inventions of the 'barbarians', demands of the Greeks: stop calling your imitations inventions![110] Later, some of the Byzantine authors gave preference to the old, 'individual' versions (the Egyptian, Syrian, Phoenician one, etc.),[111] but in Byzantium on the whole, as well as in early modern historiography, it is the 'biblical' version that determines the general view of the historical path of science: from the Jews through the Egyptians and the Babylonians to the Greeks, and from the Greeks, either directly or through the intermediary of the Arabs, to modern times.

The Arabs themselves identified Hermes, who was regarded as the forefather of science in general, as well as of its particular branches, with the biblical Enoch and with Idris, the character twice mentioned in the Koran, the founder of arts and sciences according to the Muslim tradition. The biographical histories of medicine by ibn Ǧulǧul (10th century) and ibn Abi Usaybi'a

---

106 Hecat. Abder. (*FGrHist* 264 F25); Ps.-Galen. *Intr. seu medicus*, 14, 674. See above, 298 n. 99. In Clement (*Strom.* I, 16, 75) medicine is invented in Egypt and then developed by Asclepius.

107 Aristobul. fr. 2 Denis; Goulet, R. Aristoboulos, *DPhA* I (1994) 379–380.

108 Ps.-Eupolemus (*FGrHist* 724 F 1–2); Artapan (*FGrHist* 726 F 1).

109 Clem. Alex. *Strom.* I,15.72–16.77; Euseb. *Praep. Ev.* X,1.1–7; X,4.17, etc.; Theodor. *Graec. affect. cur.* I, 12 ff.; Pythagoras and Plato studied under the Egyptians and the Jews (ibid. II, 23–26). In Roman and medieval Latin authors, the perspective was slightly different (Worstbrock, *op. cit.*, 9 ff.).

110 Tat. *Adv. Graec.*, I,1.9: ὅθεν παύσασθε τὰς μιμήσεις εὑρέσεις ἀποκαλοῦντες.

111 See e.g. *Comm. in Aratum reliquiae*, 318.20 f.; Mich. Psell. *Oratoria minora*, 18.72 f., 21.33 ff; Eustath. *Comm. in Dionys. perieg.*, 907.1–10; *Comm. ad Hom. Il.* I, 733.3–11.

(13th century) begin with Hermes and his son Asclepius.[112] Hermes is also mentioned among the Greek scientists in the bio-bibliographical encyclopaedia *Fihrist*, which also lists his astronomical works.[113] Having borrowed not only Greek science, but also the historico-scientific tradition of late Antiquity, Muslim culture successfully integrated the scheme of *inventio (translatio) artium* into the general perspective of scientific progress, whereby the legendary names and events gradually gave place to historical ones. At the turn of the 10th century, the eminent translator of Greek scientific texts Ishaq ibn Hunayn described a scholarly dispute in which one of the participants maintained that Hippocrates was the first physician and all the others derived their knowledge from him, while the other insisted that Hippocrates derived his knowledge from the ancients and his name became prominent only because he discovered many things and wrote them down systematically.[114] On the order of the vizier, who was attending the dispute, Ishaq compiled the outline of the chronology of doctors from the beginning of medicine to the present date (902); he thereby made use of the text ascribed to Yahya an-Nahwi (i.e., Joannes Philoponus).[115] This text divided the history of Greek medicine into eight periods, from Asclepius to Galen and his followers. While everything that preceded Hippocrates was sheer legend, the period from Galen to the sixth century AD was represented by historical names and texts.[116]

In a much more interesting historical perspective, *translatio artium* appears in *An Epistle to Saladin on the Revival of the Art of Healing* by ibn-Ǧumay‘ (ibn-Jami‘, d. 1198), the Jewish physician of sultan Saladin.[117] I mean here the 'Alexandria to Baghdad' complex of narratives – an account, popular in the Muslim world, of the origin, development, and decline of Greek philosophy and science and their subsequent passage to the Arabs.[118] Like Ishaq ibn Hu-

---

112 Meyerhof. Sultan Saladin's physician, 176; Ullmann, M. *Die Medizin im Islam*, Leiden 1970, 229f.

113 *Fihrist*, 634. Idris and Hermes as the inventors of astronomy: Wiedemann. Über Erfinder, 194–195.

114 Rosenthal. Ishaq b. Hunayn's Ta'rih al-attiba', 72ff.

115 There is a confusion of people known in Arabic as Yahya al-Nahwi: Meyerhof, M. Johannes Grammatikos (Philoponos) von Alexandrien und die arabische Medizin, *Mitt. des deutschen Inst. f. ägypt. Altertumskunde in Kairo* 2 (1931) 1–21; Ullmann, *op. cit.*, 27f.; Sezgin, F. *Geschichte des arabischen Schrifttums*, Vol. 3, Leiden 1973, 157ff.

116 Compiled in a similar way, though with a much shorter legendary period, is Celsus' outline of the history of medicine (see above, 285 n. 32). The brief survey of the history of medicine in Ps.-Galen (*Intr. seu medicus*, 14.674–676) starts with Asclepius, who learned it from his father Apollo and then passed it on to men, and ends with Hippocrates. Medicine existed in ancient times in Egypt as well, but Egyptian medicine does not bear comparison with the 'perfect' medicine of the Greeks.

117 Meyerhof. Sultan Saladin's physician, 169ff.

118 Strohmaier, G. 'Von Alexandrien nach Bagdad' – Eine fiktive Schultradition, *Aris-*

nayn, ibn-Ğumay' begins with Asclepius, whom he identifies with Idris and the Biblical Enoch. The legendary period, however, interests him little, and he passes immediately over to the historical one, i.e., Hippocrates. Galen, who lived 600 years later than Hippocrates, restored his doctrine by purifying it of later falsification and checking it against medical practice.

> After Galen, Christianity appeared among the Greeks and prevailed over them. The Christians considered it a fault to study intellectual subjects and their kings repudiated their cultivation and paid no heed to supporting those who sought them. So those who sough them quit taking the trouble to study them, finding that it took too long to read the books of Hippocrates and Galen. (The intellectual subjects) were in a state of crisis and their teaching fell in disorder. Then came Oribasius, after the Christian kings were firmly set in their disregard of teaching. He sought to spread (the teaching) among the common people by making it clearer and easier to follow ... He compiled compendia in which he clarified the craft and by means of which he made learning easier for them. He was followed by Paul (of Aegina) and those who came after him up to the present time.

Having told how the medical curriculum was reduced to twenty works of Hippocrates and Galen (the so-called *Summaria Alexandrinorum* of the sixth century), ibn-Ğumay' goes on:

> The instruction continued in Alexandria until the days of 'Umar ibn 'Abd-al-'Asis, for the director of instruction at that time, when 'Umar was still governor and before the caliphate was devolved upon him, converted to Islam at his hands and became his companion. After the caliphate was devolved upon 'Umar, the instruction was transferred from Alexandria to Antioch, Harran, and other places. The teaching stood on shaky ground until al-Ma'mun took over as caliph. He revived it and expanded it, and favored men of (scholarly) excellence. But for him, medicine and other sciences of the ancients would have been effaced and become extinct just as they are extinct today in the lands of the Greeks, lands in which these sciences had been cultivated the most.[119]

The summary of Greek medicine is but one of the elements of this history, which took shape, on the whole, between the reigns of caliphs 'Umar (634–644) and al-Ma'mun (813–833). In its other versions – for example in al-Mas'udi and al-Farabi – this summary is replaced by the outline of the history of Greek philosophy from Aristotle till the end of Antiquity, and could as well have been replaced by a similar outline of the history of astronomy. The key point is the decline of Greek learning with the advent of Christianity and the Byzantine rulers' disregard for it, particularly manifest when set against the flourishing of arts and sciences in the Muslim world under the wise and enlightened caliphs. D. Gutas believes these anti-Christian and anti-Byzantine polemics were part of the al-Ma'mun's official ideology, which "consisted of

---

toteles. Werk und Wirkung, Vol. 2, 380–389; Gutas. The 'Alexandria to Baghdad' complex of narratives.

119 Gutas. The 'Alexandria to Baghdad' complex of narratives, 161–166.

portraying the Christian Byzantines as benighted fanatics, who prohibited the ancient sciences, in contrast with the Muslims who welcomed and translated them".[120] What interests us now is not the propagandistic aspect of the story, but the fact that its account of the situation in the Muslim and Christian worlds was, on the whole, accurate. Al-Ma'mun went out of his way to procure Greek manuscripts, and the school of translators he founded in Baghdad (whose most important members, Hunayn ibn Ishaq and his son Ishaq ibn Hunayn, were Nestorian Christians) translated into Arabic and Syriac more than three hundred philosophical, scientific, and medical texts, so that by the end of the 9th century the Arabs appeared to be in possession of nearly everything that interested them in Greek culture.[121]

That this time *translatio artium* did take place follows from the list of Greek mathematicians and astronomers from the encyclopaedia *Fihrist*, written by ibn al-Nadim (10th century), a Baghdad bookseller and a connoisseur of literature. It includes the names of Euclid, Archimedes, Hypsicles, Apollonius, Eutocius, Menelaus, Ptolemy, Autolycus, Simplicius (as author of the introduction to the *Elements*), Theon of Alexandria, Theodosius, Pappus, Hero, Hipparchus, Diophantus, Nicomachus of Gerasa, and Aristarchus.[122] Each entry contains a vast bibliography of their scientific works, Greek and Arabic commentaries to them, and their Syriac and Arabic translations; most entries are supplied with chronological references. It is easy to see that here we are dealing with almost the entire repertoire of Greek mathematical and astronomical writings available in late Antiquity. Euclid's predecessors whose works had not survived are not mentioned, but that does not mean that they were unknown. On the contrary, it is very possible that they were known from no one else other than Eudemus, through the intermediary of some late commentators.[123]

In support of this, let us look at the refutation of astrology written by as-Samaw'al (12th century), a Jewish mathematician and physician converted to Islam.[124] An original scientist and thinker, as-Samaw'al formulates his own view of scientific progress, which substantially differs from the belief, predominant in Arabic culture, that the ancients had already discovered everything that could be known at all. Denying that the limit of perfection in science has been reached and that such a limit exists,[125] as-Samaw'al turns to historical

120 Ibid., 177.

121 Meyerhof. Sultan Saladin's physician, 177.

122 *Fihrist*, 634–644. Further on, the author lists the Indian and Arabic mathematicians up to his own time.

123 The name of Eudemus repeatedly occurs in Arabic sources, where he figures as Aristotle's pupil and a logician and physicist (Gutas, D. Eudemus in the Arabic tradition, *Eudemus of Rhodes*, 1–23). Cf. above, 167 n. 4.

124 Rosenthal. Al-Asturlabi and as-Samaw'al on scientific progress, 560ff.

125 As believed, e.g., by al-Biruni and ibn-Chaldun (Brentjes. Historiographie, 44f., 49f.).

facts, which clearly show that "in every age knowledge manifests itself in an increasing volume and with greater clarity":

> The biographies of scientists bear witness to this fact. Euclid collected the geometrical figures which were widely known in his time in a systematic work on the principles of geometry. He perfected the work by his own additions of instructive figures. The statement that before the time of Euclid, there existed no geometer or outstanding brain at all is contradicted by the testimony of history. On the other hand, the contention that Euclid knew more about geometry than the many excellent scholars who lived before his time does not necessarily imply that Euclid might not be succeeded by someone who, just as Euclid was better than his predecessors, would be better than Euclid. There is, for instance, Archimedes. His book on the *Sphere and the Prism* entitles him to such a rank (of superiority over Euclid). In his *Lemmata*, Archimedes now had to admit his inability to achieve the trisection of angles. After Archimedes, Apollonius earned greater fame than anyone else, in particular, through his discovery of the properties of conic sections.

Revealing is the confidence with which as-Samaw'al speaks of pre-Euclidean mathematics, referring to scientists' biographies and historical evidence. The historical perspective from which he sees the development of Greek mathematics implies an excellent knowledge of Greek sources, and precisely Eudemus was the principal source for the pre-Euclidean period. Yet the conclusion as-Samaw'al draws goes far beyond the limited epistemological optimism familiar to us from Greek texts;[126] the Europeans will not arrive at this conclusion until the age of the Enlightenment:

> No sage or well-informed historian will deny the fact that all the various disciplines of knowledge have manifested themselves in a process of gradual increase and ramification. This process stops at no final point and tolerates no irregularities.[127]

In Byzantium, the prospects for scientific progress looked entirely different. The time of Eutocius, Simplicius, Philoponus, and other authors of the sixth century AD, who still fully enjoyed the whole treasury of ancient tradition, is followed by the catastrophic decline of science and learning in general. Correlative with this is the absence in Byzantium of a historico-scientific tradition comparable with the Arabic one, which also, far from being an independent discipline, served the purposes of propaedeutics and theory. Here is how the situation looked through the eyes of a Greek who had an opportunity to compare Arabic science with that of Byzantium. When Stephanus the Philosopher, an astronomer and astrologer born in Persia,[128] arrived around 790 in Constantin-

---

126 Cf. above, 60 n. 62, on Seneca.

127 Rosenthal. Al-Asturlabi and as-Samaw'al on scientific progress, 563.

128 Stephanus was a disciple of another Greek, Theophilus of Edessa (ca. 695–785), the court astrologer of the Abbasids in Baghdad. See Pingree, D. From Alexandria to Baghdad to Byzantium. The transmission of astrology, *IJCT* 8 (2001) 3–37; idem. A

ople, he found that astronomy was almost extinct there. Deciding to revive it, he expounded its foundation in the work *On the Mathematical Art.*[129] At the beginning of this work, he remarks that with time and according to fate some sciences emerge and others entirely disappear – either everywhere, or in several cities only (I, 1). To demonstrate the usefulness of mathematical τέχνη for human life, Stephanus turns to its history. Discovered by the descendants of Seth (one of Adam's sons), this science passed on to the Chaldaeans, then successively to the Persians, the Greeks, the Egyptians, the Romans, and finally the Arabs. As long as these nations fostered mathematics, they remained the rulers of victorious world empires (σχεδὸν κοσμοκρατορικὰς καὶ νικητείρας εἶχον τὰς δυναστείας). That is why he, Stephanus, would like to revive and plant this science among the Christians, lest they should fall behind forever in it (I, 2). Interestingly, in Stephanus, who lived one generation before al-Ma'mun, we already find the model of *translatio artium* in its Arabic variant, schematic as it yet is. But unlike Arabic authors, who constantly emphasize that science's flourishing depends on wise and enlightened rulers,[130] Stephanus believes that the well-being of the state is conditioned by advanced science.

Stephanus' hopes for the revival of science in Byzantium did not – at least not fully – come true. The Byzantines appeared to be dependent on Arabic astronomy and mathematics and never attained their level.[131] In the course of the 9th–13th centuries, the scientific tradition of Antiquity was repeatedly broken, its achievements, if not names, forgotten.[132] But even the names partly fell into oblivion. A notion of what was known of ancient mathematicians and astronomers in the 10th century can be gained from the *Suda* lexicon by confronting the names it includes with those absent from it. Missing among the scientists of the classical period are Hippasus,[133] Oenopides, Hippocrates of Chios, Theodorus, Euctemon, and all of Eudoxus' pupils. Particularly striking is the absence

---

Greek ephemeris for 796: The work of Stephanus the Philosopher?, *Centaurus* 45 (2003) 79–82.

[129] *Catalogus Codicum Astrologorum Graecorum*, Vol. 2, ed. by F. Cumont. Brussels 1900, 181–186.

[130] Brentjes. Historiographie, 46, 48f.

[131] See e.g. Evans, J. Rec.: *An eleventh-century manual of Arabo-Byzantine astronomy*, ed. by A. Jones, Amsterdam 1987, *JHA* 20 (1989) 217–219; Saliba, G. Rec.: *The astronomical works of Gregory Chioniades*, I, ed. by D. Pingree, Amsterdam 1985, *JHA* 21 (1990) 211–215. It is revealing that in the late 13th century the astronomer Gregory Chioniades uses in his translation Persian terms even for such common notions as eclipse, latitude, or hour.

[132] Tihon, A. L'astronomie byzantine (du Ve au XVe siècle), *Byzantium* 51 (1981) 603–624; Bydén, B. *Theodore Metoichites' Stoicheiosis astronomike and the study of natural philosophy and mathematics in early Palaiologan Byzantium*, Göteborg 2003, 216ff.

[133] Hippasus is encountered in the entry on Heraclitus as the latter's teacher; his mathematical work is not mentioned.

of such Hellenistic figures as Autolycus, Euclid, Aristarchus, Archimedes,[134] Apollonius, Hypsicles, and Geminus, whose works were still accessible in Byzantium. Missing among the scientists of the Imperial age are the popularizers Nicomachus of Gerasa and Cleomedes, as well as such serious scientists as Hero, Menelaus, Diophantus, and Eutocius. No less unsatisfactory is the list of those on whom the encyclopaedia includes individual entries.[135] The classical period is represented here by Meton, Archytas, Theaetetus, and Eudoxus, the Hellenistic by Eratosthenes,[136] Hipparchus, and Theodosius, the Imperial by Ptolemy, Pappus, Theon of Alexandria, and his daughter Hypatia. The miserable state of the knowledge of ancient science is shown not only by this list of only 11 names, but also by the fact that all these mathematicians and astronomers (with the exception of Meton) figure in the *Suda* as philosophers. It is φιλόσοφος, rather than the more appropriate μαθηματικός, γεωμέτρης or ἀστρονόμος (ἀστρολόγος) that is here their principal and practically sole characteristic, independent of whether they wrote philosophical works as well as scientific ones.[137]

Since we find in the *Suda* figuring as 'philosophers' the historian Arrianus (who did, in fact, publish Epictetus), the author of a dream-book Artemidorus, and even Solon and Alcibiades, the authors of the encyclopaedia seem to understand 'philosophy' very broadly. A closer look at the use they make of such terms common in Antiquity as mathematician, geometer, astronomer, or philosopher will show, however, that the reason for this is not a particular bias they have, but their poor notion of their subject itself. Thus, Meton is presented here as μαθηματικός, ἀστρονόμος and ἰατρός, though nothing is known of his work in medicine. Theaetetus is referred to as a 'philosopher and astronomer', on which the classical tradition is silent. The only invention associated with Archytas is that of a toy rattle. Mentioned among Eudoxus' works is the *Astronomy* in verse, i.e., Aratus' *Phaenomena*. Found among the several 'astronomers' who figure in the encyclopaedia is Zoroaster, the king of the Assyrians, and Faunus (Hermes) of Egypt, who discovered gold, silver, and iron. The only author (apart from Meton) referred to as μαθηματικός appeared to be the traveler and geographer Scylax of Carianda, of whose mathematical work we

---

134 There is no special entry on Archimedes, though his name occurs in other entries in connection with various names of mechanical devices. That he was a mathematician can be inferred only from the entry on Theodosius, who wrote a commentary on Archimedes' *Method*.

135 We leave out here the names of philosophers who pursued mathematics and astronomy as well (Thales, Pythagoras, Posidonius, etc.).

136 The entry on Eratosthenes says nothing about his mathematics and astronomy; only his *Catasterismi* figures among the works on this subject.

137 Hypatia is called a philosopher by contemporary sources, including her father Theon (see Knorr. *TS*, 754 ff., 794 n. 3–4), but the *Suda* cites her scientific works only. Since her philosophical writings are unknown, Hypatia must have expounded her philosophy orally.

know nothing. Among the few characters called 'geometers' is Dicaearchus, who, like Scylax, wrote on geography, not geometry.[138] Though each of these many absurdities has causes and origins of its own, taken as a whole they demonstrate how impoverished and distorted the idea of ancient science really was among the literary men of Byzantium.

The Byzantine polymath Joannes Tzetzes (12th century) can hardly be reproached for ignorance of ancient literature. One of the cases he used to demonstrate his vast learning is connected with the Athenian astronomer Meton. Tzetzes knows Meton quite well: he dates him correctly to 432/28, cites his patronymic, and points to his discovery of the 19-year cycle. The idea that Meton was "the first of all the astronomers" found by Tzetzes in one of his contemporaries exasperates him. He gives vent to his indignation in a special chapter of his learned poem *Chiliades* (XII, 399), heaping imprecations on his opponents. Here, apparently, is an excellent occasion to show his superior knowledge of the history of Greek astronomy by citing the names of Meton's predecessors, from Thales to Oenopides. But Tzetzes, unlike as-Samaw'al, who was proving that Euclid was not the first geometer, does not seem to care for the history of science at all. What really interests him is mythical heurematography and chronology. According to him, the founder of astronomy was Atlas of Libya, who lived in the time of Osiris, Noah, and Dionysus; Hercules later borrowed this science from him. "Did not they precede Meton in time? Did not they write on astronomy?" (v. 142–143). For those who remain in doubt, Tzetzes adduces the names of the next three astronomers – Orpheus, Homer, and Hesiod. Since the youngest of them lived three hundred years before Meton, the latter can in no way be considered the author of the first work on astronomy. Despite all Tzetzes' vast learning, he gives a still more distorted perspective of the history of astronomy than the *Suda* does.

In the 14th century, the Byzantines return to the study of Ptolemy, as can be seen from the *Astronomical Elements* by Theodore Metoichites (ca. 1332) and *Astronomical Tribiblos* by Theodore of Melitene (ca. 1352).[139] In both authors we encounter extensive overviews of the history of astronomy, which culminate in Ptolemy. A chapter of Metoichites' *Semeioseis gnomikai*, entitled "That the science of mathematics was not fully developed in the beginnings", follows the scheme of *inventio–translatio artium* and starts with motifs already familiar to us from Aristotle.[140] *Mathēmata*, like many other things, "do not emerge in a perfect state at the start, but in each case develop initially from some slight beginning and yet are finally, with time, completed and established in the best

138 The other two 'geometers' figure but as the fathers of their more famous children: Nicon as the father of Galen, and Theon as the father of Hypatia. In the entry on Theon he is called a philosopher.

139 Théodore Méliténiote. *Tribiblos Astronomique*, Livre I–II, ed. by R. Leurquin, Amsterdam 1990–1993; Bydén, *op. cit.*

140 *Theodore Metoichites on ancient authors and philosophy. Semeioseis gnomikai 1–26 & 71*, ed. and transl. by K. Hult, Göteborg 2002, 135–145.

way possible and in fulfillment of their own nature." (14,1.1). The Chaldaeans studied astronomy from the beginnings, among them Abraham and Moses, whereas the Egyptians were experts in geometry, so that "both these sciences came to Greece from those people, and at a very late point in time compared to the whole of human history". Thales, Pythagoras, and later Plato personally went to Egypt and worked with mathematics; Pythagoras traveled to the Chaldaeans and even to India (14,2.1–8). Further on, Theodore discusses the achievements of Euclid, who is confused with the Socratic Euclid of Megara,[141] and Apollonius of Perga, who is named Alexander of Perga. Both of them were excellent in geometry, but in astronomy they would seem like children if compared with Ptolemy. The latter ranks Hipparchus far above all his predecessors and often pays respectful attention to him. With Ptolemy himself, astronomy comes to the point of its complete perfection:

> He has widely surpassed his predecessors, and has left his successors no opportunity to add anything to the science of astronomy, or indeed to add anything to his work, but only to spend their time going over the same ground, and labour with his results without contributing anything new unless, again, it comes from his works (14,5.6–7).

The same though more detailed historical scheme is found in Theodore of Melitene's overview "Who was the first to discover astronomy and how it came subsequently to Greece".[142] Astronomy was discovered, as Josephus Flavius says, by the descendants of Seth, one of Adam's sons, and written on two columns. After the flood, one of them fell into the hands of the Chaldaeans, who lived in Babylon. The Chaldaeans appropriated astronomy and improved it. Particularly famous among the Chaldaean astronomers were Zoroaster and Ostan, followed by Kidenas, Naburianus, Sudines, and Seleucus of Seleucia. Abraham, who traveled to Egypt, taught Chaldaean astronomy and arithmetic to Egyptian sages, who soon excelled in it, Hermes Trismegistus in particular. The sages who visited Egypt – Pythagoras, Pherecydes, Anaxagoras, Thales, Solon and above all Plato – brought astronomy over to Greece. The Greeks brought this science to perfection, surpassing by far the Chaldaeans and the Egyptians. Particularly famous among the astronomers was Hipparchus, as well as Pappus and Theon of Alexandria, the successors of the much admired Ptolemy. Theodore pays particular attention to the latter, briefly summarizing the contents of *Almagest*. In the third book, he continues his overview from the moment when the Arabs, and then the Persians, learned astronomy from Ptolemy's book, and he gives the names of a dozen Muslim astronomers of the 10th–13th centuries. According to Theodore, the Arabs and the Persians say the same as Ptolemy, and most of their books are full of things borrowed almost entirely from this author. Adding to them excellent astronomical and chronologi-

---

141 Cf. above, 4 n. 9.

142 See Balss, H. *Antike Astronomie*: Aus griech. und lat. Quellen mit Text, Übers. und Erl. dargest., Munich 1949, 201–211; *Tribiblos Astronomique* I.1, 96–100.

cal tables, they brought scientific learning to its perfection (πᾶσαν λογικὴν παιδείας εἰς ἄκρον ἐξησκημένοις).

What catches the eye when reading this outline is not even the fantastical mixture of historical legends and facts, but the total lack of references to discoveries, so typical of Eudemus' history of science. Astronomy is passed on from one people to another like a 'black box' whose contents remain unknown, with the exception of the very last period. Astronomers themselves are just listed and the question of their individual contributions to science does not even arise. The names of the three Babylonian astronomers and Seleucus, who wrote in Greek, are borrowed from Strabo,[143] and the list of the sages who traveled to Egypt from Diodorus of Sicily.[144] The whole of Greek astronomy is actually reduced to Hipparchus, mentioned by Ptolemy, Ptolemy himself, and his commentators Pappus and Theon.

Theodore's correctives to the *translatio atrium* scheme are quite revealing. His outline ends, like that of Stephanus, with the Arabs and Persians, whose works he knew in translation – not a single Byzantine astronomer is mentioned. But, for both Theodores, it is in Ptolemy that astronomy reached its ultimate summit. The Muslims, having entirely assimilated his science, perfected only the technical side of it.[145] Taken as a whole, these outlines constitute the backbone of the history of science that we find two centuries later in Petrus Ramus and in the works of his contemporaries, whose names figure on the opening pages of this book.

---

143 μέμνηνται δὲ καὶ τῶν ἀνδρῶν ἐνίων οἱ μαθηματικοί, καθάπερ Κιδηνᾶ τε καὶ Ναβουριανοῦ καὶ Σουδίνου· καὶ Σέλευκος δ' ὁ ἀπὸ τῆς Σελευκείας Χαλδαῖός ἐστι καὶ ἄλλοι πλείους ἀξιόλογοι ἄνδρες (16,1.6).

144 I,38.2–4: Thales and Anaxagoras; I,69.4: Solon and Pythagoras; I,92.2: Solon, Plato, Pythagoras; X,3.4: Pherecydes as Pythagoras' teacher. Diodorus, in turn, leaned on the book *On Egypt* by Hecataeus of Abdera (*FGrHist* 264 F 25).

145 From the standpoint of a Byzantine of the 14th century, this view of things is not totally unfounded, though now we know that many Muslim astronomers produced astronomical models different from the Ptolemaic model. See Saliba, G. *A history of Arabic astronomy: Planetary theories during the Golden Age of Islam*, New York 1994, 245 ff.; idem. Writing the history of Arabic astronomy: Problems and differing perspectives, *JAOS* 116 (1996) 709–718.

Sources[1]

Aischines von Sphettos, ed. by H. Dittmar, Berlin 1912.

Alexandri Aphrodisiensis In Aristotelis Metaphysica commentaria, ed. by M. Hayduck, Berlin 1891.

Anonymi Londinensis ex Aristotelis Iatricis et alias Medicis Eclogae, ed. by H. Diels, Berlin 1893.

Apollonii Pergaei quae Graece exstant cum commentariis antiquis (*Eutocii Commentaria in Conica*), ed. by I. L. Heiberg, Vol. 1–2, Leipzig 1891–1893.

Archimedis opera omnia cum commentariis Eutocii, ed. by I. L. Heiberg, Vol. 1–3, Leipzig 1910–1915.

Aristotelis opera. Librorum deperditorum fragmenta, ed. by O. Gigon, Berlin 1987.

Aristoxeni Elementa harmonica, ed. by R. da Rios, Rome 1954.

Asclepius Trallianus. In Nicomachi arithmeticam introductionem scholia, ed. by L. Tarán, Philadelphia 1969.

Berger, H. *Die geographischen Fragmente des Eratosthenes*, Leipzig 1880.

Claudii Ptolemaei opera quae exstant omnia (Vol. 1: *Syntaxis mathematica*. Vol. 2: *Opera astronomica minora*), ed. by I. L. Heiberg, Leipzig 1898–1903.

Comicorum Graecorum fragmenta, ed. by G. Kaibel, Vol. 1, Zurich 1999.

Commentariorum in Aratum reliquiae, ed. by E. Maass, Berlin 1898.

Davidis prolegomena et in Porphyrii Isagogen commentarium, ed. by A. Busse, Berlin 1904.

Denis, A.-M. *Fragmenta pseudepigraphorum quae supersunt Graeca* (*Pseudepigrapha veteris testamenti Graece 3*), Leiden 1970.

Diels, H. *Doxographi Graeci*, Berlin 1879.

Diels, H., Kranz, W. *Die Fragmente der Vorsokratiker*, 6th ed., Vol. 1–3, Berlin 1952.

Diogenes of Oenoanda: The Epicurean inscription, ed. by M. F. Smith, Naples 1993.

Düring, I. *Die Harmonielehre des Klaudios Ptolemaios*, Göteborg 1930.

— *Aristotle's Protrepticus: An attempt at reconstruction*, Göteborg 1961.

Edelstein, L., Kidd, I. G. *Posidonius. The fragments*, Vol. 1–3, Cambridge 1972–1988.

Eliae in Porphyrii Isagogen et Aristotelis Categorias commentaria, ed. by A. Busse, Berlin 1900.

Euclidis Elementa. After I. L. Heiberg ed. by E. S. Stamatis, Vol. 1–4, Leipzig 1969.

Euclidis Phaenomena et scripta musica, ed. by H. Menge, Leipzig 1916 (*Euclidis Opera omnia*. Vol. 8, ed. by I. L. Heiberg).

Eudoxi Ars astronomica qualis in charta Aegyptiaca superest, ed. by F. Blass, Kiel 1887 (reprint: *ZPE* 115 [1997] 79–101).

Eusebius. Praeparatio evangelica, ed. by K. Mras, Vol. 1–2, Berlin 1954–1956.

Eustathii commentarii ad Homeri Iliadem pertinentes, ed. by M. van der Valk, Vol. 1–4, Leiden 1971–1987.

Eustathii Commentarium in Dionysii periegetae orbis descriptionem, in: *Geographi Graeci minores*, ed. by K. Müller, Vol. 2, Paris 1861.

---

1 The list includes only works quoted by page and line number or name of an editor.

Gemini Elementa astronomiae, ed. by C. Manitius, Leipzig 1898.
Heronis Alexandrini opera quae supersunt omnia, ed. by W. Schmidt, H. Schöne, I. L. Heiberg, Vol. 1–5, Leipzig 1899–1914.
Hipparchi In Arati et Eudoxi Phaenomena commentariorum libri tres, ed. by C. Manitius, Leipzig 1894.
Hülser, K.–H. *Die Fragmente zur Dialektik der Stoiker*, Vol. 1–4, Stuttgart 1987.
Iamblichi de communi mathematica scientia liber, ed. by N. Festa, corr. by U. Klein, Stuttgart 1975.
Iamblichi De vita Pythagorica liber, ed. by L. Deubner, corr. by U. Klein, Stuttgart 1975.
Iamblichi In Nicomachi arithmeticam introductionem, ed. by E. Pistelli, Leipzig 1894.
Iamblichi Protrepticus, ed. by E. Pistelli, Stuttgart 1888.
Iamblichi Theologoumena arithmeticae, ed. by V. De Falco, Leipzig 1922.
Ioannis Philoponi In Aristotelis Analytica posteriora commentaria, ed. by M. Wallies, Berlin 1909.
Ioannis Philoponi in Aristotelis meteorologicorum librum primum commentarium, ed. by M. Hayduck, Berlin 1901.
Ioannis Philoponi In Aristotelis Physicorum libros commentaria, ed. by H. Vitelli, Vol. 1–2, Berlin 1887–1888.
Isnardi Parente, M. *Senocrate–Ermodoro. Frammenti*, Naples 1982.
Jacoby, F. *Die Fragmente der griechischen Historiker*, Vol. I–III, Berlin 1929–1958. – *Die Fragmente der griechischen Historiker continued.* Pt 4, ed. by G. Schepens, Leiden 1998–1999.
Kern, O. *Orphicorum fragmenta*, Berlin 1922.
Lasserre, F. *Die Fragmente des Eudoxos von Knidos*, Berlin 1966.
Luria, S. *Democritea*, Leningrad 1970.
Mensching, E. *Favorinos von Arelate*, Berlin 1963.
Merkelbach, R., West, M. L. *Fragmenta Hesiodea*, Oxford 1967.
Mette, H. J. *Die Fragmente der Tragödien des Aischylos*, Berlin 1959.
Nicomachi Geraseni Introductionis arithmeticae libri II, ed. by R. Hoche, Leipzig 1866.
Olympiodori In Aristotelis Meteora commentaria, ed. by G. Stüve, Berlin 1900.
Olympiodori In Platonis Gorgiam commentaria, ed. by W. Norvin, Leipzig 1936.
Pappi Alexandrini collectionis quae supersunt, ed. by F. Hultsch, Vol. 1–3, Berlin 1877–1878.
Platonis Timaeus: interprete Chalcidio cum eiusdem commentario, ed. by I. Wrobel, Leipzig 1876.
Poetae Comici Graeci, ed. by R. Kassel, C. Austin, Berlin 1983–.
Poetae melici Graeci, ed. by D. L. Page, Oxford 1962.
Porphyrios Kommentar zur Harmonielehre des Ptolemaios, ed. by I. Düring, Göteborg 1932.
Procli Diadochi Hypotyposis astronomicarum positionum, ed. by C. Manitius, Leipzig 1909.
Procli Diadochi In Platonis Timaeum commentaria, ed. by E. Diehl, Vol. 1–3, Leipzig 1903–1906.
Procli Diadochi In primum Euclidis elementorum librum commentarii, ed. by G. Friedlein, Leipzig 1873.

Prolegomena philosophiae Platonicae, ed. by L. G. Westerink, Paris 1990.
Radermacher, L. *Artium scriptores (Reste der voraristotelischen Rhetorik)*, Vienna 1951.
Rose, V. *Aristotelis qui ferebantur librorum fragmenta*, Leipzig 1886.
Ross, W. D. *Aristotelis fragmenta selecta*, Oxford 1955.
Scholia in Apollonium Rhodium vetera, ed. by C. Wendel, Berlin 1935.
Scholia in Dionysii periegetae orbis descriptionem, in: *Geographi Graeci minores*, ed. by K. Müller, Vol. 2, Paris 1861.
Simplicii in Aristotelis categorias commentarium, ed. by K. Kalbfleisch, Berlin 1907.
Simplicii In Aristotelis De caelo commentaria, ed. by I. L. Heiberg, Berlin 1894.
Simplicii In Aristotelis Physicorum libros commentaria, ed. by H. Diels, Vol. 1–2, Berlin 1882–1895.
Smith, A. *Porphyrii Philosophi fragmenta*, Stuttgart 1993.
Socratis et socraticorum reliquiae, ed. by G. Giannantoni, Vol. 1–5, Naples 1990.
Stoicorum veterum fragmenta, ed. by H. von Arnim, Vol. 1–4, Leipzig 1903–1924.
Suidae lexicon, ed. by A. Adler, Vol. 1–5, Leipzig 1928–1938.
Tarán, L. *Speusippus of Athens*, Leiden 1981.
Theiler, W. *Poseidonios. Die Fragmente*, Vol. 1–2, Berlin 1982.
Themistii in Aristotelis physica paraphrases, ed. by H. Schenkl, Berlin 1900.
Theonis Smyrnaei Expositio rerum mathematicarum ad legendum Platonem utilium, ed. by E. Hiller, Leipzig 1878.
Theophrastus of Eresus: Sources for his life, writings, thought, and influence, ed. by W.W. Fortenbaugh, P. M. Huby, R.W. Sharples, D. Gutas, Pt. 1–2, Leiden 1992.
Tragicorum Graecorum fragmenta, ed. by A. Nauck, Leipzig 1889.
Tragicorum Graecorum fragmenta, ed. by B. Snell, S. Radt, R. Kannicht, Vol. 1–4, Göttingen 1971–1986.
Usener, H. *Epicurea*, Leipzig 1887.
Wehrli, F. *Die Schule des Aristoteles: Texte und Kommentar*, 2nd ed., Vol. 1–10, Basel 1967–1969. – 1: Dikaiarchos; 2: Aristoxenos; 3: Klearchos; 5: Straton von Lampsakos; 7: Herakleides Pontikos; 8: Eudemos von Rhodos; 9: Phainias von Eresos, Chamaileon, Praxiphanes; 10: Hieronymos von Rhodos; Suppl. 1 (1974): Hermippos der Kallimacheer; Suppl. 2 (1978): Sotion.
Wescher, C. *Poliorcétique des Grecs: traités théoriques*, Paris 1867.

Bibliography[1]

Aaboe, A. Scientific astronomy in antiquity, *The place of astronomy in the ancient world*, ed. by F. R. Hodson, London 1974, 21–42.

Aaboe, A. et al. Saros cycle dates and related Babylonian astronomical texts, *TAPS* 81.6 (1991) 1–75.

Adam, J., ed. *The Republic of Plato*, Vol. 2, Cambridge 1902.

Allman, G. J. *Greek geometry from Thales to Euclid*, Dublin 1889.

Ancient histories of medicine. Essays in medical doxography and historiography in classical Antiquity, ed. by Ph. J. van der Eijk, Leiden 1999.

Aristotle on science. The Posterior Analytics, ed. by E. Berti, Padua 1981.

Aristotle's Metaphysics, ed. by W. D. Ross, Oxford 1924.

Aristoteles. Werk und Wirkung, ed. by J. Wiesner, Vol. 1–2, Berlin 1985–1987.

Asper, M. Mathematik, Milieu, Text. Die frühgriechische(n) Mathematik(en) und ihr Umfeld, *Sudhoffs Archiv* 87 (2003) 1–31.

Autour d'Aristote, ed. by A. Mansion, Louvain 1955.

Baldi, B. *Le vite de' matematici*, ed. by E. Nenci, Milan 1998.

Baltussen, J. *Theophrastus against the Presocratics and Plato: Peripatetic dialectic in the "De sensibus"*, Leiden 2000.

Barker, A. *Greek musical writings*, Vol. 1: *The musician and his art*; Vol. 2: *Harmonic and acoustic theory*, Cambridge 1984–1989.

— Aristoxenus' harmonics and Aristotle's theory of science, *Science and philosophy*, 188–226.

— *Scientific method in Ptolemy's* Harmonics, Cambridge 2000.

Becker, O. Die Lehre von Geraden und Ungeraden im IX. Buch der Euklidischen Elemente, *Q & St* B 3 (1934) 533–553 (= *Zur Geschichte der griechischen Mathematik*, 125–145).

— Zur Textgestaltung des Eudemischen Berichts über die Quadratur der Möndchen durch Hippokrates von Chios, *Q & St* B 3 (1936) 411–419.

— *Das mathematische Denken der Antike*, Göttingen 1957.

Berger, H. *Die geographischen Fragmente des Eratosthenes*, Leipzig 1880.

Bicknell, P. J. Anaximenes' astronomy, *AcCl* 12 (1969) 53–85.

Björnbo, A. A. *Studien über Menelaos' Sphärik. Beiträge zur Geschichte der Sphärik und Trigonometrie der Griechen*, Leipzig 1902.

— Hippokrates von Chios, *RE* 8 (1913) 1787–1801.

Blass, F. De Archytae Tarentini fragmentis mathematicis, *Mélanges Graux*, Paris 1884, 573–584.

Bleeker, C. J. *Hathor and Thoth*, Leiden 1973.

Blum, R. *Kallimachos. The Alexandrian library and the origin of bibliography*, Madison 1991.

Böker, R. Würfelverdoppelung, *RE* 9 A (1961) 1193–1223.

Boer, W. den. Progress in the Greece of Thucydides, *MKNAdW* 40.3 (1977).

[1] This bibliography includes only works which are referred to at least twice.

Boll, F. Finsternisse, *RE* 6 (1909) 2329–2364.
— Hebdomas, *RE* 7 (1912) 2547–2578.
Bollack, M. *La raison de Lucrèce*, Paris 1979.
Bonneau, D. Liber Aristotelis De inundatione Nili, *Etudes de Papyrologie* 9 (1971) 1–33.
Bowen, A. C. Menaechmus *versus* the Platonists: Two theories of science in the early Academy, *AncPhil* 3 (1983) 13–29.
— Eudemus' history of early Greek astronomy, *Eudemus of Rhodes*, 307–322.
— Goldstein, B. R. Meton of Athens and astronomy in the late 5th century BC, *A scientific humanist: Studies in memory of A. Sachs*, ed. by E. Leichty et al., Philadelphia 1988, 39–81.
Brentjes, S. Historiographie der Mathematik im islamischen Mittelalter, *AIHS* 42 (1992) 27–63.
Bretschneider, A. C. *Die Geometrie und die Geometer vor Euklides*, Berlin 1870.
Burk, A. *Die Pädagogik des Isokrates*, Würzburg 1923.
Burkert, W. *Lore and science in ancient Pythagoreanism*, Cambridge, Mass. 1972.
— Philodems Arbeitstext zur Geschichte der Akademie, *ZPE* 97 (1993) 87–94.
Burnyeat, M. F. Plato on why mathematics is good for the soul, *Mathematics and necessity. Proc. of British Academy* 103 (2000) 1–81.
Bydén, B. *Theodore Metochites' Stoicheiosis astronomike and the study of natural philosophy and mathematics in early Palaiologan Byzantium*, Göteborg 2003.
Cantor, M. *Vorlesungen über die Geschichte der Mathematik*, 2nd ed., Vol. 1–4, Leipzig 1901–1908.
Cambiano, G. Archimede meccanico e la meccanica di Archita, *Elenchos* 19 (1998) 291–324.
— Philosophy, science and medicine, *The Cambridge history of Hellenistic philosophy*, ed. by K. Algra et al., Cambridge 1999, 585–613.
Cherniss, H. *Aristotle's criticism of Presocratic philosophy*, Baltimore 1935.
— Plato as mathematician, *Rev. Met.* 4 (1951) 395–425 (= Cherniss, H. *Selected papers*, 222–252).
— *Selected papers*, ed. by L. Tarán, Leiden 1977.
Classen, C. J. Thales, *RE Suppl.* 10 (1965) 930–947.
— Anaximandros, *RE Suppl.* 12 (1970) 30–69.
Cole, T. *Democritus and the sources of Greek anthropology*, Ann Arbour 1967.
The commentary of Pappus on book X of Euclid's Elements, transl. by G. Junge, W. Thomson, London 1930.
Cornford, F. M. Mathematics and dialectic in the *Republic* VI–VII (1932), *Studies in Plato's metaphysics*, ed. by R. E. Allen, London 1965, 61–95.
Cuomo, S. *Pappus of Alexandria and the mathematics of late Antiquity*, Cambridge 2000.
De Ste. Croix, G. E. M. Aristotle on history and poetry (1975), *Essays on Aristotle's Poetics*, ed. by A. O. Rorty, Princeton 1992, 23–32.
Demandt, A. Verformungstendenzen in der Überlieferung antiker Sonnen- und Mondfinsternisse, *AAWM* no. 7 (1970).
Dicaearchus of Messana, ed. by W. W. Fortenbaugh, E. Schütrumpf, New Brunswick 2001 (Rutgers University Studies in Classical Humanities, Vol. 10).

Dicks, D. R. Thales, *CQ* 9 (1959) 294–309.
— Solstices, equinoxes, and the Presocratics, *JHS* 86 (1966) 26–40.
— *Early Greek astronomy*, London 1970.
Diels, H. Über die Excerpte von Menons Iatrika in dem Londoner Papyrus 137, *Hermes* 28 (1893) 487–494.
Diocles. On burning mirrors, ed. and transl. by G. J. Toomer, New York 1976.
Dorandi, T. *Platone e l'Academia (PHerc. 1021 e 164)/Filodemo,* Naples 1991.
Ducatillon, J. *Polémiques dans la Collection hippocratique*, Lille 1977.
Dörrie, H. *Der Platonismus in der Antike*, Vol. 1, Stuttgart 1987.
Düring, I. *Ptolemaios und Porphyrios über die Musik*, Göteborg 1934.
— *Aristoteles*, Heidelberg 1966.
Edelstein, L. *The idea of progress in classical Antiquity*, Baltimore 1967.
Effe, B. *Studien zur Kosmologie und Theologie der Aristotelischen Schrift "Über die Philosophie"*, Munich 1970.
Eggers Lan, C. Eudemo y el "catálogo de geómetras" de Proclo, *Emerita* 53 (1985) 127–157.
Eichholtz, P. *De scriptoribus* Περὶ εὑρημάτων (Diss.), Halle 1867.
Eucken, C. *Isokrates*, Berlin 1983.
Eudemus of Rhodes, ed. by I. Bodnár, W. W. Fortenbaugh, New Brunswick 2002 (Rutgers University Studies in Classical Humanities, Vol. 11).
Festugière, A.-J. *La révélation d'Hermès Trismégiste*, Vol. 1–3, Paris 1949.
Fiedler, W. *Analogiemodelle bei Aristoteles: Untersuchungen zu den Vergleichen zwischen den einzelnen Wissenschaften und Künsten*, Amsterdam 1978.
Fragmentsammlungen philosophischer Texte der Antike, ed. by W. Burkert et al., Göttingen 1998.
Franciosi, F. *Le origini scientifiche dell' astronomia greca*, Rome 1990.
Fritz, K. von. Oinopides, *RE* 17 (1937) 2258–2272.
— Die Bedeutung des Aristoteles für die Geschichtsschreibung, *Entretiens Fondation Hardt* IV (1956) 83–128.
— *Grundprobleme der Geschichte der antiken Wissenschaft*, Berlin 1971.
Froidefond, C. *Le Mirage égyptienne dans la littérature grecque d'Homère à Aristote*, Paris 1971.
Fuchs, H. *Enkyklios paideia*, *RLAC* 5 (1962) 365–398.
Gaiser, K. *Platons ungeschriebene Lehre*, Stuttgart 1963.
— *Philodems Academica*, Stuttgart 1988.
Geschichtliche Grundbegriffe, ed. by O. Brunner et al., Vol. 2, Stuttgart 1975.
Geus, K. *Eratosthenes von Kyrene*, Munich 2002.
Gigon, O. *Der Ursprung der griechischen Philosophie*, Basel 1945.
— Die ἀρχαί der Vorsokratiker bei Theophrast und Aristoteles, *Naturphilosophie bei Aristoteles und Theophrast*, ed. by I. Düring, Heidelberg 1969, 114–123.
Ginzel, F. K. *Spezieller Kanon der Sonnen- und Mondfinsternisse*, Berlin 1899.
Gottschalk, H. Eudemus and the Peripatos, *Eudemus of Rhodes*, 25–37.
Grensemann, H. *Knidische Medizin*, T. 1, Berlin 1975.
Gundel, H. Zodiakos, *RE* 10 A (1972) 462–709.
Gutas, D. Paul the Persian on the classification of the parts of Aristotle's philosophy: A milestone between Alexandria and Baghdad, *Islam* 60 (1983) 231–267.

Gutas, D. The 'Alexandria to Baghdad' complex of narratives. A contribution to the study of philosophical and medical historiography among the Arabs, *Documenti e studi sulla tradizione filosofica medievale* 10 (1999) 155–193.

Guthrie, W. K. C. *A history of Greek philosophy*, Vol. 1, Cambridge 1971.

Hartner, W. Eclipse periods and Thales' prediction of a solar eclipse, *Centaurus* 14 (1969) 60–71.

Heath, T. L. *Apollonius of Perga*, Oxford 1896.

— *Aristarchus of Samos*, Oxford 1913.

— *A history of Greek mathematics*, Vol. 1–2, Oxford 1921.

— *The thirteen books of Euclid's Elements*, Vol. 1–3, Oxford 1927.

— *Mathematics in Aristotle*, Oxford 1949.

Heiberg, J. L. Jahresberichte, *Philologus* 43 (1884) 467–522.

— *Mathematisches zu Aristoteles*, Leipzig 1904.

Heinimann, F. Eine vorplatonische Theorie der τέχνη, *Mus. Helv.* 18 (1961) 105–130.

Heitsch, E. *Platon. Phaidros*, Göttingen 1993.

Herter, H. Die kulturhistorische Theorie der hippokratischen Schrift von der Alten Medizin, *Maia* (1963) 464–483.

Hippocrate. De L'Art, ed. by J. Jouanna, Paris 1988.

— *L'ancienne médicine*, ed. by J. Jouanna, Paris 1990.

Hippokratische Medizin und antike Philosophie, ed. by R. Wittern, P. Pellegrin, Zurich 1996.

History and the disciplines: The reclassification of knowledge in early modern Europe, ed. by D. R. Kelley, Rochester 1997.

Huffman, C. A. *Philolaus of Croton. Pythagorean and Presocratic*, Cambridge 1993.

— *Archytas of Tarentum: Pythagorean, philosopher and mathematician king*, Cambridge 2005.

Hultsch, F. Astronomie, *RE* 2 (1896) 1828–1862.

Huxley, G. On Aristotle's historical methods, *GRBS* 13 (1972) 157–169.

Infinity and continuity in ancient and medieval thought, ed. by N. Kretzmann, Ithaca 1982.

Jaeger, W. *Aristotle: Fundamentals of the history of his development*, 2nd ed., Oxford 1948.

Janko, R. *Homer, Hesiod and the hymns*, Cambridge 1982.

Jones, A. Geminus and the Isia, *HSCPh* 99 (1999) 255–267.

Joos, P. ΤΥΧΗ, ΦΥΣΙΣ, ΤΗΧΝΗ: *Studien zur Thematik frühgriechischer Lebensbetrachtung* (Diss.), Winterthur 1955.

Kahn, C. H. On early Greek astronomy, *JHS* 90 (1970) 99–116.

Kienle, W. von. *Die Berichte über die Sukzessionen der Philosophen in der hellenistischen und spätantiken Literatur* (Diss.), Berlin 1961.

Kienzle, E. *Der Lobpreis von Städten und Ländern in der älteren griechischen Dichtung* (Diss.), Kallmünz 1936.

Kirk, G. S., Raven, J. E., Schofield, M. *The Presocratic philosophers*, 2nd ed., Cambridge 1983.

Kleingünther, A. ΠΡΩΤΟΣ ΕΥΡΕΤΗΣ, Leipzig 1933.

Knorr, W. R. *The evolution of the Euclidean Elements*, Dordrecht 1975.

— *The ancient tradition of geometric problems*, Boston 1986.

— *Textual studies in ancient and medieval geometry*, Boston 1989.
— Plato and Eudoxus on the planetary motions, *JHA* 21 (1990) 313–329.
Koselleck, R. Fortschritt, *Geschichtliche Grundbegriffe*, 351–421.
Krämer, H. J. Die Ältere Akademie, *Die Philosophie der Antike*, Vol. 3, 1–165.
Krafft, F. Der Mathematikos und der Physikos. Bemerkungen zu der angeblichen Platonischen Aufgabe, die Phänomene zu retten, *BGWT* 5 (1965) 5–24.
— *Dynamische und statische Betrachtungsweise in der antiken Mechanik*, Wiesbaden 1970.
Kramer, S. N. *Sumerian mythology*, Philadelphia 1944.
Kremmer, M. *De catalogis heurematum* (Diss.), Leipzig 1890.
Kühnert, F. *Allgemeinbildung und Fachbildung in der Antike*, Berlin 1961.
Kullmann, W. *Wissenschaft und Methode. Interpretationen zur Aristotelischen Theorie der Naturwissenschaft*, Berlin 1974.
Kurth, D. Thoth, *Lexikon der Ägyptologie* 6 (1986) 497–523.
Lanata, G. *Poetica pre-platonica,* Florence 1963.
Lasserre, F. *The birth of mathematics in the age of Plato*, London 1964.
— *De Léodamas de Thasos à Philippe d'Oponte*, Naples 1987.
Lebedev, A.V. Aristarchus of Samos on Thales' theory of eclipses, *Apeiron* 23 (1990) 77–85.
Leo, F. *Die griechisch-römische Biographie nach ihrer litterarischen Form*, Leipzig 1901.
Lloyd, A. B. *Herodotus book II. Commentary 1–98*, Leiden 1976.
— *Herodotus book II. Commentary 99–182*, Leiden 1988.
Lloyd, G. E. R. Saving the appearances, *CQ* 28 (1978) 202–222.
— *Methods and problems in Greek science*, Cambridge 1991.
Louis, P. Le mot ἱστορία chez Aristote, *RPh* 29 (1955) 39–44.
Manetti, D. 'Aristotle' and the role of doxography in the Anonymus Londiniensis, *Ancient histories of medicine*, 95–142.
Mansfeld, J. *Studies in historiography of Greek philosophy*, Assen 1990.
— Aristotle and the others on Thales, or the beginning of natural philosophy (= Mansfeld. *Studies*, 126–146).
— Aristotle, Plato, and the Preplatonic doxography and chronography (= Mansfeld. *Studies*, 22–83).
— Doxographical studies, Quellenforschung, tabular presentation and other varieties of comparativism, *Fragmentsammlungen philosophischer Texte*, 16–40.
— Doxography and dialectic: *The Sitz im Leben* of the 'Placita', *ANRW* II 36.4 (1990) 3057–3231.
— *Prolegomena mathematica: From Apollonius of Perga to the late Neoplatonists*, Leiden 1998.
— Cosmic distances: Aëtius 2. 31 Diels and some related texts, *Le style de la pensée. Recueil de textes en hommage à J. Brunschwig*, ed. by M. Canto-Sperber, P. Pellegrin, Paris 2002, 429–463.
Mansfeld, J., Runia, D. *Aëtiana: The method and intellectual context of a doxographer*, Vol. 1, Dordrecht 1997.
Marsden, E.W. *Greek and Roman artillery. Technical treatises*, Oxford 1971.
Mathieu, B. Archytas de Tarent pythagoricien et ami de Platon, *BAGB* (1987) 239–255.

McDiarmid, J. B. Theophrastus on Presocratic causes, *HSCPh* 61 (1953) 85–156.
McKirahan, R. Aristotle's subordinate sciences, *BJHS* 11 (1978) 197–220.
Meier, C. 'Fortschritt' in der Antike, *Geschichtliche Grundbegriffe*, 353–363.
— Ein antikes Äquivalent des Fortschrittsgedankens: das "Könnens-Bewusstsein" des 5. Jahrhunderts v. Chr., *HZ* 226 (1978) 265–316.
Mendell, H. The trouble with Eudoxus, *Ancient and medieval traditions in the exact sciences. Essays in memory of Willbur Knorr*, ed. by P. Suppes et al., Stanford 2000, 59–138.
Mensching, E. *Favorinos von Arelate*, Berlin 1963.
Merlan, P. *From Platonism to Neoplatonism*, The Hague 1960.
Meyerhof, M. Sultan Saladin's physician on the transmission of Greek medicine to the Arabs, *BHM* 18 (1945) 169–178.
Mittelstraß, J. *Die Rettung der Phänomene*, Berlin 1963.
Moraux, P. *Les listes anciennes des ouvrages d'Aristote*, Louvain 1951.
— *Der Aristotelismus bei den Griechen*, Vol. 2, Berlin 1984.
Mueller, I. Aristotle and the quadrature of the circle, *Infinity and continuity*, 146–164.
— Remarks on Euclid's *Elements* I, 32 and the parallel postulate, *Science in Context* 16 (2003) 287–297.
Netz, R. Greek mathematics: A group picture, *Science and mathematics*, 196–216.
— Eudemus of Rhodes, Hippocrates of Chios and the earliest form of a Greek mathematical text, *Centaurus* 46 (2004) 243–286.
Neuenschwander, E. Die ersten vier Bücher der Elemente Euklids, *AHES* 9 (1973) 325–380.
Neugebauer, O. *The exact sciences in Antiquity*, 2nd ed., New York 1962.
— *A history of ancient mathematical astronomy*, Pt. 1–3, Berlin 1975.
Nicomachus of Gerasa. *Introduction to arithmetic*, transl. by M. L. D'Ooge, New York 1926.
Palmer, J. Aristotle on the ancient theologians, *Apeiron* 33 (2000) 181–205.
Panchenko, D. Ὅμοιος and ὁμοιότης in Thales and Anaximander, *Hyperboreus* 1 (1994) 28–55.
— Thales's prediction of a solar eclipse, *JHA* 25 (1994) 275–288.
Pasquali, G. Doxographica aus Basiliusscholien, *NGWG* (1910) 194–228.
Partsch, J. Des Aristoteles Buch "Über das Steigen des Nil", *ASGW* 27 (1909) 553–600.
Patzer, A. *Der Sophist Hippias als Philosophiehistoriker*, Freiburg 1986.
Pesch, J. G. van. *De Procli fontibus* (Diss.), Leiden 1900.
Philipp, H. *Tektonon Daidala. Der bildende Künstler und sein Werk im vorplatonischen Schrifttum*, Berlin 1968.
Die Philosophie der Antike, ed. by H. Flashar, Vol. 3: *Ältere Akademie, Aristoteles, Peripatos*, 2nd ed., Basel 2004; Vol. 4: *Hellenistische Philosophie*, Basel 1994.
Pinault, J. R. *Hippocratic lives and legends*, Leiden 1992.
Pritchett, W. K., Waerden, B. L. van der. Thucydidean time-reckoning and Euctemon's seasonal calendar, *BCH* 85 (1961) 17–52.
Proclus. – Lecteur et interprète des Anciens, ed. by J. Pépin, H. D. Saffrey, Paris 1987.
Regenbogen, O. Theophrastos, *RE Suppl.* 7 (1940) 1354–1562.
Rehm, A. Das Parapegma des Euktemon, *SHAW* no. 3 (1913).
— Kallippos, *RE Suppl.* 4 (1924) 1431.

— *Parapegmastudien*, *ABAW* no. 19 (1941).

Riginos, A. E. *Platonica. The anecdotes concerning the life and writings of Plato*, Leiden 1976.

Robins, I. Mathematics and the conversion of the mind, *Republic* vii 522c 1–531e 3, *AncPhil* 15 (1995) 359–391.

Rose, P. L. *The Italian Renaissance of mathematics*, Geneva 1975.

Rosenthal, F. Al-Asturlabi and as-Samaw'al on scientific progress, *Osiris* 9 (1945) 555–564 (= Rosenthal, F. *Science and medicine in Islam*, Aldershot 1991).

— Ishaq b. Hunayn's Ta'rih al-attiba', *Oriens* 7 (1954) 55–80 (= Rosenthal, F. *Science and medicine in Islam*, Aldershot 1991).

Runia, D. The *placita* ascribed to doctors in Aëtius' doxography on physics, *Ancient histories of medicine*, 189–250.

— What is doxography?, *Ancient histories of medicine*, 33–55.

Rusch, A. Thoth, *RE* 7 A (1936) 351–362.

Sachs, E. *Die fünf platonischen Körper*, Berlin 1917.

Samuel, A. *Greek and Roman chronology*, Munich 1972.

Schaerer, R. ΕΠΙΣΤΗΜΗ et ΤΕΧΝΗ. *Etude sur les notions de connaissance et d'art d'Homère à Platon*, Mâcon 1930.

Schlachter, A. *Der Globus*, Berlin 1927.

Schneider, H. *Das griechische Technikverständnis*, Darmstadt 1989.

Schmidt, M. Die Fragmente des Mathematikers Menaechmus, *Philologus* 42 (1884) 72–81.

— Philologische Beiträge zu den griechischen Mathematikern, *Philologus* 45 (1886) 63–81, 278–320.

Schramm, M. *Ibn al-Haythams Weg zur Physik*, Wiesbaden 1961.

Schürmann, A. *Griechische Mechanik und antike Gesellschaft*, Stuttgart 1991.

Science and mathematics in ancient Greek culture, ed. by C. J. Tuplin, T. E. Rihll, Oxford 2002.

Science and philosophy in classical Greece, ed. by A. C. Bowen et al., New York 1991.

Sedley, D. Epicurus and the mathematicians of Cyzicus, *CErc* 6 (1976) 23–54.

Sharples, R. W. *Theophrastus of Eresus. Commentary*, Vol. 3.1: *Sources on physics*, Leiden 1998.

Snell, B. *Die Ausdrücke für den Begriff des Wissens in der vorplatonischen Philosophie*, Berlin 1924.

Spadafora, D. *The idea of progress in England*, New Haven 1990.

Spoerri, W. Kulturgeschichtliches im Alpha der aristotelischen "Metaphysik", *Catalepton. Festschrift B. Wyss*, Basel 1985, 45–68.

Steele, J. M. Eclipse predictions in Mesopotamia, *AHES* 54 (2000) 421–454.

Steinmetz, P. *Die Physik des Theophrastos von Eresos*, Bad Homburg 1964.

Stephenson, F. R., Fatoohi, L. J. Thales's prediction of a solar eclipse, *JHA* 28 (1997) 279–282.

Svenson-Evers, H. *Die griechischen Architekten archaischer und klassischer Zeit*, Frankfurt 1996.

Szabó, A., Maula, E. ΕΓΚΛΙΜΑ. *Untersuchungen zur Frühgeschichte der antiken griechischen Astronomie, Geographie und Sehnentafeln*, Athens 1982.

Tannery, P. *La géométrie grecque*, Paris 1887.

— *Pour l'histoire de la science hellène*, Paris 1887.
— *Recherches sur l'histoire de l'astronomie ancienne*, Paris 1893.
— Sur les fragments d'Eudème de Rhodes relatifs á l'histoire des mathématiques (1882), *Mémoires scientifiques*, T. 1, Toulouse 1912, 168–177.
Tarán, L. *Academica: Plato, Philip of Opus and the Pseudo-Platonic Epinomis*, Philadelphia 1975.
— Proclus on the Old Academy, *Proclus*, 227–276.
Taylor, A. E. *A commentary on Plato's Timaeus*, Oxford 1928.
Taylor, C. C. W. Plato and the mathematicians, *PhilosQ* 17 (1968) 193–203.
The Cambridge companion to the Stoics, ed. by B. Inwood, Cambridge 2003.
The Fihrist of al-Nadim, transl. by B. Dodge, Vol. 1–2, New York 1970.
Thraede, K. Das Lob des Erfinders. Bemerkungen zur Analyse der Heuremata-Kataloge, *RhM* 105 (1962) 158–186.
— Erfinder, *RLAC* 5 (1962) 1191–1278.
— Fortschritt, *RLAC* 7 (1965) 139–182.
Tittel, K. Geminos, *RE* 7 (1912) 1026–1050.
Trampedach, K. *Platon, die Akademie und die zeitgenössische Politik*, Stuttgart 1994.
Ullmann, M. *Die Medizin im Islam*, Leiden 1970.
Usener, H. Analecta Theophrastea (1858), *Kleine Schriften*, Vol. 1, Leipzig 1912, 71–87.
Uxkull-Gyllenband, W. *Griechische Kultur-Entstehungslehren*, Berlin 1924.
Vogt, J. Herodot in Ägypten, *Genethliakon W. Schmid*, Stuttgart 1929, 97–137.
Waerden, B. L. van der. *Erwachende Wissenschaft*, 2nd ed., Stuttgart 1966.
— Die Postulate und Konstruktionen in der frühgriechischen Geometrie, *AHES* 18 (1978) 343–357.
— *Die Pythagoreer: Religiöse Bruderschaft und Schule der Wissenschaft*, Zurich 1979.
— *Die Astronomie der Griechen*, Darmstadt 1988.
Waschkies, H.-J. *Von Eudoxos zu Aristoteles*, Amsterdam 1977.
Weidler, J. F. *Historia astronomiae, sive de ortu et progressu astronomiae liber singularis*, Wittenberg 1741.
Weil, R. *Aristote et l'histoire*, Paris 1960.
— Aristotle's view of history, *Articles on Aristotle 2. Ethics and politics*, ed. by J. Barnes et al., London 1977, 202–217.
Wending, E. *De Peplo Aristotelico* (Diss.), Strassburg 1891.
West, M. L. *The east face of Helicon. West Asiatic elements in Greek poetry and myth*, Oxford 1997.
Wiedemann, E. *Aufsätze zur arabischen Wissenschaftsgeschichte*, Vol. 1–2, Hildesheim 1970.
— Über Erfinder nach arabischen Angaben, *Gesammelte Schriften zur arabisch-islamischen Wissenschaftsgeschichte*, Vol. 2, Frankfurt 1984, 848–850.
Wilms, H. *Techne und Paideia bei Xenophon und Isokrates*, Stuttgart 1995.
Wöhrle, G. Wer entdeckte die Quelle des Mondlichts?, *Hermes* 123 (1995) 244–247.
Wolfer, E. P. *Eratosthenes von Kyrene als Mathematiker und Philosoph*, Groningen 1954.
Worstbrock, F. J. Translatio artium. Über Herkunft und Entwicklung einer kulturhistorischen Theorie, *ArKult* 47 (1965) 1–22.

Zaicev, A. *Das Griechische Wunder. Die Entstehung der griechischen Zivilisation*, Konstanz 1993.

Zeller, E. *Die Philosophie der Griechen*, 6th ed., Vol. 1–3, Leipzig 1919.

Zhmud, L. *Wissenschaft, Philosophie und Religion im frühen Pythagoreismus*, Berlin 1997.

— Some notes on Philolaus and the Pythagoreans, *Hyperboreus* 4 (1998) 243–270.

— Revising doxography: Hermann Diels and his critics, *Philologus* 145 (2001) 219–243.

— Eudemus' history of mathematics, *Eudemus of Rhodes*, 263–306.

— Doxographie in ihrer Beziehung zu den anderen Genres der antiken Philosophiegeschichte, *Die Philosophie der Antike*, Vol. 1: *Frühgriechische Philosophie*, ed. by H. Flashar et al., Basel 2007 (*forthcoming*).

Zur Geschichte der griechischen Mathematik, ed. by O. Becker, Darmstadt 1965.

General index

Bei Fragen zur Produktsicherheit wenden Sie sich bitte an:
If you have any questions regarding product safety,
please contact:

Walter de Gruyter GmbH
Genthiner Straße 13
10785 Berlin
productsafety@degruyterbrill.com